World Sustainability Series

Series Editor

Walter Leal Filho, European School of Sustainability Science
and Research, Research and Transfer Centre "Sustainable Development and
Climate Change Management", Hamburg University of Applied Sciences,
Hamburg, Germany

Due to its scope and nature, sustainable development is a matter which is very interdisciplinary, and draws from knowledge and inputs from the social sciences and environmental sciences on the one hand, but also from physical sciences and arts on the other. As such, there is a perceived need to foster integrative approaches, whereby the combination of inputs from various fields may contribute to a better understanding of what sustainability is, and means to people. But despite the need for and the relevance of integrative approaches towards sustainable development, there is a paucity of literature which address matters related to sustainability in an integrated way.

Notes on the quality assurance and peer review of this publication

Prior to publication, the works published in this book are initially assessed and reviewed by an in-house editor. If suitable for publication, manuscripts are sent for further review, which includes a combined effort by the editorial board and appointed subject experts, who provide independent peer-review. The feedback obtained in this way was communicated to authors, and with manuscripts checked upon return before finally accepted. The peer-reviewed nature of the books in the "World Sustainability Series" means that contributions to them have, over many years, been officially accepted for tenure and promotion purposes.

Mourade Azrour · Jamal Mabrouki ·
Azidine Guezzaz
Editors

Sustainable and Green Technologies for Water and Environmental Management

Springer

Editors
Mourade Azrour
Faculty of Sciences and Techniques
Moulay Ismail University of Meknes
Errachidia, Morocco

Jamal Mabrouki
Faculty of Sciences
Mohammed V University
Rabat, Morocco

Azidine Guezzaz
Technology Higher School Essaouira
Cadi Ayyad University
Ghazoua, Morocco

ISSN 2199-7373 ISSN 2199-7381 (electronic)
World Sustainability Series
ISBN 978-3-031-52418-9 ISBN 978-3-031-52419-6 (eBook)
https://doi.org/10.1007/978-3-031-52419-6

This Springer imprint is published by the registered company Springer Nature Switzerland AG
The registered company address is: Gewerbestrasse 11, 6330 Cham, Switzerland

Contents

State-Of-The-Art Methods for Dynamic Texture Classification: A Comprehensive Review

Manal Benzyane, Mourade Azrour, Imad Zeroual, and Said Agoujil

Abstract Dynamic texture classification has become an increasingly important research area due to the growing availability of video data and the need for efficient video analysis. Recently, deep learning models have demonstrated remarkable success in automatically classifying dynamic textures. This review provides a comprehensive and concise overview of recent advances in dynamic texture classification, with a particular focus on deep learning-based approaches. We survey the most popular deep learning architectures used for this task, highlighting key findings based on existing deep learning models and offering future research directions. Our review covers network architecture, evaluation criteria, datasets used in video classification, and provides an overview of deep learning methods.

Keywords Dynamic texture · Classification · Deep learning · Video

1 Introduction

Video classification is an essential field of study within computer vision, as it provides the ability to categorize videos into different classes automatically. In the modern era of digital advancements, the amount of available video data has grown tremendously, making it more challenging to manage, analyze and extract valuable insights from this data. Efficient video analysis has, therefore, become increasingly important in various applications, including surveillance, entertainment, education, and more.

The process of video classification involves assigning one or more predefined labels to a given video segment based on its content. This can be accomplished using

M. Benzyane · S. Agoujil
Faculty of Sciences and Techniques, MMIS, MAIS, Moulay Ismail University of Meknes, Meknes, Morocco

M. Azrour (✉) · I. Zeroual
Faculty of Sciences and Techniques, STI, IDMS, Moulay Ismail University of Meknes, Meknes, Morocco
e-mail: mo.azrour@umi.ac.ma

M. Azrour et al. (eds.), *Sustainable and Green Technologies for Water and Environmental Management*, World Sustainability Series,
https://doi.org/10.1007/978-3-031-52419-6_1

various methods, including conventional methods relying on handcrafted features and more recent approaches based on deep learning. With the rise of deep learning models, researchers have achieved significant progress in video classification, outperforming traditional methods by a considerable margin.

Deep learning models have the ability to automatically learning relevant features and representations from raw video data, rendering them highly effective for this task. Usually, these models undergo training using extensive video datasets in order to learn and capture the fundamental attributes of the data. This training process enables them to generalize well to unseen data, thereby improving their classification accuracy.

In this overview, our attention will be directed toward the latest developments in video classification, with a particular emphasis on deep learning-based approaches. We will survey the most popular deep learning architectures used for this task and discuss the datasets and commonly employed evaluation metrics for assessing the effectiveness of video classification models.

In summary, the aim of this review is to present a comprehensive overview of the state-of-the-art in video classification and its future prospects, which will enable researchers to better understand the challenges and opportunities in this field and to identify promising directions for further research.

2 Data Set

The success of any video classification task greatly depends on the careful choice of a suitable dataset. Video datasets consist of videos that have been annotated with labels or tags, serving the purpose of facilitating numerous computer vision tasks like video classification, object detection, action recognition, and more. The choice of a dataset can impact the performance and accuracy of video classification models. The availability of large-scale video datasets has facilitated the development and evaluation of deep learning-based video classification models. In recent years, several video datasets have been released, covering a wide range of action categories, video modalities, and complexities. This has led to significant advancements in video classification research, with the latest models achieving human-level performance on some datasets. In this context, it is important for researchers to be familiar with the different video datasets available and their characteristics, so they can choose the most appropriate dataset for their specific task and evaluate their models against the state-of-the-art. In this article, we present a summary of the most used video datasets in research related to video classification, their properties, and their suitability for different types of video classification tasks.

Various datasets are employed for video classification, each tailored to assess the efficacy of video classification models across diverse video categories. A summary of several significant datasets is provided in Table 1.

Table 1 Synopsis of video task dataset

Name of the dataset	Year	Number of video	Number of class
UCF50	2012	6618	50
UCF101	2012	13,320	101
HMDB51	2013	6766	51
Sports-1 M	2014	1,133,158	487
Youtube-8 M	2016	6.1 M	3862
Kinetics	2017	300,000	400

3 Performance Metrics

Performance metrics in video classification can be used to evaluate the effectiveness of a model in accurately predicting the class labels for videos. When evaluating the performance of a binary classifier, it is important to consider the four possible outcomes: true positives (TP), true negatives (TN), false positives (FP), and false negatives (FN) (Studying Employee Absenteeism Due to Health-Related Factors: A Data-Science Approach - ProQuest 2023; Krstinić et al. 2020). These outcomes are used to calculate metrics like precision, recall, F1-measure, and accuracy, which provide different perspectives on the classifier's effectiveness. Achieving high values for TP and TN and low values for FN and FP is the goal, indicating a highly accurate classifier. Understanding these concepts is essential for evaluating and improving the performance of binary classifiers in practical applications.

Some commonly used performance metrics in video classification include:

3.1 *Accuracy*

The proportion of videos correctly classified out of the total videos present in the dataset (Krstinić et al. 2020; Kapadia and Mehta 2023).

$$Accuracy = \frac{TP + TN}{TP + FP + FN + TN} \tag{1}$$

3.2 *Precision*

The ratio of correctly predicted videos (true positives) to the total number of predicted videos for a specific class (positive predictions)(Krstinić et al. 2020) (Kapadia and Mehta 2023).

$$Precision = \frac{TP}{TP + FP} \quad (2)$$

3.3 Recall

The proportion of correctly predicted positive instances to the overall count of actual positive videos within the dataset (Krstinić et al. 2020).

$$Recall = \frac{TP}{TP + FP} \quad (3)$$

3.4 F1-Score

The harmonic mean of precision and recall, which provides a balanced measure of both precision and recall (Krstinić et al. 2020).

$$F1 - Score = \frac{2 * (Recall * Precision)}{(Recall + Precision)} \quad (4)$$

3.5 Confusion Matrix

A table that provides a summary of the predicted class labels in comparison to the actual class labels., which can be used to calculate various performance metrics such as accuracy, precision, and recall (Pawan_Dubey, arvindpdmn 2023).

Table 2 presents a compilation of recent research that has employed these metrics.

4 Applications of Video Classification

Video classification involves automatically classifying video content into predetermined categories using visual and/or audio attributes. Its applications span diverse fields like entertainment, surveillance, healthcare, education, and social media.

Surveillance (Amrutha and Jyotsna 2021): Video classification is used in security and surveillance systems to detect and track objects and people, identify suspicious behavior, and alert security personnel in real-time. This is done by analyzing the video

Table 2 Evaluation metric

Evaluation metric	Reference	Year of publication
Accuracy	Vrskova et al. (2020)	2020
	Vrskova et al. (2022)	2022
	Benzyane et al. (2022)	2023
Precision	Vrskova et al. (2020)	2020
	Vrskova et al. (2022)	2022
	Benzyane et al. (2022)	2023
Recall	Vrskova et al. (2022)	2022
	Benzyane et al. (2022)	2023
F1 score	Vrskova et al. (2022)	2022
	Benzyane et al. (2022)	2023
Confusion matrix	Vrskova et al. (2020)	2020
	Benzyane et al. (2022)	2023

stream and identifying objects and people based on their appearance and motion patterns. Video classification is also used to recognize and track specific individuals or objects over time, making it easier to identify potential threats or criminal activity.

Healthcare (Chen et al. 2023; Fadil et al.): Video classification is used in medical imaging to diagnose and detect diseases such as lung cancer. This is done by analyzing medical images and extracting features such as the shape and texture of the lung nodules. Video classification is also used in telemedicine to monitor patient symptoms and diagnose diseases remotely. This is done by analyzing video of the patient and identifying symptoms such as shortness of breath, coughing, and wheezing.

Social Media (Trzcinski 2018): Video classification is used to detect and remove hate speech, nudity, and other inappropriate content on social media platforms. This is done by analyzing the content of the videos and identifying inappropriate or offensive language and images. Video classification is also used to classify and tag user-generated content for easy search and retrieval, making it easier for users to find and share content.

Education (Gautam et al. 2023): Video classification is used to classify and tag educational videos, making it easier for students to search and find the content they need. This is done by analyzing the content of the videos and extracting features such as the topic, the difficulty level, and the intended audience. Video classification is also used to track student engagement and provide personalized learning experiences. This is done by analyzing video of the student and identifying their level of engagement and attention.

5 Video Classification Models

There are many different video classification models, and the choice of model depends on the application and the specific requirements of the task. Some popular video classification models include:

5.1 3D CNNs

3D CNNs represent advanced deep-learning models utilized in video classification (Muhammad et al. 2021). They possess the capability to simultaneously analyze multiple frames of a video, enabling the extraction of both spatial and temporal features from the video data. This distinctive trait facilitates their proficiency in comprehending the intricacies of video content (Studying Employee Absenteeism Due to Health-Related Factors: A Data-Science Approach—ProQuest 2023; Muhammad et al. 2021).

Over the past several years, utilizing neural network architecture that is founded on 3D convolution layers has become a commonly utilized technique for classifying video data. This method is especially adept at examining the temporal alignment of objects within dynamic 3D visuals, such as medical images (Vrskova et al. 2022). The architecture of 3DCNN produces a 3D activation map as a result of the convolution process, which holds significant importance not just in terms of data analysis, but also for considering temporal and volumetric context. With the aim of determining the low-level representation of elements, a three-dimensional filter is employed to execute 3D convolution across the dataset. The kernel moves along three axes (x, y, z), as depicted in Fig. 1 (Vrskova et al. 2022).

The value at each position within the layer's feature map is determined using a specific equation (Vrskova et al. 2022).

$$v_{ij}^{xyz} = \tanh < b_{ij} + \sum_{m} \sum_{p=0}^{P_i-1} \sum_{q=0}^{Q_i-1} \sum_{r=0}^{R_i-1} w_{ijm}^{pqr} v_{\langle i-1 \rangle m}^{\langle x+p \rangle \langle y+q \rangle \langle z+r \rangle}$$

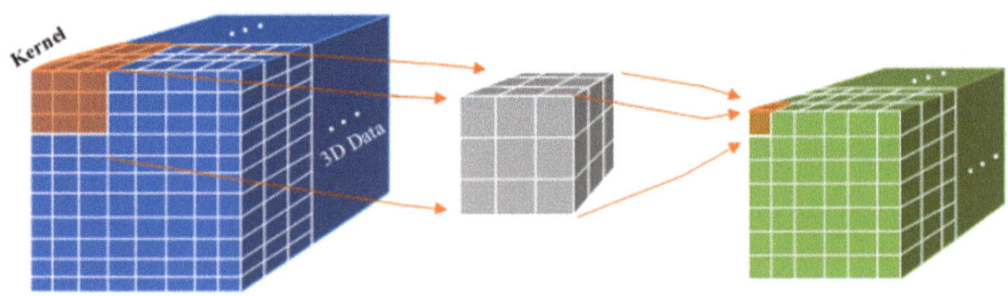

Fig. 1 Convolution operation in the 3DCNN architecture (Vrskova et al. 2022)

where R_i is the size of the 3Dkernel and w_{ijm}^{pqr} is the value of the kernel linked to the feature map in the preceding layer (Vrskova et al. 2022).

Recently, there has been a notable increase in the adoption of 3DCNN-based approaches, which harness 3D convolutional layers to extract features from video data (Liu et al. 2009). Hence, we utilized the three-dimensional (3D) convolution layer within the 3DCNN architecture. This 3D convolution layer functions by utilizing a moving 3D convolution window across the input data, enabling the identification of diverse filters. Each filter is dedicated to capturing distinctive patterns (Krstinić et al. 2020) (Fadil et al.). Importantly, these 3D filters are employed across all three dimensions, adeptly encompassing both spatial and temporal features present in the video data (Vrskova et al. 2022).

5.2 CNN with LSTM's (LRCN)

CNNs (Convolutional Neural Networks) are commonly used in video classification because of their ability to analyze spatiotemporal features. In video classification tasks, CNNs are typically applied to 3D data representations that incorporate both spatial and temporal dimensions (Ratan 2023). This methods is commonly used in various domains including security (Sciences and|Free Full-Text|FSCB-IDS: Feature Selection and Minority Class Balancing for Attacks Detection in VANETs, 2023; Khan et al. 2023), agriculture (IoT-Enabled Smart Agriculture: Security Issues and Applications|SpringerLink), energy (Machine Learning and Deep Learning Applications for Solar Radiation Predictions Review: Morocco as a Case of Study | SpringerLink 2023; System simulation and modulation of solar thermal domestic hot water production systems|SpringerLink 2023).

The typical CNNs architecture designed for video classification involves a series of convolutional layers, which work together to extract spatial features, Subsequently, there are pooling layers which serve to decrease the spatial dimensionality of the data (Ratan 2023). These layers are succeeded by one or more fully connected layers responsible for carrying out the classification process (Ratan 2023). CNN architectures as depicted in Fig. 2.

Long Short-Term Memory (LSTM) represents a form of recurrent neural network (RNN) architecture strategically crafted to tackle the challenge of vanishing gradients encountered in conventional RNNs (Onan 2023). This challenge often obstructs the network's ability to effectively grasp extended dependencies in sequential data (Onan 2023). The LSTM design incorporates a memory cell and a series of gates to manage the passage of information in and out of the cell (Muhammad et al. 2021; Onan 2023).

The memory cell allows the network to selectively store and retrieve information over long periods of time, while the gates regulate the amount of information that is allowed to flow through the cell (Muhammad et al. 2021). This makes LSTMs well-suited for tasks that involve sequential data with long-term dependencies, such

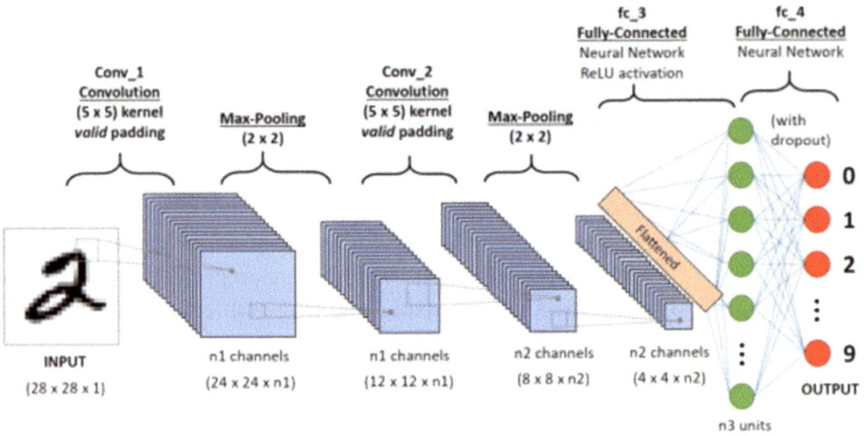

Fig. 2 Convolutional neural network architectures (Ratan 2023)

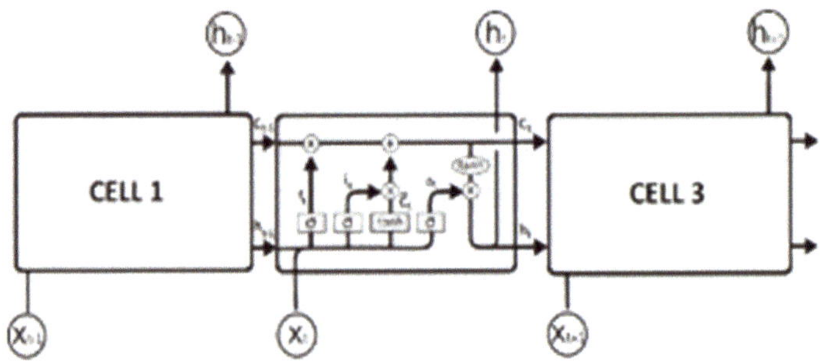

Fig. 3 Chained LSTM structure (Shivam et al. 2020)

as speech recognition, machine translation, and video analysis (Muhammad et al. 2021). The chained structure can be seen in Fig. 3.

5.3 ConvLSTM

ConvLSTM (Convolutional Long Short-Term Memory) is a variant of the LSTM recurrent neural network that introduces convolutional layers (Ye et al. 2019). It is widely employed in video classification tasks to effectively model both spatial and temporal dependencies within video sequences (Liu et al. 2009; Ye et al. 2019).

Within the ConvLSTM framework, the conventional fully-connected gates of LSTM are substituted with convolutional gates. This structural adjustment replaces the matrix multiplication processes with convolution operations at each gate (Ye et al. 2019). By doing so, ConvLSTM integrates spatial information into the LSTM

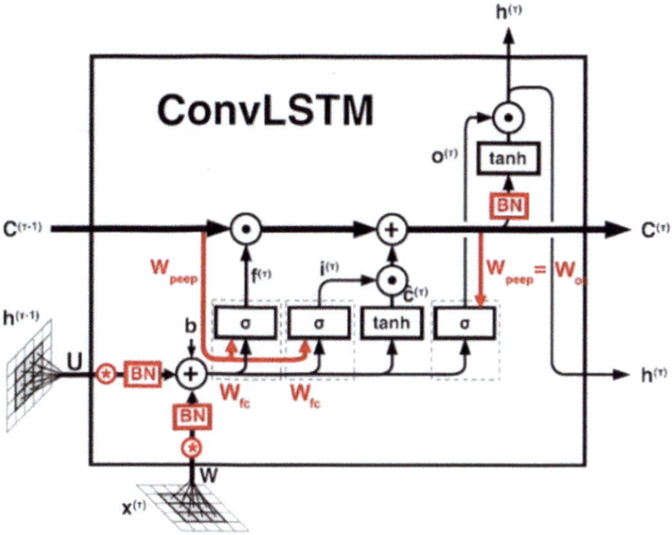

Fig. 4 Architecture of a ConvLSTM cell (Benzyane et al. 2022)

framework, allowing it to learn spatial representations and capture long-term dependencies more effectively. This combination of convolutional and LSTM layers makes ConvLSTM well-suited for video analysis tasks, as it can simultaneously handle spatial and temporal aspects of the data (Ye et al. 2019) (Fig. 4).

5.4 *Optical Flow and CNN's*

Optical flow constitutes a fundamental notion applied within the realm of video analysis and video classification endeavors. It denotes the observable movement pattern of objects across successive frames within a given video sequence (Akpınar and Alpaslan 2014). Optical flow provides information about the movement and motion patterns within a video (Akpınar and Alpaslan 2014).

In video classification, optical flow can be utilized as an additional input alongside the raw frames of the video. It captures the motion information between frames and helps in understanding the dynamic aspects of the video. By performing optical flow calculations, it becomes possible to ascertain both the direction and magnitude of movement for every pixel present within the image (Akpınar and Alpaslan 2014) (Fig. 5).

In this approach, a dual-stream architecture is employed, consisting of two parallel streams of convolutional networks. The top stream, known as the Spatial Stream, processes individual frames from the video. It applies a bunch of CNN kernels to extract spatial information and features from each frame. Based on these

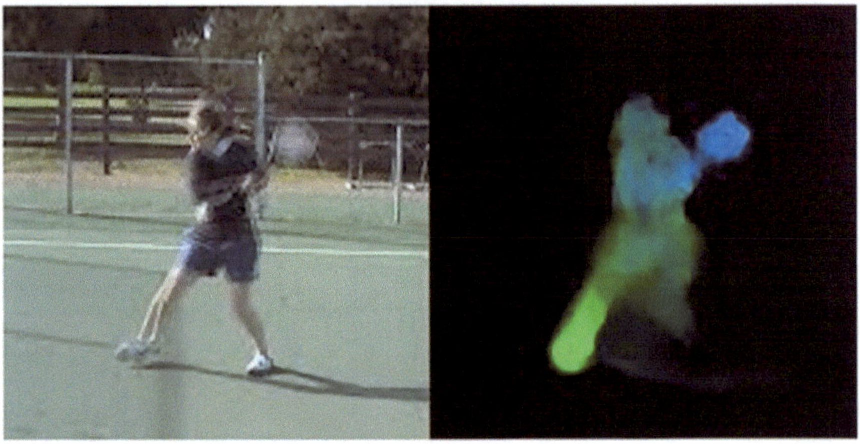

Fig. 5 Optical flow application (Brox and Malik 2011)

spatial features, the Spatial Stream makes predictions or performs classification tasks (Sargano et al. 2017).

6 Comparing Video Classification Approaches on UCF101 and HMDB51 Datasets

The performance of video classification algorithms can be significantly influenced by the quality and size of the dataset. Larger and more diverse datasets tend to yield better performance, as they offer the model a broader range of examples and enhance its ability to generalize across various classes. It's essential to understand that the effectiveness of video classification algorithms is significantly impacted by the quality and quantity of the dataset used (Table 3).

Table 3 Comparing the accuracy of video classification methods on UCF101 and HMDB51

Method	HMDB51 (%)	UCF101 (%)
Two-stream CNNs (Simonyan and Zisserman 2023)	59.4	88
LSTM (Ng et al. 2015)	–	88.6
ConvLSTM (Uzzaman et al. 2022; Savran Kızıltepe et al. 2021)	67.67	82.04
LRCN (Uzzaman et al. 2022; Donahue et al. 2016)	71.55	82.9

7 Conclusion

In conclusion, this article presents a thorough overview of deep learning techniques employed in the realm of video classification. Notable studies in the field have been summarized, emphasizing their key findings. The aim of reporting these findings is to support the research community in advancing the creation of novel deep learning models tailored specifically for the task of video classification.

References

Akpınar S, Alpaslan FN (2014) Video action recognition using an optical flow based representation

Amrutha CV, Jyotsna C (2021) A robust system for video classification: identification and tracking of suspicious individuals from surveillance videos. In: Reddy VS, Prasad VK, Wang J, Reddy KTV (eds) Soft computing and signal processing. Springer, Singapore, pp 263–271. https://doi.org/10.1007/978-981-33-6912-2_24

Applied Sciences|Free Full-Text|FSCB-IDS: Feature Selection and Minority Class Balancing for Attacks Detection in VANETs. https://www.mdpi.com/2076-3417/13/13/7488. Last accessed 18 Aug 2023

Benzyane M, Zeroual I, Azrour M, Agoujil S (2022) Convolutional long short-term memory network model for dynamic texture classification: a case study. In: International conference on advanced intelligent systems for sustainable development, J. Kacprzyk, M. Ezziyyani, V. E. Balas (eds) in Lecture Notes in Networks and Systems. Cham: Springer Nature Switzerland, 2023, pp 383–395. https://doi.org/10.1007/978-3-031-26384-2_33

Brox T, Malik J (2011) Large displacement optical flow: descriptor matching in variational motion estimation. IEEE Trans Pattern Anal Mach Intell 33:500–513. https://doi.org/10.1109/TPAMI.2010.143

Chen J, Wee L, Dekker A, Bermejo I (2023) Using 3D deep features from CT scans for cancer prognosis based on a video classification model: A multi-dataset feasibility study. Medical Physics. n/a. https://doi.org/10.1002/mp.16430

Donahue J, Hendricks LA, Rohrbach M, Venugopalan S, Guadarrama S, Saenko K, Darrell T (2016) Long-term recurrent convolutional networks for visual recognition and description. http://arxiv.org/abs/1411.4389

Fadil O., Lemhadri I, Zheng C Deep learning based endoscopy video classification intermediary report

Gautam A, Hazra S, Verma R, Maji P, Balabantaray BK (2023) ED-NET: Educational teaching video classification network. In: Tistarelli M, Dubey SR, Singh SK, Jiang X (eds) Computer vision and machine intelligence. Springer Nature, Singapore, pp. 141–151. https://doi.org/10.1007/978-981-19-7867-8_12

IoT-Enabled Smart Agriculture: Security Issues and Applications. SpringerLink, https://link.springer.com/chapter/https://doi.org/10.1007/978-3-031-26254-8_82. Last accessed 19 Aug 2023

Kapadia N, Mehta R (2023) Dynamic route optimization for IoT based intelligent waste collection vehicle routing system. Intell Decis Technol 17:751–772. https://doi.org/10.3233/IDT-230032

Khan S, Siddiqui T, Mourade A, Alabduallah BI, Alajlan SA, Almjally A, Albahlal BM, Alfaifi A (2023) Manufacturing industry based on dynamic soft sensors in integrated with feature representation and classification using fuzzy logic and deep learning architecture. Int J Adv Manuf Technol. https://doi.org/10.1007/s00170-023-11602-y

Krstinić D, Braović M, Šerić L, Božić-Štulić D (2020) Multi-label classifier performance evaluation with confusion matrix. In: Computer Science & Information Technology. AIRCC Publishing Corporation, pp. 01–14. https://doi.org/10.5121/csit.2020.100801

Liu J, Luo J, Shah M (2009) Recognizing realistic actions from videos "in the wild." In: 2009 IEEE conference on computer vision and pattern recognition. IEEE, Miami, FL, pp 1996–2003. https://doi.org/10.1109/CVPR.2009.5206744

Machine Learning and Deep Learning Applications for Solar Radiation Predictions Review: Morocco as a Case of Study|SpringerLink. https://link.springer.com/chapter/https://doi.org/10.1007/978-3-031-05258-3_6. Last accessed 19 Aug 2023

Muhammad K, Mustaqeem, Ullah A, Imran AS, Sajjad M, Kiran MS, Sannino G, de Albuquerque VHC (2021) Human action recognition using attention based LSTM network with dilated CNN features. Futur Gener Comput Syst 125:820–830. https://doi.org/10.1016/j.future.2021.06.045

Ng JY-H, Hausknecht M, Vijayanarasimhan S, Vinyals O, Monga R, Toderici G (2015) Beyond short snippets: deep networks for video classification. http://arxiv.org/abs/1503.08909

Onan A (2023) Leveraging synonyms and antonyms for data augmentation in sarcasm identification. In: Silhavy R, Silhavy P (eds) Artificial intelligence application in networks and systems. Springer International Publishing, Cham, pp 703–713. https://doi.org/10.1007/978-3-031-35314-7_59

Pawan_Dubey, arvindpdmn: Confusion Matrix. https://devopedia.org/confusion-matrix. Last accessed 18 Aug 2023

Ratan P What is the Convolutional neural network architecture? https://www.analyticsvidhya.com/blog/2020/10/what-is-the-convolutional-neural-network-architecture/. Last accessed 14 May 2023

Sargano AB, Angelov P, Habib Z (2017) A comprehensive review on handcrafted and learning-based action representation approaches for human activity recognition. Appl Sci 7:110. https://doi.org/10.3390/app7010110

Savran Kızıltepe R, Gan JQ, Escobar JJ (2021) A novel keyframe extraction method for video classification using deep neural networks. Neural Comput & Applic. https://doi.org/10.1007/s00521-021-06322-x

Shen J, Huang Y, Wen M, Zhang C (2019) Towards an efficient deep pipelined template-based architecture for accelerating the entire 2D and 3D CNNs on FPGA. IEEE Trans Comput Aided Des Integr Circuits Syst. https://doi.org/10.1109/TCAD.2019.2912894

Shivam K, Tzou J-C, Wu S-C (2020) Multi-step short-term wind speed prediction using a residual dilated causal convolutional network with nonlinear attention. Energies 13:1772. https://doi.org/10.3390/en13071772

Simonyan K, Zisserman A Two-stream convolutional networks for action recognition in videos. https://arxiv.org/abs/1406.2199v2. Last accessed 03 April 2023

Study, simulation and modulation of solar thermal domestic hot water production systems|SpringerLink. https://link.springer.com/article/https://doi.org/10.1007/s40808-021-01200-w. Last accessed 19 Aug 2023

Studying Employee Absenteeism Due to Health-Related Factors: A Data-Science Approach—ProQuest. https://www.proquest.com/openview/8061a8ec5e2a302425ff1ded2df659f6/1?pq-origsite=gscholar&cbl=18750&diss=y. Last accessed 18 July 2023

Trzcinski T (2018) Multimodal social media video classification with deep neural networks. In: Photonics applications in astronomy, communications, industry, and high-energy physics experiments 2018. SPIE, pp. 879–886. https://doi.org/10.1117/12.2501679

Uzzaman MS, Debnath C, Uddin DMA, Islam MM, Talukder MA, Parvez S (2022) LRCN based human activity recognition from video data. https://papers.ssrn.com/abstract=4173741. https://doi.org/10.2139/ssrn.4173741

Vrskova R, Hudec R, Kamencay P, Sykora P (2022) Human activity classification using the 3DCNN architecture. Appl Sci 12:931. https://doi.org/10.3390/app12020931

Vrskova R, Hudec R, Sykora P, Kamencay P, Radilova M (2020) Education of video classification based by neural networks. In: 2020 18th International Conference on Emerging eLearning Technologies and Applications (ICETA). IEEE, Košice, Slovenia, pp 762–767. https://doi.org/10.1109/ICETA51985.2020.9379190

Ye W, Cheng J, Yang F, Xu Y (2019) Two-stream convolutional network for improving activity recognition using convolutional long short-term memory networks. IEEE Access. 7:67772–67780. https://doi.org/10.1109/ACCESS.2019.2918808

Intelligent Real-Time Monitoring System for Wastewater Management Using Artificial Neural Network

Fouad Essahlaoui, Nourddine Elhajrat, Mohammed Halimi, Mourade Azrour, Zeyneb Kılıç, and Ahmed El Abbassi

Abstract The importance of real-time wastewater quality monitoring in addressing the increasing demand for water and the impact of climate change on wastewater treatment systems is discussed in this work. It highlights the potential of Artificial Intelligence (AI)-based solutions, specifically Artificial Neural Networks (ANNs), to accurately monitor water quality and manage wastewater effectively. Emphasis is placed on the integration of ANNs with the Kalman filter, a technique renowned for enhancing accuracy. This work presents the architecture of a wireless sensor network-based system designed to monitor wastewater quality. Sensor nodes strategically positioned and equipped with microcontrollers, sensors, and transceivers collect data, which undergoes processing using the Kalman filter to estimate and rectify any inaccuracies before transmission to the base station. The system employs multi-sensor data collection and ANN analysis to improve the accuracy and reliability of water quality monitoring. The system's performance is evaluated in terms of recognition and false alarm rates. It achieves a high recognition rate of 95.60% and a low false alarm rate, significantly increasing wastewater utilization efficiency. The ANN decision-making process is efficient, taking only 15 μs. The evaluation also

F. Essahlaoui (✉) · A. E. Abbassi
LRETTI: Renewable Energies, Treatment and Transmission of Information Laboratory, Faculty of Sciences and Techniques, Moulay Ismail University of Meknes, Meknes, Morocco
e-mail: f.essahlaoui@edu.umi.ac.ma

N. Elhajrat
LOTEA: Optoelectronics and Applied Energetic Techniques Laboratory, Faculty of Sciences and Techniques, Moulay Ismail University of Meknes, Meknes, Morocco

M. Halimi
OETA: Optoelectronics and Energetic Techniques Applied Team, Faculty of Sciences and Techniques, Moulay Ismail University of Meknes, Meknes, Morocco

M. Azrour
STI Laboratory, IDMS Team, Faculty of Sciences and Techniques, Moulay Ismail University of Meknes, Meknes, Morocco

Z. Kılıç
Engineering Faculty, Civil Engineering Hydraulic Department, Adıyaman University, Adıyaman, Türkiye

M. Azrour et al. (eds.), *Sustainable and Green Technologies for Water and Environmental Management*, World Sustainability Series,
https://doi.org/10.1007/978-3-031-52419-6_2

includes the F1 score, demonstrating the system's classification abilities and overall performance. In conclusion, the proposed intelligent system, which integrates ANN and the Kalman filter, represents a notable advancement in wastewater management. It offers superior performance, accurate classification, and rapid response, making it a practical and effective solution for the management of water resources.

Keywords Wastewater monitoring · Artificial neural networks · Kalman filter · Water quality management · Real-time

1 Introduction

The increasing water demand greatly influences the management of wastewater. A study published in (Elbasiouny et al. 2023; Lima et al. 2021) predicts that nearly 6 billion people could face water scarcity by 2050 due to factors such as population growth, economic development, and shifts in consumption patterns. Furthermore, another study emphasizes the considerable impact of climate change on wastewater treatment systems (Mohy-Eddine et al. 2023; Mabrouki et al. 2023, 2022, 2021a, 2021b; Fattah et al. 2023; Anouzla et al. 2022). To address this problem, real-time wastewater quality monitoring is suggested as a potential solution, as mentioned in a study by Xuexiu Jia and Fouad (Jia et al. 2022). Artificial intelligence (AI)-based solutions can enable real-time water quality monitoring, leak detection, and source identification, leading to more effective wastewater management while reducing environmental and health risks. The deep autoencoder-based approach to anomaly identification is suggested in the work by Russo et al. (2020). From the monitoring data of wastewater treatment systems, the authors employ an autoencoder to extract the key features. The ability to identify anomalies in the challenging time series data of wastewater treatment system monitoring is demonstrated by this method.

On the other hands, Artificial Neural Networks (ANNs) is an extremely successful scientific discipline, and its latest advancement and development have earned AI widespread recognition in a variety of sectors like security (Mohy-eddine et al. 2023, 2022; Hazman et al. 2023a, 2023b, 2023c, 2022; Dargaoui et al. 2023; Attou et al. 2023; Amaouche et al. 2023; Guezzaz et al. 2022a, 2022b; Douiba et al. 2022a, 2022b; Chaganti et al. 2022), energy (Mabrouki et al. 2021a; Hissou et al. 2023; Boutahir et al. 2022a, 2022b), and so ones. The aim of this technology is to train machines in order to learn from failures and accomplish new tasks more effectively in difficult situations.

The monitoring of effluent quality in wastewater treatment plants is the main topic of the study by Sharghi et al. (2019). Using various data clustering techniques, they create three models of artificial neural networks. The best results come from Model, which uses the self-organizing map (SOM) clustering technique. This method shows how artificial neural networks and clustering techniques can improve effluent quality monitoring in wastewater treatment facilities. Integrating ANNs is a successful approach for wastewater quality classification, as indicated in a study published in PubMed (Zhang et al. 2023). ANN models can be trained with historical data to categorize wastewater quality based on different parameters. However, including another powerful technique, the Kalman filter, is essential to improve the accuracy of water quality assessments. In relation to the paper "Contribution à la modélisation de la qualité des eaux par réseaux de neurones" published in (Heddam 2006; Essahlaoui et al. 2017), ANN can be used for real-time water quality monitoring. Software sensors for water quality monitoring can be developed using artificial neural networks. The effectiveness of ANN and Extreme Learning Machines (ELM) approaches has been evaluated in developing these software sensors.

The efficacy and efficiency of ANN in evaluating wastewater quality have been demonstrated in various studies, as highlighted in (Sharghi et al. 2019; Dadar et al. 2020; Essahlaoui et al. 2020). Using the capabilities of artificial neural networks, the monitoring system can analyse data from multiple sensors and make informed decisions about wastewater quality. This combination of multisensor data collection and artificial neural network processing improves the accuracy and reliability of the monitoring system, providing valuable insights for efficient water management in agricultural settings. AI-driven systems can detect changes in water quality, alert authorities to potential sources of water pollution, and identify water leaks, among other issues, as emphasized in (Bui et al. 2021; Hejabi et al. 2021). Using AI for water quality monitoring, it is possible to improve the effectiveness of water management while minimising associated environmental and health risks.

By combining the Kalman filter with real-time measurements of various water quality parameters and the estimations provided by ANN, it is possible to accurately assess water quality at any given time. Considering both ANN predictions and noisy sensor measurements, the Kalman filter improves the estimation and prediction of water quality, enabling more accurate monitoring and faster response to changes and potential problems. The use of the Kalman filter in conjunction with ANN in wastewater management offers the opportunity to monitor and analyze water quality in real-time more accurately. This strategy enables more effective wastewater management, reduces environmental and public health risks, and meets the growing demand for water.

2 Overall System Architecture

The system's overall architecture is based on the deployment and integration of various hardware and software components for a wireless sensor network. The system was specifically developed using a wireless sensor network to monitor wastewater quality. Sensor nodes are strategically positioned near the treatment tank outlet and along the pipelines to agricultural fields. Each sensor node consists of a platform equipped with a microcontroller, sensors, and a radiofrequency transceiver. Additionally, the sensor nodes are equipped with a Kalman filter to estimate and correct any inaccuracies in the data before transmitting it to the base station, also known as the "sink." The sink node serves as a coordinator and/or gateway using the RF 433 module, while the other sensor nodes use modules as routers. This collaborative operation of the RF modules creates a unified network topology using optimal routing protocols for efficient data detection.

As shown in Fig. 1, the system employs a multisensor approach combined with artificial neural networks with Kalman filtering for wastewater monitoring. The integration of ANN has consistently demonstrated the effectiveness and efficiency of this non-parametric technology for assessing wastewater quality. Using the capabilities of artificial neural networks, the monitoring system can analyse data collected from multiple sensors and make informed decisions about wastewater quality. This combination of multisensor data collection and analysis by ANNs enhances the accuracy and reliability of the monitoring system, providing valuable insights for effective water management in agricultural environments.

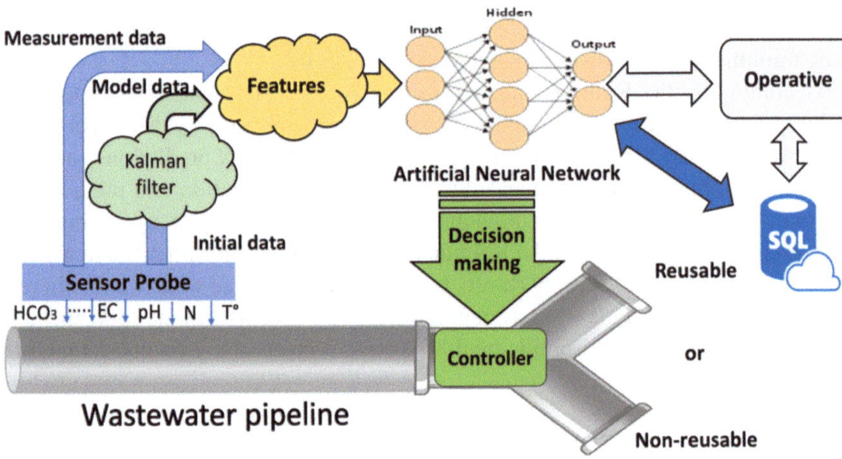

Fig. 1 Block diagram of the multi-sensors wastewater monitoring system

To facilitate remote configuration and management of sensor networks, a gateway application has been developed. It enables monitoring of dispersed sensor node performance and data retrieval. The data undergo a preprocessing phase and are fed into an artificial neural network decision-making system for learning and decision-making processes before being stored on the server. Additionally, on the base station, an application has been created using the Apache web server and embedded Linux operating system to allow external access to sensor nodes and data as quickly as possible, all on a cost-effective and simplified deployment platform (Raspberry Pi). A Web application has been developed to provide users with a simple means of interacting with the system, allowing them to remotely monitor sensor data and configure deployed sensor nodes via the local network or any device connected to the Internet.

3 Results and Discussion

The system architecture involves deploying and integrating various hardware and software components for a wireless sensor network. A gateway application enables remote configuration and management of the sensor network, allowing monitoring of dispersed sensor node performance and data retrieval. The collected data undergoes preprocessing and is fed into an artificial neural network decision-making system. The resulting decisions are stored on the server. An example of the database collected by the sensor network deployed in their system is represented in Table 1. Include decisions made by experts to assess wastewater quality. This valuable information is incorporated into the system during the learning phase, enabling informed decision-making processes that converge or diverge from expert decisions based on validation and testing modes. Using this extensive database significantly enhances the system's efficiency in analysing and evaluating water quality. Additionally, a web application on the base station, using the Apache Web server and the embedded Linux operating system, allows external access to sensor nodes and data, facilitating remote monitoring and configuration. Real descriptor vectors from a database are utilized as input for the neural network, and the decision outcomes from the database and simulation are represented in Fig. 2. The blue graph in Fig. 1 shows the decisions made by the new artificial neural network-controlled surveillance system during a 350-min period, while the green portion indicates the previous decision state. The system achieves a recognition rate of 95.60%, with a low percentage of false alarms. This improvement leads to a 96% increase in wastewater utilization rate, significantly enhancing the efficiency of the wastewater treatment plant. The decision-making process of the network takes only 15 μs, emphasizing its efficiency. Further modifications are necessary to address sensitivity to location and unforeseen threats, reducing associated risks. The dynamic processing capability of the multi-sensor system in irrigation channels plays a vital role in effective water management.

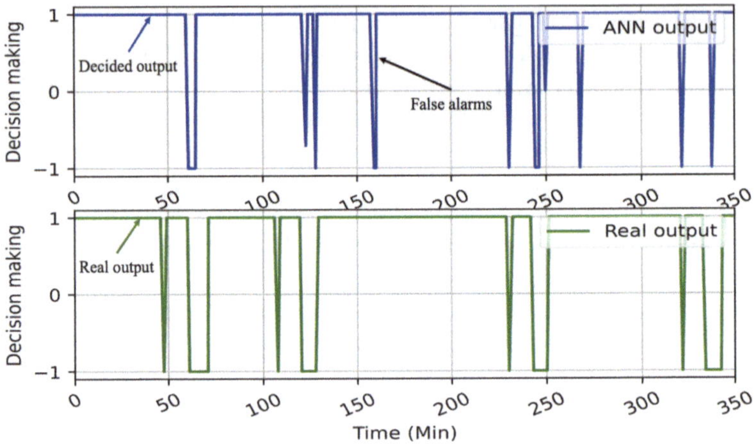

Fig. 2 Graphs comparing real and ANN decision-making

A thorough system performance evaluation was performed using the F1 score to quantify the recognition rate and the false alarm rate. The F1 score was selected as a criterion to assess the precision and efficiency of the system for classifying wastewater. Recall gauges how well the system can locate all pertinent instances, while precision gauges the percentage of correctly categorised occurrences. The F1 score, which accounts for false positives and false negatives, gives a balanced measurement of precision and recall, yielding a 96.70% score. These metrics allowed for a thorough analysis of the system's categorization abilities and overall performance.

In conclusion, based on an artificial neural network and a Kalman filter, the proposed intelligent system represents a significant advancement in controlling wastewater irrigation systems. It offers superior performance, accurate classification, and rapid response, making it a practical and effective solution for the management of water resources. The system is based on an artificial neural network (ANN), which uses machine learning to make decisions and adapt to new configurations. Once the ANN has trained, it can make real-time decisions using newly provided input data .

This means that the system can easily adjust to new configurations or environments by utilizing the knowledge gained during training. When new sensors are added to the system, the data from these sensors can be used as inputs to the ANN. By adjusting the weights and parameters of the network, the ANN can learn to incorporate the additional information provided by the new sensors. Furthermore, the use of the Kalman filter enables the system to utilize less sensitive sensors, as the filter can resolve this issue along with problems related to missing information. This not only addresses the sensitivity concern but also contributes to cost reduction, making the overall system more economically viable. As a result, the system has the potential to be widely applicable and can be deployed in various operational environments without significant modifications, making it a practical solution for the management of water resources.

4 Conclusion

The study emphasizes the significance of real-time wastewater quality monitoring in addressing water demand and climate change impacts. It highlights the potential of AI-based solutions, particularly artificial neural networks (ANNs) and the Kalman filter, as a significant advance in wastewater management. These solutions offer superior performance, accurate classification, and rapid response, making them practical and effective for water resource management. The system's adaptability to new configurations and environments through machine learning, as well as its potential for widespread application, make it a valuable tool for decision makers and practitioners in wastewater management. It enables accurate water quality monitoring and effective wastewater management. The system consists of a wireless sensor network-based architecture, with sensor nodes equipped with microcontrollers, sensors, and transceivers that collect data. The collected data are processed using a Kalman filter to estimate and correct inaccuracies before being transmitted to the base station. The system utilizes multi-sensor data collection and ANN analysis to enhance the accuracy and reliability of water quality monitoring. It achieves a high recognition rate of 95.60%, and the F1 score evaluation demonstrates the system's classification abilities and overall performance of approximately 96.70%, resulting in increased wastewater utilization efficiency. The ANN decision making process is efficient, taking only 15 μs (Table 1).

Table 1 A brief illustration of the collected data from a Moroccan wastewater treatment plant

Turbidity (NTU)	Trihalomethanes (µg/L)	Organic carbon (ppm)	Conductivity (µS/cm)	Sulfate (mg/L)	Chloramines (ppm)	Solids (ppm	Hardness (mg/L)	pH	Reusability
3,70	65,34	17,16	522,79	339,05	6,5	28544,2	208,09	7,45	0
3,98	73,76	14,72	451,14	300,12	8,02	19859,8	128,1	9,02	1
3,93	66,73	16,56	411,3	378,53	5,92	22833,1	252,22	4,78	0
3,89	77,58	8,65	592,66	310,56	9,44	13168,55	145,83	9,45	1
3,2	32,14	16,32	347,98	355,48	6,16	27948,56	216,73	9,42	0
4,1	68,47	16,84	351,19	325,25	7,21	20418,86	187,37	7,81	0
2,75	59,45	14,18	512,43	323,23	9,22	27357,44	98,74	9,83	1
3,43	70,54	15,59	297,61	374,22	6,13	26849,14	280,02	5,2	1
4,13	99,12	11,29	503,17	333,38	4,27	15156,75	265,04	9,38	0
3,56	62,51	13,16	417,84	308,42	8,86	15976,92	215,24	6,2	1
4,38	77,63	15,56	481,48	278,71	7,46	16392,12	217,65	4,82	1

References

Amaouche S, Guezzaz A, Benkirane S, et al (2023) FSCB-IDS: Feature selection and minority class balancing for attacks detection in VANETS. Applied sciences

Anouzla A, Kastali M, Azoulay K, et al (2022) Multi-response optimization of coagulation–flocculation process for stabilized landfill leachate treatment using a coagulant based on an industrial effluent. Desalination Water Treat 10

Attou H, Guezzaz A, Benkirane S et al (2023) Cloud-based intrusion detection approach using machine learning techniques. Big Data Min Anal 6:311–320

Boutahir MK, Farhaoui Y, Azrour M, et al (2022a) Effect of feature selection on the prediction of direct normal irradiance. Big Data Min Anal 5:309–317. https://doi.org/10.26599/BDMA.2022.9020003

Boutahir MK, Farhaoui Y, Azrour M (2022b) Machine learning and deep learning applications for solar radiation predictions review: morocco as a case of study. In: Digital economy, business analytics, and big data analytics applications. Springer, pp 55–67

Bui HM, Bui HN, Le TM, Karri RR (2021) Chapter 5—Application of artificial neural networks on water and wastewater prediction: a review. In: Karri RR, Ravindran G, Dehghani MH (eds) Soft computing techniques in solid waste and wastewater management. Elsevier, pp 95–109

Chaganti R, Azrour M, Vinayakumar R, et al (2022) A particle swarm optimization and deep learning approach for intrusion detection system in internet of medical things. Sustainability 14:12828

Dadar S, Pezeshki A, Đurin B, Dogančić D (2020) Qualitative evaluation of wastewater treatment plant performance by a neural network model optimized by genetic algorithm. 1. https://doi.org/10.3390/ECWS-5-08047

Dargaoui S, Azrour M, El Allaoui A et al (2023) An Overview of the Security Challenges in IoT Environment. In: Mabrouki J, Mourade A, Irshad A, Chaudhry SA (eds) Advanced technology for smart environment and energy. Springer International Publishing, Cham, pp 151–160

Douiba M, Benkirane S, Guezzaz A, Azrour M (2022a) Anomaly detection model based on gradient boosting and decision tree for IoT environments security. J Reliab Intell Environ:1–12

Douiba M, Benkirane S, Guezzaz A, Azrour M (2022b) An improved anomaly detection model for IoT security using decision tree and gradient boosting. J Supercomput:1–20

Elbasiouny H, El-Ramady H, Elbehiry F (2023) Sustainable and green management of wastewater under climate change conditions. In: Nasr M, Negm AM (eds) Cost-efficient wastewater treatment technologies: engineered systems. Springer International Publishing, Cham, pp 443–461

Essahlaoui F, Elhajrat N, Abbassi AE et al (2020) Modeling design and implementation of an embeds system real time over a network of wireless sensors to environmental monitoring. Mod Appl Sci 14:10. https://doi.org/10.5539/mas.v14n1p41

Essahlaoui F, Ouadoudi N, Abbassi A, Skouri R (2017) Emulated artificial arduino neural network

Fattah G, Elouardi M, Benchrifa M, et al (2023) Modeling of the coagulation system for treatment of real water rejects. In: Advanced technology for smart environment and energy. Springer, pp 161–171

Guezzaz A, Benkirane S, Azrour M (2022a) A novel anomaly network intrusion detection system for internet of things security. In: IoT and smart devices for sustainable environment. Springer, pp 129–138

Guezzaz A, Azrour M, Benkirane S, et al (2022b) A lightweight hybrid intrusion detection framework using machine learning for edge-based IIoT security. Int Arab J Inf Technol 19

Hazman C, Guezzaz A, Benkirane S, Azrour M (2022) lIDS-SIoEL: intrusion detection framework for IoT-based smart environments security using ensemble learning. Clust Comput:1–15

Hazman C, Guezzaz A, Benkirane S, Azrour M (2023a) Toward an intrusion detection model for IoT-based smart environments. Multimed Tools Appl. https://doi.org/10.1007/s11042-023-16436-0

Hazman C, Benkirane S, Guezzaz A, et al (2023b) Intrusion detection framework for IoT-based smart environments security. In: Artificial intelligence and smart environment: ICAISE'2022. Springer, pp 546–552

Hazman C, Benkirane S, Guezzaz A, et al (2023c) Building an intelligent anomaly detection model with ensemble learning for IoT-based smart cities. In: Advanced technology for smart environment and energy. Springer, pp 287–299

Heddam S (2006) Contribution à la modélisation de la qualité des eaux par les réseaux des neurones. Thesis, INA

Hejabi N, Saghebian SM, Aalami MT, Nourani V (2021) Evaluation of the effluent quality parameters of wastewater treatment plant based on uncertainty analysis and post-processing approaches (case study). Water Sci Technol 83:1633–1648. https://doi.org/10.2166/wst.2021.067

Hissou H, Benkirane S, Guezzaz A et al (2023) A novel machine learning approach for solar radiation estimation. Sustainability 15:10609. https://doi.org/10.3390/su151310609

Jia X, Shahzad K, Klemeš JJ, Jia X (2022) Changes in water use and wastewater generation influenced by the COVID-19 pandemic: A case study of China. J Environ Manage 314:115024. https://doi.org/10.1016/j.jenvman.2022.115024

Lima A, Abreu T, Figueiredo S (2021) Water and wastewater optimization in a food processing industry using water pinch technology. Sustain Water Resour Manag 7:82. https://doi.org/10.1007/s40899-021-00560-6

Mabrouki J, Fattah G, Al-Jadabi N et al (2021a) Study, simulation and modulation of solar thermal domestic hot water production systems. Model Earth Syst Environ. https://doi.org/10.1007/s40808-021-01200-w

Mabrouki J, Azrour M, Hajjaji SE (2021b) Use of internet of things for monitoring and evaluating water's quality: a comparative study. Int J Cloud Comput 10:633–644

Mabrouki J, Fattah G, Kherraf S, et al (2022) Artificial intelligence system for intelligent monitoring and management of water treatment plants. In: Emerging Real-World Applications of Internet of Things. CRC Press, pp 69–87

Mabrouki J, Benchrifa M, Ennouhi M, et al (2023) Geographic information system for the study of water resources in Chaâba El Hamra, Mohammedia (Morocco). In: Artificial intelligence and smart environment: ICAISE'2022. Springer, pp 469–474

Mohy-eddine M, Guezzaz A, Benkirane S, Azrour M (2023) An efficient network intrusion detection model for IoT security using K-NN classifier and feature selection. Multimed Tools Appl. https://doi.org/10.1007/s11042-023-14795-2

Mohy-Eddine M, Azrour M, Mabrouki J et al (2023) Embedded web server implementation for real-time water monitoring. In: Mabrouki J, Mourade A, Irshad A, Chaudhry SA (eds) Advanced Technology for Smart Environment and Energy. Springer International Publishing, Cham, pp 301–311

Mohy-eddine M, Guezzaz A, Benkirane S, Azrour M (2022) An effective intrusion detection approach based on ensemble learning for IIoT edge computing. J Comput Virol Hacking Tech:1–13

Russo S, Disch A, Blumensaat F, Villez K (2020) Anomaly Detection using deep autoencoders for in-situ wastewater systems monitoring data

Sharghi E, Nourani V, Ashrafi AA, Gökçekuş H (2019) Monitoring effluent quality of wastewater treatment plant by clustering based artificial neural network method. DWT 164:86–97. https://doi.org/10.5004/dwt.2019.24385

Zhang S, Jin Y, Chen W et al (2023) Artificial intelligence in wastewater treatment: A data-driven analysis of status and trends. Chemosphere 336:139163. https://doi.org/10.1016/j.chemosphere.2023.139163

Fumes and Smoke Car Detection Using YOLOv8

Ali Omari Alaoui, Omaima El Bahi, and Ahmad El Allaoui

Abstract Reducing vehicle emissions is crucial for achieving environmental sustainability and mitigating the detrimental impact of air pollution. This scientific article explores the utilization of the YOLOv8 object detection algorithm to detect and mitigate smoke emissions from cars, contributing to the overall reduction of vehicle emissions. By training the YOLOv8 model using a dataset obtained from RoboFlow, consisting of annotated images of smoke-emitting cars, a robust and accurate detection system is developed. The article delves into the process of dataset preparation, including the annotation of images and the creation of corresponding text files with bounding box coordinates around smoke-emitting vehicles. Through iterative model training, which involves optimizing the model's parameters and loss function using backpropagation, the YOLOv8 model becomes adept at detecting and classifying smoke-emitting cars. Once trained, the model can be deployed in real-time scenarios to identify vehicles emitting excessive smoke. Upon detection, measures can be implemented to minimize the smoke percentage in the air. These actions may encompass notifying the driver, suggesting maintenance actions, or even automatically adjusting factors such as vehicle speed and fuel mixture. The integration of fumes and smoke car detection using YOLOv8 into sustainable technologies offers several advantages, including proactive identification of smoke-emitting vehicles and targeted interventions to reduce emissions. This approach fosters responsible vehicle maintenance practices and encourages eco-friendly behaviors among drivers, ultimately contributing to a greener and healthier environment. By embracing these

A. Omari Alaoui (✉) · A. El Allaoui
Faculty of Sciences and Techniques, L-STI, T-IDMS, Moulay Ismail University of Meknes, Meknes, Morocco
e-mail: al.omarialaoui@edu.umi.ac.ma

A. El Allaoui
e-mail: a.elallaoui@umi.ac.ma

O. El Bahi
Faculty of Sciences and Techniques, L-MAIS, T-MIS, Moulay Ismail University of Meknes, Meknes, Morocco
e-mail: omaima.elbahi@edu.umi.ac.ma

M. Azrour et al. (eds.), *Sustainable and Green Technologies for Water and Environmental Management*, World Sustainability Series,
https://doi.org/10.1007/978-3-031-52419-6_3

advancements and reducing the smoke percentage in the air, we can take significant strides towards a more sustainable future for all.

Keywords Fumes and smock detection · Image classification · Computer vision · YOLOv8

1 Introduction

As concerns over environmental sustainability and the impact of vehicle emissions continue to grow, innovative approaches are being sought to reduce air pollution and promote greener technologies. Among the significant contributors to air pollution are fumes and smoke emitted by cars, which not only harm the environment but also pose a threat to public health. In response to this pressing issue, advanced computer vision techniques have emerged as a promising solution to detect and mitigate vehicle emissions. This article focuses on the application of the YOLOv8 object detection algorithm (Jocher et al. 2023), in conjunction with a dataset from universe.roboflow.com, to address the challenge of fumes and smoke car detection. By leveraging YOLOv8's exceptional capabilities in real-time object detection, we can identify vehicles emitting excessive smoke and implement measures to minimize the smoke percentage in the air. This sustainable approach not only helps combat air pollution but also promotes green technologies by encouraging responsible vehicle maintenance and reducing harmful emissions. By exploring the intricacies of training the YOLOv8 model on a smoke-emitting car dataset and discussing the potential actions to repair smoke-emitting vehicles, we delve into a vital aspect of achieving a cleaner and more environmentally friendly future.

2 Related Works

A significant body of research has been conducted on the detection of smoke and vehicles, employing diverse methodologies and techniques. Tao and Lu have made notable contributions by proposing multi-feature fusion and ensemble neural networks (Tao and Lu 2018a), a three-stage framework for smoky vehicle detection (Tao et al. 2020) and leveraging features such as multi-scale block Tamura features (Tao and Lu 2018b), range filtering on three orthogonal planes, and motion orientation histogram (Tao and Lu 2018c). Peng et al. have developed a sophisticated coarse-to-fine framework for video-based smoky vehicle detection (Peng et al. 2022) and introduced innovative cascaded vehicle matching and short-term spatial–temporal networks (Peng et al. 2023). In the domain of video-based smoke detection, Islam et al. have focused on smoke object segmentation and the dynamic growth feature model (Islam et al. 2020). Li et al. have presented a range of approaches, including united video dehazing and detection (Li et al. 2017a), an all-in-one network for dehazing and beyond (Li et al. 2017b), and vehicle emission detection through deep learning models (Tao and Lu 2020).

In the realm of smoke detection, Chen and Peng proposed DB-Net, a network specifically designed for detecting vehicle smoke using deep block networks (Chen and Peng 2023). Cao and Lu delved into the exploration of learning spatial–temporal representation for smoke vehicle detection (Cao and Lu 2019). Frizzi et al. made significant strides by developing a deep dual-channel neural network tailored for image-based smoke detection (Gu et al. 2020). In the context of video-based smoke detection, Yin et al. introduced a recurrent convolutional network specifically designed for this purpose (Yin et al. 2018), while Filonenko et al. conducted a comprehensive comparative study of modern convolutional neural networks for smoke detection on image data (Filonenko et al. 2017). Moreover, Li et al. presented a noteworthy contribution with their 3D parallel fully convolutional network, which enables real-time detection of smoke in video footage of wildfire scenarios (Li et al. 2020).

Furthermore, Toreyin et al. contributed to the field by focusing on real-time smoke and flame detection in video, enabling timely identification of smoke and fire occurrences (Toreyin et al. 2006). Yuan et al. proposed a novel approach using high-order local ternary patterns with locality preserving projection for effective smoke detection and image classification (Yuan et al. 2016).

Detecting car fumes is vital for sustainable and green technologies, as vehicle emissions significantly impact the environment and public health. Car fumes contain harmful pollutants that degrade air quality and pose health risks. Accurate detection enables the identification of smoke-emitting vehicles and facilitates measures to minimize emissions, reducing air pollution and fostering a healthier future.

3 Methodology

This section outlines the methodology used to detect and mitigate smoke emissions from vehicles using the YOLOv8 object detection algorithm. The process involves dataset acquisition and preparation, model training, validation, and fine-tuning, as well as smoke detection and minimization actions.

3.1 Dataset Acquisition and Preparation

Obtain a dataset from RoboFlow (Smoke vehicles Object Detection Dataset (v5, 2022–10–11 4:24pm) by alpha ai), that includes annotated images and corresponding text files. These annotations contain bounding box coordinates around smoke-emitting cars. Organize the dataset into appropriate formats for training. Figure 1 shows an example of images used in the training.

Fig. 1 Example of dataset

3.2 *Model Training*

By employing the YOLOv8 algorithm, the model was trained on the prepared dataset for 20 epochs using a batch size of 8. The model's parameters were fine-tuned through backpropagation, and the loss function was optimized to enhance the accuracy of smoke detection. A confidence threshold of 0.5 was set, ensuring robust detection, and an image size of 640 was utilized during the training process. These adjustments contribute to refining the model's performance and improving the accuracy of smoke detection in the implemented system. The results obtained are as follows, in Figs. 2 and 3.

3.3 *Validation and Fine-Tuning*

Evaluate the trained model using a separate validation dataset. Fine-tune the model's hyperparameters based on validation results to enhance its performance and generalization capabilities.

3.4 *Smoke Detection*

The trained YOLOv8 model is deployed for real-time smoke detection. It identifies smoke-emitting vehicles by generating bounding box predictions around them. This

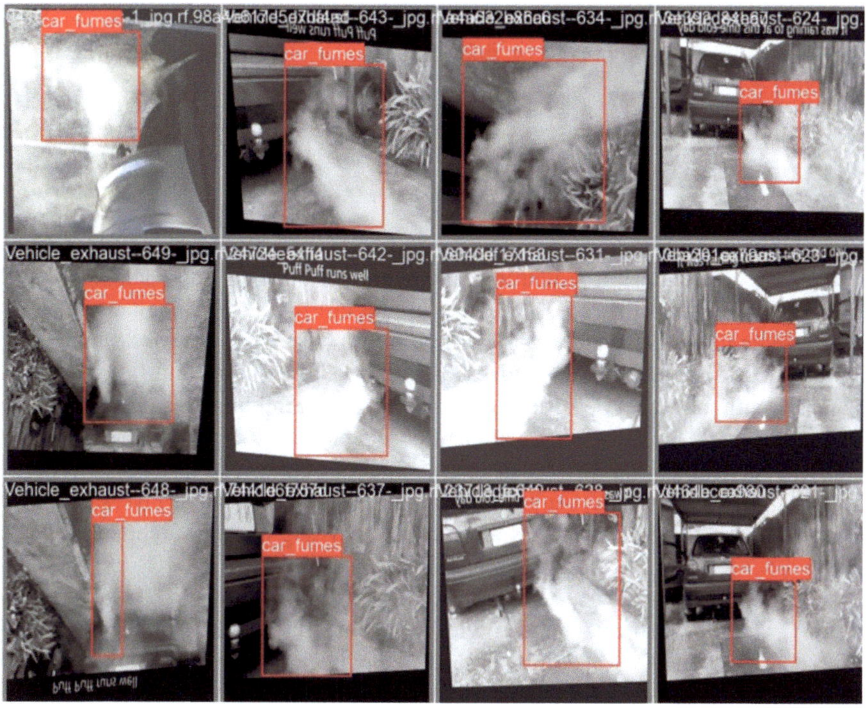

Fig. 2 Training results

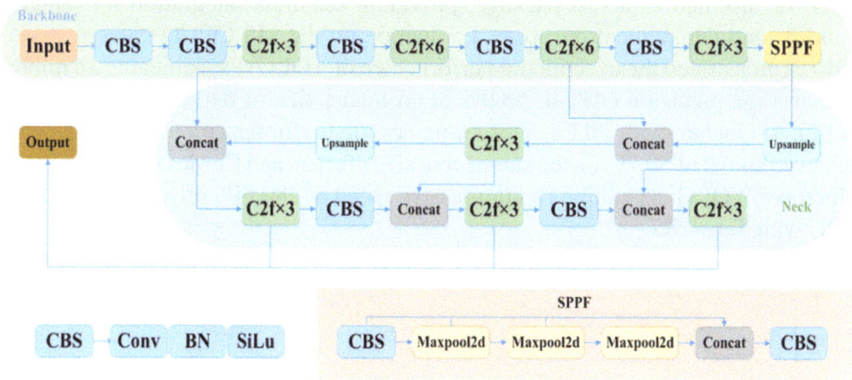

Fig. 3 YOLOv8 architecture (Lou et al. 2023)

detection process enables the localization and identification of vehicles emitting excessive smoke within images or video frames.

4 Architecure (YOLOv8)

YOLOv8 represents an innovative convolutional neural network (CNN) that excels in real-time object detection with remarkable precision. This approach leverages a single neural network to process the entire image, effectively dividing it into multiple components to predict accurate bounding boxes and probabilities for each element. By estimating these bounding boxes based on expected probabilities, YOLOv8 achieves optimal object detection performance by examining the image only once through a single forward propagation. Subsequently, the algorithm applies non-maximal suppression to ensure each object is identified uniquely, resulting in efficient detection. These distinct components can be perceived as a set of layers, acting as a functional unit, which draws inspiration from existing network architectures with subtle modifications to enhance its capabilities.

YOLOv8 incorporates a CSPDarknet53 feature extractor as its backbone architecture, which is further enhanced with a C2f module in place of the traditional YOLO neck architecture. Additionally, the model integrates two segmentation heads to predict semantic segmentation masks for input images. Similar to YOLOv8, the detection heads consist of five detection modules and a prediction layer (Terven and Cordova-Esparza 2023). YOLOv8 can be executed via the command line interface (CLI) or installed as a PIP package, providing seamless integration for labeling, training, and deployment purposes. Evaluation on the MS COCO dataset test-dev 2017 demonstrated the exceptional performance of YOLOv8x, achieving an impressive average precision (AP) of 53.9% at an image size of 640 pixels (as depicted in Fig. 4) (Jocher et al. 2023). This surpasses the performance of YOLOv5, which achieved an AP of 50.7% on the same input size (Terven and Cordova-Esparza 2023). Moreover, YOLOv8x exhibits an impressive speed of 280 FPS on an NVIDIA A100 GPU with TensorRT.

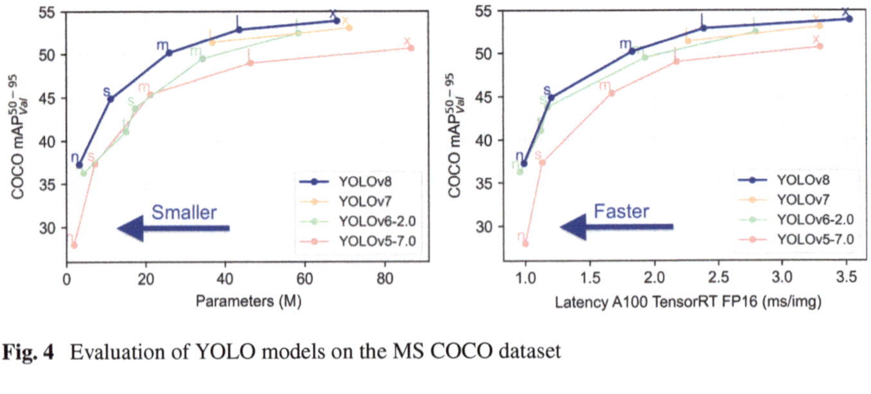

Fig. 4 Evaluation of YOLO models on the MS COCO dataset

YOLOv8 is available in five distinct models n, s, m, l, and x, each tailored to deliver varying levels of accuracy and detection performance. These models exhibit different capabilities and trade-offs, enabling users to choose the one that best suits their specific requirements. Figure 4 illustrates the comparative analysis of YOLO's models, providing insights into their respective strengths and performance characteristics.

5 Results

The performance evaluation of the smoke detection system employing the YOLOv8 model was carried out on a comprehensive test dataset comprising 2184 images. Among these images, 1875 contained fumes, while 309 were without fumes or smoke. The quantitative results of the evaluation are presented in Tables 1 and 2, providing a summarized overview of the system's performance.

In Table 1, the smoke detection algorithm successfully identified 1350 smoke-emitting vehicles from the 2184 images in the dataset. However, it also produced 525 false positive detections, indicating instances where smoke was incorrectly identified in non-smoke-emitting vehicles or false alarms (Fig. 5).

Table 2 presents the evaluation metrics of the smoke detection system. The precision, which measures the accuracy of positive detections, achieved a value of 0.821. The recall, indicating the percentage of actual smoke-emitting vehicles correctly identified, attained a value of 0.717. The mAP50, a harmonic mean of average precision was calculated as 0.811. Additionally, the overall mAP50-95 of the smoke detection system was determined to be 0.485. These results provide valuable insights into the performance of the trained YOLOv8 model for smoke detection. The achieved precision, recall, and mAP50 demonstrate a reasonable level of effectiveness in identifying smoke-emitting vehicles. However, further optimizations and fine-tuning may be necessary to improve these metrics and reduce the occurrence of false positives.

These results signify the potential of the developed smoke detection system, serving as a foundation for further research and advancements in sustainable technologies to reduce vehicle emissions and mitigate the environmental impact of smoke-emitting vehicles.

Table 1 Smoke detection results

Dataset size	Smoke-emitting vehicles detected	False positives
2184 images	1350	525

Table 2 Evaluation metrics

Metrics	Precision	Recall	mAP50	mAP50-95
Results	0.821	0.717	0.811	0.485

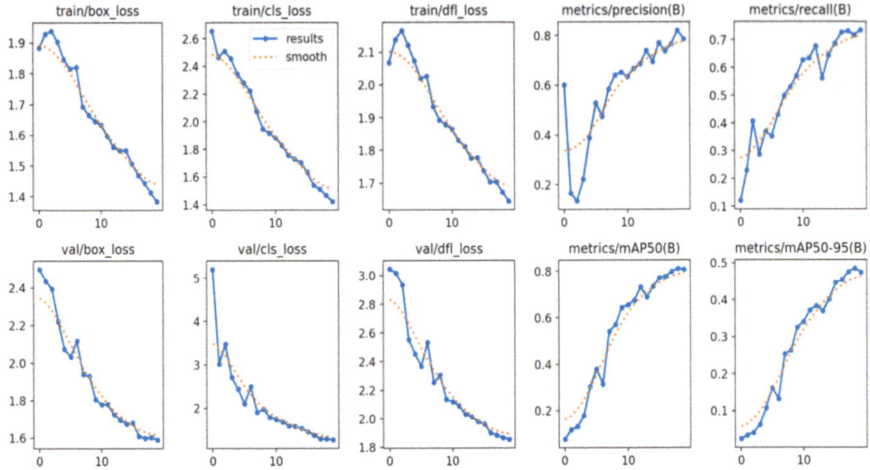

Fig. 5 Training results

6 Conclusion

The development and evaluation of a smoke detection system using the YOLOv8 model have showcased promising results and significant potential for addressing the challenge of identifying smoke-emitting vehicles. The system successfully detected smoke emissions in a test dataset, demonstrating the effectiveness of the trained model in identifying instances of smoke emission accurately. Although some false positive detections were observed, the evaluation metrics indicated reasonable precision, recall, and overall accuracy.

The results highlight the significance of advanced computer vision techniques, such as the YOLOv8 algorithm, in enabling real-time smoke detection and its potential for promoting sustainable and green technologies. By identifying smoke-emitting vehicles, targeted interventions can be initiated to minimize emissions and reduce the smoke percentage in the air. These actions may include notifying drivers, suggesting maintenance measures, or implementing automated mechanisms to mitigate emissions.

Further advancements and refinements are necessary to improve the performance of the smoke detection system. Optimization of model parameters, continued training on diverse datasets, and rigorous evaluation in real-world scenarios are crucial for enhancing the system's accuracy, robustness, and generalization capabilities. Additionally, incorporating additional sensor data and environmental factors could further enhance the precision and reliability of the system.

References

Cao Y, Lu X (2019) Learning spatial-temporal representation for smoke vehicle detection. Multimed Tools Appl 78:27871–27889. https://doi.org/10.1007/s11042-019-07926-1

Chen J, Peng X (2023) DB-Net: detecting vehicle smoke with deep block networks. Appl Sci 13:4941. https://doi.org/10.3390/app13084941

Fan Q, Brown LM, Smith JR (2016) A closer look at Faster R-CNN for vehicle detection. https://doi.org/10.1109/ivs.2016.7535375

Filonenko A, Kurnianggoro L, Jo K-H (2017) Comparative study of modern convolutional neural networks for smoke detection on image data. https://doi.org/10.1109/hsi.2017.8004998

Frizzi S, Kaabi R, Bouchouicha M, Ginoux J-M, Moreau E, Fnaiech F (2016) Convolutional neural network for video fire and smoke detection. https://doi.org/10.1109/iecon.2016.7793196

Gu K, Xia Z, Qiao J, Lin W (2020) Deep dual-channel neural network for image-based smoke detection. IEEE Trans Multimedia 22:311–323. https://doi.org/10.1109/tmm.2019.2929009

He Z, Ye G, Jiang H, Fu Y (2020) Vehicle emission detection in data-driven methods. Math Probl Eng 2020:1–13. https://doi.org/10.1155/2020/4875310

Islam MR, Amiruzzaman, Nasim S, Shin J (2020) Smoke object segmentation and the dynamic growth feature model for video-based smoke detection systems. Symmetry 12, 1075 (2020). https://doi.org/10.3390/sym12071075

Jocher G, Chaurasia A, Qiu J (2023) YOLO by Ultralytics. https://github.com/ultralytics/ultralytics

Li X-Q, Chen Z, Wu Q, Liu C (2020) 3D parallel fully convolutional networks for real-time video wildfire smoke detection. IEEE Trans Circuits Syst Video Technol 30:89–103. https://doi.org/10.1109/tcsvt.2018.2889193

Li B, Peng X, Wang Z, Jizheng X, Feng D (2017a) End-to-end united video dehazing and detection. arXiv (Cornell University). https://doi.org/10.48550/arxiv.1709.03919

Li B, Peng X, Wang Z, Jizheng X, Feng D (2017b) An all-in-one network for dehazing and beyond. arXiv (Cornell University). https://doi.org/10.48550/arxiv.1707.06543

Lou H, Duan X, Guo J, Liu H, Gu J, Bi L, Chen H (2023) DC-YOLOV8: Small-size object detection algorithm based on camera sensor. Electronics 12:2323. https://doi.org/10.3390/electronics12102323

Peng X, Fan X, Wu Q, Zhao J, Gao P (2023) Cascaded vehicle matching and short-term spatial-temporal network for smoky vehicle detection. Appl Sci 13:4841. https://doi.org/10.3390/app13084841

Peng X, Fan X, Wu Q, Zhao J, Gao P (2022) Video-based Smoky Vehicle Detection with A Coarse-to-Fine Framework. arXiv (Cornell University). https://doi.org/10.48550/arxiv.2207.03708

Smoke vehicles Object Detection Dataset (v5, 2022–10–11 4:24pm) by alpha ai. https://universe.roboflow.com/alpha-ai-nwrrb/smoke-vehicles/dataset/5

Tao H, Lu X (2018a) Smoky vehicle detection based on multi-feature fusion and ensemble neural networks. Multimed Tools Appl. https://doi.org/10.1007/s11042-018-6248-2

Tao H, Lu X (2018b) Smoky vehicle detection based on multi-scale block Tamura features. SIViP 12:1061–1068. https://doi.org/10.1007/s11760-018-1254-4

Tao H, Lu X (2018c) Smoky vehicle detection based on range filtering on three orthogonal planes and motion orientation histogram. IEEE Access. https://doi.org/10.1109/access.2018.2873757

Tao H, Lu X (2019a) Smoke vehicle detection based on robust codebook model and robust volume local binary count patterns. Image vis Comput 86:17–27. https://doi.org/10.1016/j.imavis.2019.03.008

Tao H, Lu X (2019b) Smoke vehicle detection based on multi-feature fusion and hidden Markov model. J Real-Time Image Proc 17:745–758. https://doi.org/10.1007/s11554-019-00856-z

Tao H, Lu X (2020) Smoke vehicle detection based on spatiotemporal bag-of-features and professional convolutional neural network. IEEE Trans Circuits Syst Video Technol 30:3301–3316. https://doi.org/10.1109/tcsvt.2019.2920657

Tao H, Zheng P, Xie C, Lu X (2020) A three-stage framework for smoky vehicle detection in traffic surveillance videos. Inf Sci 522:17–34. https://doi.org/10.1016/j.ins.2020.02.053

Terven J, Cordova-Esparza D (2023) A comprehensive review of Yolo: From Yolov1 and beyond. arXiv (Cornell University). https://doi.org/10.48550/arxiv.2304.00501

Toreyin BU, Dedeoğlu Y, Cetin AE (2006). Real-time smoke and flame detection in video. https://doi.org/10.1109/siu.2005.1567780

Yin M, Lang C, Li Z, Feng S, Wang T (2018) Recurrent convolutional network for video-based smoke detection. Multimed Tools Appl 78:237–256. https://doi.org/10.1007/s11042-017-5561-5

Yuan F, Shi J, Xia X, Fang Y, Fang Z, Mei T (2016) High-order local ternary patterns with locality preserving projection for smoke detection and image classification. Inf Sci 372:225–240. https://doi.org/10.1016/j.ins.2016.08.040

Enhancing the Reliability and Efficiency of Solar Systems Through Fault Detection in Solar Cells Using Electroluminescence (EL) Images and YOLO Version 5.0 Algorithm

Naima El yanboiy, Mohamed Khala, Ismail Elabbassi, Nourddine Elhajrat, Omar Eloutassi, Youssef El hassouani, Choukri Messaoudi, and Ali Omari Alaoui

Abstract The importance of solar energy as a renewable power source has led to increased adoption of solar modules for electricity generation. However, faults in solar cells can significantly impact their performance and efficiency. Manual defect detection is time-consuming and subjective, hence the need for an intelligent and efficient detection solution. In this study, we propose a method for detecting defective solar cells in electroluminescence imaging using an advanced object detection algorithm, specifically YOLO5 version. An important step in the algorithm is to formulate the detection problem in terms of real-time detection of defects. We evaluate our method on a dataset of different types of solar modules containing a total of 240 solar cells with various defects, including finger interruptions, microcracks, electrically separated or degraded cell parts and material defects. Experimental evaluation on solar cell images extracted from high-resolution electroluminescence images of photovoltaic modules datasets reveals that the proposed framework successfully

---

N. El yanboiy (✉) · M. Khala · I. Elabbassi · N. Elhajrat · C. Messaoudi
Optoelectronics and Applied Energy Techniques, Faculty of Science and Technology, Moulay Ismail University of Meknes, Errachidia, Morocco
e-mail: naima.elyanboiy@gmail.com

O. Eloutassi
Materials and Modelling Laboratory, Department of Physics, Faculty of Science, Moulay Ismail University of Meknes, Meknes, Morocco

Y. El hassouani
New Energies and Materials Engineering, Faculty of Science and Technology, Moulay Ismail University of Meknes, Errachidia, Morocco

A. Omari Alaoui
Decisional Computing and Systems Modelling Team, Engineering Sciences and Techniques Laboratory, Faculty of Sciences and Techniques Errachidia, Moulay Ismail University of Meknes, Meknes, Morocco

M. Azrour et al. (eds.), *Sustainable and Green Technologies for Water and Environmental Management*, World Sustainability Series,
https://doi.org/10.1007/978-3-031-52419-6_4

mitigates the influence of defect image degradation. The precision and recall confidence curves indicate a moderate performance, suggesting that the framework shows promising capabilities in detecting and localizing defects. This research contributes to the widespread adoption and sustainable utilization of solar energy, ensuring the optimal performance and longevity of solar cells.

Keywords Solar energy system · Smart detection · Electroluminescence (EL) imaging · YOLOv5

1 Introduction

The adoption of solar energy as a sustainable and renewable power source has gained significant importance in recent years (Rabaia et al. 2020). Solar modules, which are composed of interconnected solar cells play a crucial role in converting sunlight into electricity (Shubbak 2019). However, as with any complex system, solar modules are subject to various faults that can compromise their performance and longevity (Aghaei et al. 2022). Among the various faults that can affect photovoltaic systems are cracks, hot spots and broken connections. Detecting these faults is crucial to ensuring optimal operation of solar modules and optimizing their energy production (Dhanraj et al. 2021). Traditional fault detection methods often rely on manual inspection, which is time-consuming, subjective, and prone to human error. In this context, the concept of smart detection has emerged as a promising solution, leveraging advancements in computer vision and deep learning algorithms to automate the fault detection process. Many research studies have focused on this particular topic and have generally used a dual neural network approach to solve the problems of efficiency and accuracy. In the context of solar panel fault detection, many researchers have used convolutional neural networks as the main method for object detection. In the paper (Abbas and Zhang 2021) an intelligent PV fault detection system is presented, utilizing ANFIS methodology. ANFIS model is trained using Grid Partition (GP) and Subtractive Clustering (SC) strategies, with SC outperforming GP in fault detection and classification. The trained model demonstrates excellent performance on unseen data, with high values for R correlation. In (Chen et al. 2018) the Random forest (RF) ensemble learning algorithm is utilized for early fault detection and diagnosis of PV arrays. The optimized RF model achieves high detection and diagnosis performance, surpassing decision tree models. Fault experiments on simulated and laboratory PV systems validate the effectiveness of the proposed RF-based method. It offers an efficient and practical solution for PV array fault detection and classification. The researchers of papers (Chen et al. 2017) present an intelligent fault diagnosis method for photovoltaic arrays using an optimized kernel extreme learning machine (KELM) and I–V characteristics. By proposing a seven-dimensional feature vector derived from key points and model parameters of monitored I–V curves, the fault diagnosis model achieves improved accuracy and reliability. In paper (Han et al. 2021) a Deep learning (DL) approach is proposed for fault detection in solar panels using a UAV equipped with a thermal camera and GPS. The improved YOLOv3-tiny DL model is employed for fault detection, with information transmitted via LTE to a remote server. The proposed model outperforms the default YOLOv3-tiny model, achieving high performance in fault detection. In the context of detecting faults in solar panels, it is crucial to balance detection accuracy and processing speed. To address this challenge, this study proposes a deep DL approach that leverages an enhanced version of the You Only Look Once (YOLO) algorithm. Specifically, versions 5 of YOLO is utilized to achieve both accurate and efficient fault detection in solar cells. Compared to previous methodologies that faced challenges in real-time detection and processing time, the enhanced version of YOLO overcomes these issues. The experimental evaluation of the trained YOLOv5 model confirms its effectiveness in detecting and localizing faults in solar cells. Additionally, With an inference time of 9 ms. This low inference demonstrates its ability to process images rapidly and makes it well-suited for real-time or near-real-time fault detection in solar panels, ultimately enhancing the overall efficiency of photovoltaic module production. The rest of the document is organized as follows: Sect. 2 presents the various faults that can affect solar panels. In Sect. 3 the proposed approach for defect detection in solar panels. Section 4 details the experimental results. Finally, Sect. 5 concludes paper.

2 Faults that Can Affect the Solar Panel

Solar panels can be affected by various types of defect, which can have a significant impact on their performance and efficiency. Here are some common types of faults that can occur in solar panels (Madeti and Singh 2017; Hwang et al. 2021):

Hot spots: Hot spots are areas of a solar panel that experience high temperatures due to shading, dirt or debris on the panel surface, or damaged cells. Hot spots can reduce the efficiency of the panel and lead to permanent damage if not untreated.
Cracks or breaks in cells: Cracks or breaks in the solar cells can reduce the power output of the panel and can also cause hot spots.
Faulty bypass diodes: Bypass diodes are used to prevent hot spots by diverting cur-rent around shaded or damaged cells. Faulty bypass diodes can lead to hot spots and reduced efficiency.
Degradation or corrosion of materials: The materials used in solar panels can de-grade or corrode over time, reducing the efficiency and lifespan of the panel.
Potential-Induced Degradation (PID): Occurs when a voltage potential is present between the solar panel and the ground, causing a flow of current that can degrade the panel over time.
Soiling: Refers to the accumulation of dirt, dust, and debris on the surface of solar panels. When the panels are heavily soiled, it can create uneven shading on the cells. The shaded cells generate less current compared to the unshaded cells,

leading to a mismatch in the current flow. This mismatch can result in localized heating and the formation of hot spots at the shaded areas.

Shadow: Happens when objects like buildings, trees, or structures cast shadows on solar panels. This shading creates a disparity in current generation between shaded and unshaded cells. The resulting difference in current flow can cause the shaded cells to heat up more than the rest of the panel, leading to hot spots in the shaded areas.

Detecting and addressing these faults in a timely manner is crucial for maintaining the efficiency and lifespan of solar panels.

3 Proposed Method

3.1 The Architecture of YOLOv5

The YOLOv5 framework plays a crucial role in the efficient detection of faults in solar panels. Its importance lies in its ability to provide accurate and efficient fault detection, addressing the challenges of real-time detection and processing speed. The architecture of YOLOv5 can be summarized as follows as illustrated in Fig. 1.

Backbone Network: YOLOv5 utilizes a modified variant of the CSPDarknet53 architecture as its backbone network. CSPDarknet53 is a deep convolutional neural network that extracts features from the input image.

Neck: YOLOv5 introduces a "neck" module called PANet (Path Aggregation Network) to fuse features from different layers with different scales. PANet enhances the model's ability to detect objects at various sizes and resolutions.

Head: The head of YOLOv5 consists of multiple convolutional layers followed by a prediction layer. The prediction layer predicts bounding boxes, confidence scores, and class probabilities. Each grid cell in the output predicts a fixed number of bounding boxes along with their associated confidence scores and class probabilities.

Training: YOLOv5 uses the "cross-entropy" loss function to train the network. It also implements data augmentation techniques, such as random scaling, cropping, and flipping, to improve generalization and robustness.

Inference: During inference, non-maximum suppression (NMS) is applied to eliminate redundant and overlapping bounding boxes. A confidence threshold is used to filter out low-confidence predictions.

Output: The final output of YOLOv5 is a list of bounding boxes, each associated with a class label and a confidence score (Zhang et al. 2023).

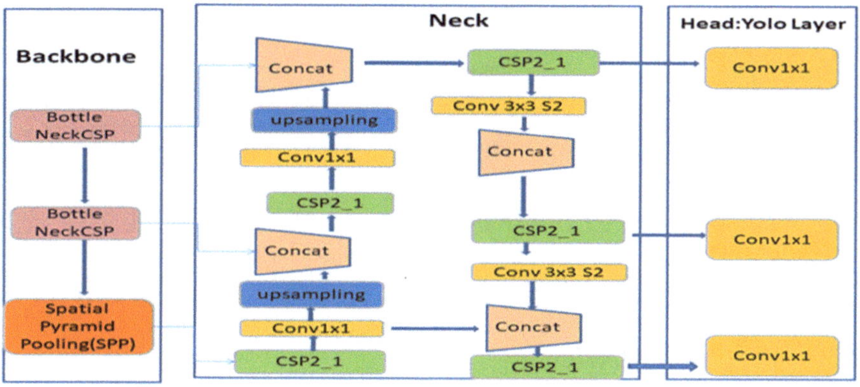

Fig. 1 YOLOv5 architecture

3.2 Dataset

In this study we exploit a publicly available dataset comprising solar cells obtained from high-resolution electroluminescence (EL) images of monocrystalline and poly-crystalline PV modules (Açikgöz and Korkmaz 2022). The dataset includes various images of defective solar panels as shown in Fig. 2, each with a resolution of 300 × 300 pixels. The collected images were annotated using the make Sence software (https://www.makesense.ai/). Annotations included rectangular frames representing the identified affected areas, along with associated category labels and classification probabilities (Sager et al. 2021). The identified rectangles were labeled in red and pink to indicate the different categories. These annotations provided valuable information about the defects detected in the solar panel images. After completing the labeling of the data, it was required to divide the dataset into distinct sets. The data set was divided into training and validation sets. The data set was then fed into the YOLOv5 algorithm to undergo further processing and analysis for defect detection as indicated in Fig. 3.

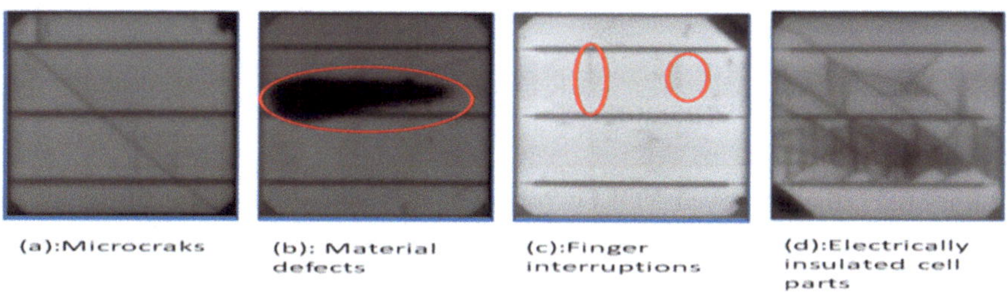

Fig. 2 Data set of different types of solar cell defects: **a** solar cell with microcrack, **b** typical material defect, **c** finger interruptions in circled areas, **d** electrically separated cell parts and material defects

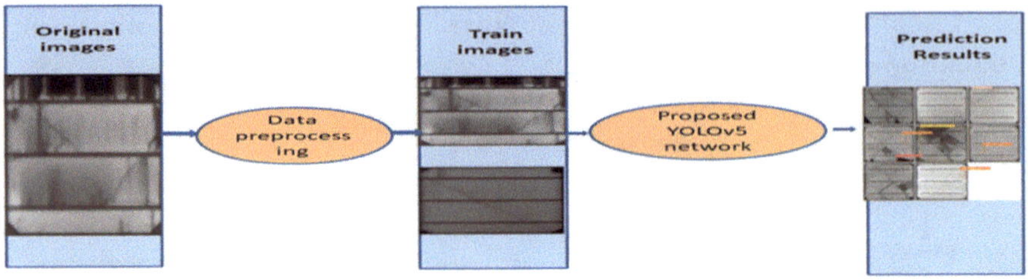

Fig. 3 General object detection framework

3.3 Evaluation Metrics

For the YOLO models, several common evaluation metrics can be used to assess their performance in object detection tasks (Malta et al. 2021). These metrics include:

Precision: Precision measures the proportion of correctly predicted objects out of all the objects detected by the model. It quantifies the accuracy of the model's predictions.
Recall: Recall, also known as sensitivity or true positive rate, measures the proportion of correctly detected objects out of all the ground truth objects. It quantifies the model's ability to detect the relevant objects.
Mean Average Precision (mAP): mAP is a widely used metric that calculates the average precision across different object categories or classes. It provides an overall assessment of the model's detection accuracy by taking into account both precision and recall.
Speed: Speed refers to the inference time required for the model to process an image and make object detection predictions. Faster inference times indicate higher efficiency and real-time or near-real-time performance.

These evaluation metrics help assess the accuracy, efficiency, and overall performance of the YOLO models, in object detection tasks. They provide insights into the model's ability to detect and localize objects accurately while considering computational efficiency and real-time applicability.

4 Results and Discussion

In this study, the improved YOLOv5 model was trained for 45 epochs using the Electroluminescent (EL) Image Dataset of PV Modules. The results and performance indicators were evaluated on both the training set and validation set, as shown in Fig. 4. The training progress of the model was observed through three loss curves: box loss, object loss, and classification loss. These curves consistently decreased during the training epochs, indicating enhanced model fitting, stability, and accuracy.

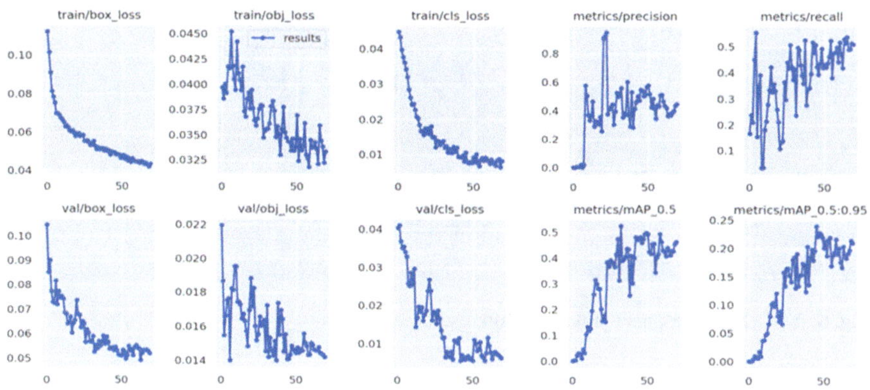

Fig. 4 Performance values for the YOLOv5 model

The PR curves in the last two columns of Fig. 4 represent precision and recall, evaluating the object-detection performance across different confidence level thresholds. Higher curve values correspond to increased model confidence. The visual representation in Fig. 4 further confirm the effectiveness of the proposed improved YOLOv5 model in achieving reliable and accurate object detection.

Figure 5 shows the PR curves for the proposed model. It is observed that the precision and recall confidence curves indicate a moderate performance. However, further analysis is required to assess the overall effectiveness of the proposed model.

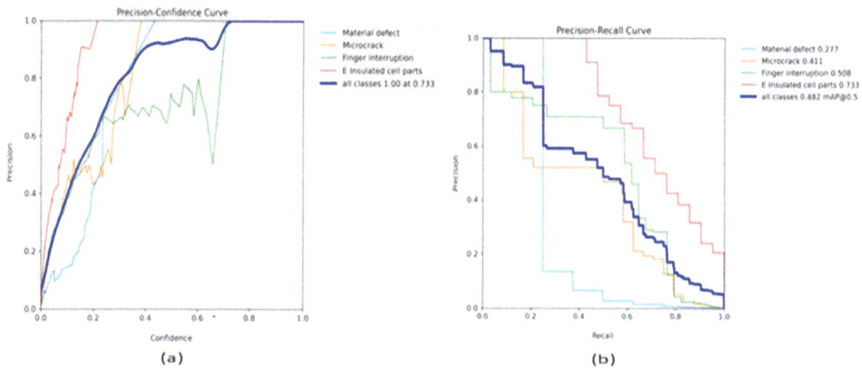

Fig. 5 PR curves for the suggested model

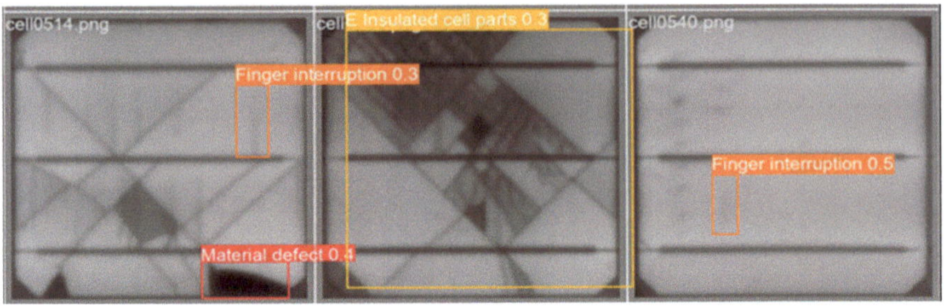

Fig. 6 YOLOv5 solar panel fault detection results

By analyzing the validation batch prediction, the bounding boxes provided by the model indicate the predicted location and size of detected defects, such as microcracks, material defects, and finger interruptions as shown in Fig. 6. Hence the ability of YOLOv5 to detect and classify these defects.

5 Conclusion

Smart detection plays a important role in various industries and applications, and its significance in fault detection within solar panels. The ability to accurately and efficiently detect faults is crucial for maintaining the optimal performance, reliability, and safety of solar energy systems. Based on advanced technologies such as YOLOv5, which offers real-time and accurate object detection capabilities, the detection of faults in solar panels becomes more effective. The proposed approach makes it able to detect solar panel faults such as micro-cracks, material defects and finger interruptions, enabling rapid maintenance and necessary repairs. However, it is essential to acknowledge that further improvements are still required. These include expanding the dataset size and exploring alternative techniques or versions of YOLO to enhance the fault detection accuracy and broaden its capabilities within the solar energy domain. Continual advancements and research in smart detection techniques hold immense potential for improving the efficiency and sustainability of solar energy systems.

References

Abbas MN, Zhang D (2021) A smart fault detection approach for PV modules using adaptive neuro-fuzzy inference framework. Energy Rep

Açikgöz H, Korkmaz D (2022) Automatic classification of defective photovoltaic module cells in electroluminescence images. Fırat Üniversitesi Mühendislik Bilimleri Dergisi

Aghaei M et al (2022) Review of degradation and failure phenomena in photovoltaic modules. Renew Sustain Energy Rev

Chen Z et al (2017) Intelligent fault diagnosis of photovoltaic arrays based on optimized kernel extreme learning machine and I–V characteristics. Appl Energy

Chen Z et al (2018) Random forest based intelligent fault diagnosis for PV arrays using array voltage and string currents. Energy Conv Manag

Dhanraj JA et al (2021) An effective evaluation on fault detection in solar panels. Energies

Han SH et al (2021) Detection of faults in solar panels using deep learning. In: 2021 international conference on electronics, information, and communication (ICEIC)

Hwang M-H et al (2021) A study on the improvement of efficiency by detection solar module faults in deteriorated photovoltaic power plants. Appl Sci

Madeti SR, Singh S (2017) A comprehensive study on different types of faults and detection techniques for solar photovoltaic system. Solar Energy

Makesense. https://www.makesense.ai/. Accessed 13 July 2023

Malta A et al (2021) Augmented reality maintenance assistant using YOLOv5. Appl Sci

Rabaia MK, Hussien et al (2020) Environmental impacts of solar energy systems: a review. Sci Total Environ

Sager C et al (2021) A survey of image labelling for computer vision applications. J Bus Anal

Shubbak MH (2019) Advances in solar photovoltaics: technology review and patent trends. Renew Sustain Energy Rev

Zhang J et al (2023) An improved YOLOv5-based underwater object-detection framework. Sensor

Deep Multi-temporal Matching of Satellite Images for Agricultural Dams

Omaima El Bahi, Ali Omari Alaoui, Youssef Qaraai, and Ahmad El Allaoui

Abstract Accurate multi-temporal satellite image matching plays a crucial role in monitoring agricultural areas, particularly in the context of water resource management. This study presents a feature-based approach for multi-temporal satellite image matching, using the VGG16 model. More specifically, our research focuses on the application of this approach to satellite images of agricultural dams in order to assess the reduction in water quantity caused by drought. Using the discriminant features extracted by the VGG16 model, we establish points of correspondence between images captured at different times, enabling precise alignment. Experimental results demonstrate the effectiveness of our approach in achieving accurate alignment of agricultural dam images, this study contributes to the advancement of feature-based matching techniques and their application in satellite image analysis for agricultural monitoring.

Keywords Remote sensing · Multi-temporal image matching · VGG16 · Agricultural dams · Satellite images

O. El Bahi (✉) · Y. Qaraai
Mathematical and Computer Modelling of Systems Team, Applied Mathematics, Computing and Systems Laboratory, Faculty of Sciences and Techniques Errachidia, Moulay Ismail University of Meknes, Meknes, Morocco
e-mail: omaima.elbahi@edu.umi.ac.ma

Y. Qaraai
e-mail: y.qaraai@fste.umi.ac.ma

A. Omari Alaoui · A. El Allaoui
Engineering Sciences and Techniques Laboratory, Faculty of Sciences and Techniques Errachidia, Decisional Computing and Systems Modelling Team, Moulay Ismail University of Meknes, Meknes, Morocco
e-mail: al.omarialaoui@edu.umi.ac.ma

A. El Allaoui
e-mail: a.elallaoui@umi.ac.ma

M. Azrour et al. (eds.), *Sustainable and Green Technologies for Water and Environmental Management*, World Sustainability Series,
https://doi.org/10.1007/978-3-031-52419-6_5

1 Introduction

Monitoring agricultural areas and detecting changes in water resources play a crucial role in sustainable resource management (Corwin 2020; Kumar and Singh 2021). Satellite imagery is useful due to its wide coverage, frequent revisits and ability to capture the Earth's surface in different spectral bands.

In this study, we propose a deep feature-based approach for multi-temporal non-rigid image matching using the VGG16 model. Leveraging the powerful feature extraction capabilities of the VGG16 model, our approach captures discriminative features from satellite images. These features are then used to identify points of correspondence between images captured at different times, enabling precise alignment and matching. Our main objective is to demonstrate the effectiveness of the proposed approach in matching satellite images of agricultural dams.

The remainder of this chapter is organized as follows: Sect. 2 provides an overview of related work in the field of multi-temporal image matching and its applications in agricultural monitoring. Section 3 presents the methodology and outlines the detailed steps involved in our proposed deep feature-based matching approach. In Sect. 4, we describe the data acquisition and experiment. This is followed by a discussion of the findings and their implications in Sect. 5. Finally, in Sect. 6, we conclude the chapter by summarizing the key contributions of our work and highlighting potential future research directions.

2 Literature Review

Monitoring agricultural areas through the analysis of multi-temporal satellite images has gained significant attention in recent years (Roy and Inamdar 2019). In this section, we provide an overview of related work in the field of multi-temporal image matching and its applications in agricultural monitoring.

Several studies have focused on the registration and matching of satellite images to enable accurate monitoring of agricultural regions. Traditional registration techniques (Dalmiya and Dharun 2015), such as feature-based methods and geometric transformations, have been widely explored. These methods aim to align images by finding corresponding points and estimating transformation parameters. However, they often struggle to handle non-rigid deformations, which are common in agricultural areas due to changes in land cover and varying environmental conditions.

To address these limitations, recent research has turned towards deep learning-based approaches for image matching. Deep neural networks, particularly convolutional neural networks (CNNs) (Rezaee et al. 2018; Zhang et al. 2019), have shown remarkable capabilities in feature extraction and matching tasks. The VGG16 model (Mascarenhas and Agarwal 2021), with its deep architecture and strong representation learning abilities, has been successfully applied in various computer vision tasks.

In the context of agricultural monitoring, Yang et al. (2018) proposed a deep feature-based matching method using CNNs for multi-temporal image registration. They used the VGG16 model to extract discriminative features and achieved accurate registration of a multi-temporal satellite image and multi-temporal UAV image datasets to improve the robustness of feature point registration. Similarly, Song et al. (2018) presented a multi-scale feature-based land cover change detection in mountainous terrain using multi-temporal and multi-sensor remote sensing images. By generating a multi-scale feature description using layers formed via a pretrained VGG network.

Furthermore, advances in deep learning have led to the development of Siamese networks (Melekhov et al. 2016) and metric learning (Liong et al. 2017) techniques. These methods aim to learn similarity measures directly from the data and have shown promising results in image matching tasks.

3 Methodology

3.1 Feature Extraction with VGG16 Model

We leverage the VGG16 model, shown in Fig. 1, a deep convolutional neural network (CNN), for feature extraction from the preprocessed satellite images. The VGG16 model, pre-trained on a large dataset, has demonstrated exceptional performance in capturing high-level features. By passing the preprocessed images through the VGG16 model, we extract deep discriminative features that effectively represent the image content. The selected VGG16 model offers several advantages: it exhibits impressive image classification performance, has a concise structure composed of convolution, pooling, and fully connected layers, and is a deep model trained on diverse images, enabling its convolution filters to identify universal patterns and achieve robust generalization. The use of VGG16 for feature extraction is common in computer vision applications, including object detection and super-resolution solutions.

3.2 Correspondence Estimation

To identify matching points between the reference and target images in our proposed deep feature-based matching approach, we use the Euclidean distance metric. After extracting deep features from the satellite images using the VGG16 model, each feature vector represents a specific location in the image. To find corresponding points, we compute the Euclidean distance between feature vectors in the reference and target images.

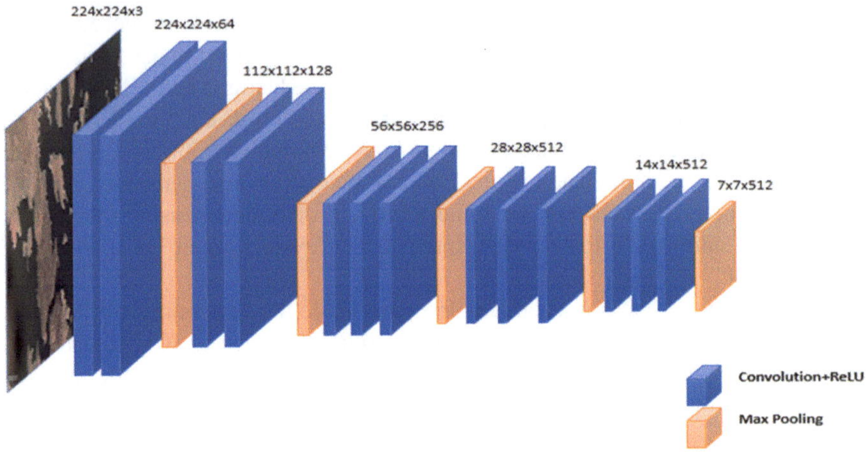

Fig. 1 VGG-16 network architecture

$$d(x; y) = \text{Euclidean-distance}\,(Di(x); Di(y)) \tag{1}$$

The Euclidean distance serves as a measure of dissimilarity between the feature vectors, allowing us to identify points with similar features. By considering points with the smallest Euclidean distances, we can establish correspondences between the reference and target images, enabling precise alignment and matching. The use of Euclidean distance as a matching metric provides a straightforward and effective approach for identifying similar features and aligning multi-temporal satellite images.

4 Data Acquisition and Experiment

4.1 Data Acquisition and Preprocessing

In this study, multitemporal satellite images were acquired from Google Earth. The dataset consisted of 9 pairs of images, each captured at different time instances between 2009 and 2023. The images were selected to represent a range of agricultural dams. The size of the images in the dataset varied, ranging from 600 × 400 to 1920 × 1080 pixels. These variations in image size were due to differences in satellite image resolutions and the specific areas covered by each image pair (Fig. 2).

Fig. 2 Illustrative examples from the dataset

4.2 Experiment Setting

The experimental setup involved using Google Colab with GPU support and Python as the programming language. We utilized the TensorFlow framework, leveraging the pre-trained VGG16 model for feature extraction. The image processing and registration algorithms were implemented using libraries such as OpenCV and scikit-image. This setup enabled efficient processing and analysis of the satellite images.

The performance of our methodology is evaluated using our multi-temporal satellite image dataset. To assess the effectiveness of our approach, we compare our feature descriptor with the widely used SIFT (Scale-Invariant Feature Transform) method (Chang et al. 2019). For each pair of test images, we extract and pre-match feature points using both methods. From these matches, we measure precision to evaluate the performance and robustness of our feature descriptor in achieving accurate and reliable matching results.

5 Results and Discussion

5.1 Results

In this section, we present the results of our feature matching precision test on our dataset. The numerical results are summarized in Table 1 and Fig. 3.

The results clearly demonstrate that our method outperformed the SIFT method in terms of precision. By leveraging the discriminative features extracted from the VGG16 model, we achieved a significantly higher number of accurately matched points compared to the SIFT method. This improvement can be attributed to the VGG16 model's ability to capture more complex and informative features from the satellite images.

Table 1 Feature matching precision test. Unit: %

	Index	VGG16	SIFT
Multi-temporal satellite image	Avg	94.14	62.74
	Max	100.00	73.73
	Min	87.09	39.32
	Std	5.08	12.09

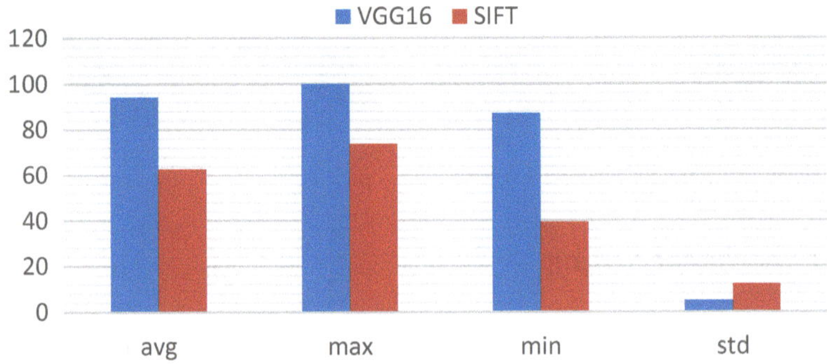

Fig. 3 Precision statistics

Overall, these results validate the superiority of our approach in terms of feature matching precision. The higher number of correct correspondences and better distribution of feature points contribute to the overall accuracy and reliability of multi-temporal matching for satellite images.

5.2 *Discussion*

Our deep feature-based matching approach outperforms the widely used SIFT method in accuracy, thanks to VGG16, which captures more useful information from satellite images. It is important to note that our evaluation focused on precision, but future work could explore other performance measures such as recall, and computational efficiency to provide a comprehensive evaluation of the proposed approach. The results confirm that deep feature matching is effective for monitoring agricultural dams. It improves accuracy over the SIFT method, validating the use of the VGG16 model. Deep learning methods have indeed demonstrated significant potential for accurately matching satellite images in agriculture.

6 Conclusion

In conclusion, our study presented an efficient deep feature-based matching approach for agricultural dam monitoring using multi-temporal satellite imagery. Relying on the VGG16 model, we succeeded in extracting discriminating features and obtaining accurate matching between images captured at different times. The results demonstrated higher accuracy than traditional methods such as SIFT, highlighting the ability of deep learning techniques to capture complex patterns and deal with non-rigid deformations in agricultural images.

For the future, there are promising avenues to explore for further improvements. Although our approach used the VGG16 model, future research could investigate the performance of other deep learning architectures, such as VGG19 (Mateen et al. 2018), ResNet (Sarwinda et al. 2021), or EfficientNet (Marques et al. 2020), in the context of satellite image matching. Exploring these models could provide additional insights and potentially improve the accuracy and efficiency of the matching process.

Furthermore, our study focused on matching, and it is possible to extend our work to the registration of satellite data. Registration involves aligning images from different sensors, resolutions or viewpoints, enabling comprehensive analysis and monitoring. Studying satellite data registration techniques would contribute to a more comprehensive understanding of agricultural areas, including dam monitoring, and enable more advanced applications in precision agriculture, land-use planning and environmental management.

References

Chang H-H, Wu G-L, Chiang M-H (2019) Remote sensing image registration based on modified SIFT and feature slope grouping. IEEE Geosci Remote Sens Lett 16:1363–1367

Corwin DL (2020) Climate change impacts on soil salinity in agricultural areas. Eur J Soil Sci 72:842–862

Dalmiya CP, Dharun VS (2015) A survey of registration techniques in remote sensing images. Indian J Sci Technol 8

Kumar S, Singh R (2021) Geospatial applications in land use/land cover change detection for sustainable regional development: the case of central Haryana, India. Geomat Environ Eng 15:81–98

Liong VE, Lu J, Tan Y-P, Zhou J (2017) Deep coupled metric learning for cross-modal matching. IEEE Trans Multimedia 19:1234–1244

Marques G, Agarwal D, De La Torre Díez I (2020) Automated medical diagnosis of COVID-19 through EfficientNet convolutional neural network. Appl Soft Comput 96:106691

Mascarenhas S, Agarwal M (2021) A comparison between VGG16, VGG19 and ResNet50 architecture frameworks for image classification

Mateen M, Wen J, Nasrullah, Song SO, Huang Z (2018) Fundus image classification using VGG-19 architecture with PCA and SVD. Symmetry 11(1)

Melekhov I, Kannala J, Rahtu E (2016) Siamese network features for image matching

Rezaee MA, Mahdianpari M, Zhang Y, Salehi B (2018) Deep convolutional neural network for complex wetland classification using optical remote sensing imagery. IEEE J Sel Top Appl Earth Observ Remote Sens 11:3030–3039

Roy A, Inamdar AB (2019) Multi-temporal land use land cover (LULC) change analysis of a dry semi-arid river basin in western India following a robust multi-sensor satellite image calibration strategy. Heliyon 5(4)

Sarwinda D, Paradisa RH, Bustamam A, Anggia P (2021) Deep learning in image classification using residual network (ResNet) variants for detection of colorectal cancer. Procedia Comput Sci 179:423–431

Song F, Yang Z, Gao X, Dan T, Yang Y, Zhao W, Yu R (2018) Multi-scale feature based land cover change detection in mountainous terrain using multi-temporal and multi-sensor remote sensing images. IEEE Access 6:77494–77508

Yang Z, Dan T, Yang Y (2018) Multi-temporal remote sensing image registration using deep convolutional features. IEEE Access 6:38544–38555

Zhang H, Ni W, Yan W, Xiang D, Wu J, Yang X, Bian H (2019) Registration of multimodal remote sensing image based on deep fully convolutional neural network. IEEE J Sel Top Appl Earth Observ Remote Sens 12:3028–3042

Blockchain Technology and Customs Clearance Procedures: Facilitating and Smoothing Products Importation in Morocco's Customs and Excises Administration

Mohamed Elkoutour, Hajar Raji, and Mohcine Bakhat

Abstract This research intends to unveil the positive implications of using blockchain-based technology in international trade, specifically in importation customs clearance activities. The application of blockchain technology in customs clearance offers several benefits, such as simplifying customs procedures, improving global supply chain management, preventing customs-related risks, and enabling real-time product verification in terms of quantity, quality, and origin. Through qualitative analysis using Nvivo software, this research examines the implementation of the blockchain platform PortNet in Morocco and its impact on import customs procedures. The findings suggest that the use of blockchain technology through PortNet has effectively facilitated and streamlined the import customs process. By utilizing blockchain technology, this study highlights the potential for increased efficiency, transparency, and reliability in international commercial transactions, which can contribute to improved trade facilitation and economic growth.

Keywords Blockchain · International trade · Customs clearance · Qualitative analysis

M. Elkoutour
Inspector at the Moroccan Customs and Excise Administration, Casablanca, Morocco

H. Raji (✉)
Laboratory of Studies and Research in Economics and Management Sciences, Institute of Technology, Sultan Moulay Slimane University, Beni Mellal, Morocco
e-mail: h.raji@usms.ma

M. Bakhat
Laboratory of Studies and Research in Economics and Management Sciences, Faculty of Legal, Economic and Social Sciences, Hassan II University, Casablanca, Morocco

M. Azrour et al. (eds.), *Sustainable and Green Technologies for Water and Environmental Management*, World Sustainability Series,
https://doi.org/10.1007/978-3-031-52419-6_6

1 Introduction

Foreign trade involves a complex procedure with the integration of third-party logistics and administrative agencies such as logistics companies, customs authorities, transporters, freight forwarders, and banking institutions. The ever-increasing interconnection between all these parties of global supply chain forces them to adopt and improve internet-based solutions. Indeed, new technologies have been acknowledged as original methods to tackle these issues such as the Internet of Things (Suppatvech et al. 2019; Dargaoui et al. 2023; Mohy-eddine et al. 2023a, b; Hazman et al. 2023a; Mabrouki et al. 2021), blockchain, and artificial intelligence (Benzidia et al. 2021; Mabrouki et al. 2019, 2022; Mabrouki and Jamal, 2021; Hissou et al. 2023; Khan et al. 2023; Hazman et al. 2023b). Hastig and Sodhi (Hastig and Sodhi 2020) along with (Li 2023) have outlined that the major issues of applying blockchain-based technologies in supply chains are to ensure information sharing, facilitate traceability, and integrate multiple entities. Blockchain has been recognized as a technology that facilitates security, improves transparency, and increases transaction speed in international trade and, mainly, in customs procedures (Sunny et al. 2020; Belu 2019).

Lately, blockchain-centered technology has attracted the lion's share of research attention in international trade and particularly customs clearance processes. For instance, in an influential chapter on the potential of blockchain for customs, Okazaki (2018) emphasized the disruptive capabilities and dynamic potential of blockchain in revolutionizing customs procedures. The author has stated that this technology could facilitate customs procedures, reduce paperwork, and improve data usage related to clearance forms, invoices, transporters, buyers, and suppliers. Further, in the supply chain context, blockchain may tackle multiple issues related to transparency and responsibility (Kshetri 2018), security (Rahmanzadeh et al. 2020; Guezzaz et al. 2020, 2022; Douiba et al. 2022; Hazman et al. 2023c), uncertainty (Kim and Laskowski 2017), and cost-reducing (Roeck et al. 2020).

Research has underscored the practical implications of blockchain technology for customs administrations. This distributed ledger-based technology can mitigate various risks related to information sharing such as information delays, lack of information transparency, and information security (Tseng et al. 1055) and, therefore, ensuring an immutable, encrypted, distributed, and auditable chain of data ledger. Thus, implementing blockchain in customs agencies can facilitate risk assessment of imported or exported goods, prevent fraudulent claims, and enhance targeting measures of clearance forms (Tyagi and Goyal 2021). To identify the blockchain implications for customs procedures, we adopted an exploratory qualitative study to empirically investigate the extent to which blockchain technology can facilitate customs clearance processes. Through a qualitative analysis, semi-structured interviews were used in order to seize the opportunity of interacting with interviewees' answers with spontaneously formulated additional questions (Qu and Dumay 2011). The semi-structured interviews were mainly conducted in French and analyzed using Sphinx software. It is worth mentioning that the analysis method adoption is prompted by the fact that this study, to our knowledge, has not been conducted in Morocco's Customs Administration and Excises, and blockchain-based technology is still a prominent research topic, mainly its implementation within customs clearance processes. Ultimately, the selection of interviewees was based on the following criteria: interviewees had to be an employee of the Provincial Directorate of Nouasseur of Customs and Excises Administration, or an employee of an aeronautics company installed in Nouasseur, or an A-grade forwarding agent operating in Nouasseur Directorate.

2 Literature Review

Over the last few decades, the advancement of free trade has resulted in the growth of global value networks that cross boundaries and connect several countries worldwide. Supply networks have become more complicated as a result of globalization and technological improvements (Aitken et al. 2016). These supply chains include a vast number of actors and participants from various corners of the globe, including raw material suppliers, manufacturers, shipping and transportation companies, government authorities (especially customs and regulators), banks, insurance companies, and consumers. The extensive involvement of diverse stakeholders in these supply chains creates an unprecedented need for coordination and cooperation among parties that may not inherently trust each other (Chen et al. 2019).

Indeed, the growing complexity of global supply chains increases the operating expenses involved with the transportation of commodities and services among industries and worldwide economies. According to a report by the Global Alliance for Trade Facilitation, a joint initiative of the World Bank Group and the International Finance Corporation (Gnata), supply chain operating costs account for two-thirds of the final cost of traded goods, while documentation costs absorb 7% of the overall trade value. Thus, stakeholders' integration has become crucial in supply chains (Wang et al. 2020). In fact, Saberi et al. (2019) have shown that implementing blockchain technology to improve supply chain management entails a combination of various approaches involving these technologies. Wang et al. (2019a), for example, undertook a comprehensive examination to unveil how blockchain can impact supply chain activities and legislation.

A growing body of research to date demonstrates that blockchain has an accumulative effect on supply chains. For instance, Saberi et al. (2019) found that blockchain technology can improve supply chain sustainability, make tracking sustainable habits easier, and have an impact on the economy, society, and the environment. In the same vein, Tapscott and Kaplan (2019) emphasized that blockchain technology has the potential to cut transaction costs, boost efficiency, and lead to new business models. Other research has shown that this technology can also lower the expenses and delays connected with customs operations, giving businesses a robust competitive advantage (Narayanan et al. 2016). Therefore, the utilization of blockchain-based technologies in supply chains can be a critical step toward establishing a more productive and sustainable economy mainly.

From this perspective, blockchain technology is important in the field of supply chain management because it provides a solution for traceability, transparency, and security at various levels of the supply chain (Hastig and Sodhi 2020). This technology enables the immutable and transparent storage of logistics data, facilitating product management, lowering costs and delays, and improving product quality. Furthermore, blockchain can be used to automate operations and payments, improving confidence among chain actors and enabling more effective and speedier supply chain management (Agarwal et al. 2022).

The integration of blockchain technology in supply chain management is continually growing and has a significant impact on organizations and sectors globally. According to a Deloitte (Wang et al. 2019b) study, 55% of surveyed organizations have already embraced blockchain technology or aim to do so within the next two years to improve supply chain management. Blockchain allows more effective inventory management, improved product visibility, and lower expenses connected with product tracking activities. To summarize, blockchain-based technology's role in supply chains has become critical for firms wanting to improve their competitiveness and efficiency.

Drawing on the foundation of supply chain literature, blockchain-based technology can indeed facilitate and improve customs clearance operations. As part of international trade, customs procedures depend largely on information systems embedded mainly in blockchain technologies. Indeed, blockchain has a prominent role in reducing inefficiencies and increasing visibility of customs operations (Ioannou and Demirel 2022). Thus, the implementation of blockchain technology can streamline this operation and cut costs for firms while assuring data integrity and transparency in trade exchanges.

3 Introduction the Use of Blockchain in Import Customs Procedures

Generally speaking, logistical proceedings involve all importation, exportation, and transportation-related activities manifested mainly by products' packaging, containers' loading,, goods pre-transportation (i.e., moving products to the airport or port of embarkation, or to the processing platforms), customs clearance procedures, international trade insurance, post-transportation, and products' unloading from all means of transportation (Belu 2019).

Various research suggests that blockchain-based technology can simplify and decrease product delivery while ensuring transparency and security of international trade participants. Due to these advantages, many businesses are increasingly interested in implementing this technology in their operations. Indeed, specific commercial records related to international transactions can be stored on the blockchain,

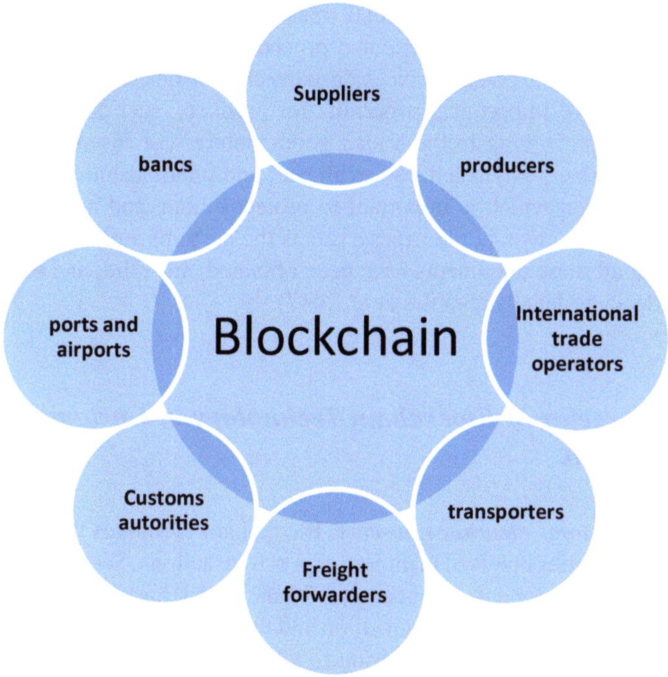

Fig. 1 Blockchain participants in an import and export process (Medium 2018)

enabling online access and third-parties verification such as the supplier, client, corresponding bank, customs agencies, and others. Utilizing blockchain-based technology could reduce the enormous number of participants associated with import and export processes (refer to Fig. 1) which, in turn, eliminates the need for intermediaries and third parties, thereby reducing time, costs, and risk management (Wouda and Opdenakker 2019).

Following the same reasoning, Dobrovnik et al. (2018) emphasized the positive implications of implementing blockchain-centered technology in customs clearance procedures by stressing that this technology could enhance physical product checking, end-to-end tracking, and paperless trading. The author also highlighted the role of blockchain in preventing information alteration of trade documents as this technology doesn't allow information modification.

In international trade, Distributed Ledger Technology-based (DLT) applications can eliminate massive volumes of documentation and various governmental interventions. Customs agencies can greatly improve risk management and facilitate international transactions. According to Medium (2018), blockchain-based solutions in supply chain management enable parties to frequently update information, resulting in correct data processing that is required to carry out tasks ranging from raw material sourcing to final goods delivery.

Nowadays, international commercial transactions are heavily impacted by the costs associated with customs clearance procedures, as well as the long waiting duration to get import/export clearances. In many circumstances, international traders are required to give import/export permits for products, such as licenses, as well as other permits according to the traded goods nature. For instance, plant-related products necessitate a phytosanitary certificate, and pharmaceutical products need health authorization which is important to protect human and animal health. The ultimate arbiter in a cross-border transaction is the customs authority, whose role is to ensure that all these documents have been obtained, are valid, and that the goods have been declared legally (Staples et al. 2017).

3.1 The Benefits of Blockchain Technology in Customs Operations

Accelerating customs clearance process. Blockchain facilitates the verification of participants' identities involved in import-export transactions. Scrutinizing the implications of blockchain for customs agencies, Ganne (2018) has shown that many organizations like World Customs Organization and the Centre for Trade Facilitation and Electronic Business have been exploring to which extent this technology could be implemented in customs information systems and authorization processes. Customs authorities would gain a more comprehensive understanding of international trade transactions that are undergoing customs clearance, including information on the nature, route, and participants involved. This enables more effective risk management in customs clearance, directly impacting the time required for goods clearance. For instance, International Trade Data System implementation in the United States intends to facilitate clearance operations by adopting a unique platform for capturing and managing all transportation data and documents. Various international organization like World Customs Organization and the Association of Southeast Asian Nations has been advocating for the concept of a "single window" in international trade. Chang et al. (2020), on the other hand, outlined that the focus of a "single window" strategy has mainly revolved around giving international operators a single-entry point for submitting standardized digital information to customs authorities.

Streamlining international trade. Integrating blockchain-based technology, customs agencies and other authorities associated with import–export activities can enhance their ability for analysis and inspection, thereby promoting international trade. Blockchain technology has the potential to combine customs agencies into a unique platform that connects all economic operators, facilitating greater information exchange among all stakeholders involved (Chang et al. 2020). Thus, customs authorities can access details involving export/import activities such as supplier, client, transporters, origin, methods of payments, insurance, etc., and monitor the real-time location and status of these products.

Compliance with security and sanitary standards. To ensure imported product safety and quality, a set of documents, including permits for importation and exportation, health authorizations, and phytosanitary permits have become due. Blockchain-based technology allows a shared system accessible to industrialists, logistics partners, clients, and government authorities, offering complete transparency regarding product origins, certification tests, and the authorization of import/export licenses.

By utilizing blockchain, electronic certificates can be reliably issued and authenticated by authorized regulatory agencies, mitigating the risks of content alteration, falsification, or unauthorized manipulation. Customs authorities integrated into the blockchain network can automatically process goods' clearance if they have previously been verified in the customs registry, streamlining the declaration process and potentially eliminating the need for physical inspections. This enhanced efficiency and traceability contribute to strengthening import controls and ensuring the integrity and safety of products entering the market.

Facilitation of goods' origin verification procedures. Drawing on blockchain-based technology advantages, the customs agency of the importing country can utilize tracking mechanisms to trace the origin of imported goods, thus ensuring their authenticity and preventing the entry of counterfeit products. In the same vein, Medium reported that by harnessing this innovative technology, comprehensive information is stored throughout the entire lifecycle of traded goods, encompassing customs clearance, manufacturing, controlling, and transportation. He argued that each unique item is assigned a unique identification code, leading to a substantial enhancement of logistics activities transparency. For instance, with the collaboration of IBM, Walmart in 2016 managed to promote food traceability, conducting initial pilot tests to digitally track the movement of pork from small Chinese farms to Chinese stores and international movements between Latin American production sites to the United States stores. All of these test data recorded critical information regarding shipping information, manufacturing details, and more all have been securely included the blockchain platform. This data will allows Walmart to trace the product's origin and react correspondingly in case of good outbreak (Hackius and Petersen 2017).

Minimization of tax evasion. A significant hurdle faced by tax authorities is bridging the gap between projected value-added tax (VAT) revenues and the actual amount gathered. Blockchain technology, with its enhanced transparency and traceability, offers a potential solution to this issue by facilitating the easy detection of fraud and errors. Chang et al. (2020) hypothesized that real-time exchange of information among customs offices, exporters, importers, and other relevant stakeholders empowers customs authorities to swiftly uncover fraudulent activities.

Management of customs risks. Customs agencies will be educated and prepared to address special hazards connected with import–export activities inside a blockchain network. Chang et al. (2020) reported that Illicit profits can be generated, for instance, by indicating/declaring bogus routes, by overvaluing or undervaluing traded products, or by transshipment sites. Supporting this assumption, Okazaki (2018) added that blockchain can help to encourage lawful trade by reducing bureaucracy and improving customs operations.

More blockchain-based platforms, including customs authorities, will be built in the future. One potential stumbling block could be the difficulties in integrating all customs agencies so that they can easily retrieve data held on these systems. Accordingly, it is expected that public blockchains will be used for improved accessibility for the multiple parties involved in supply chain management, which may reduce trust among network members.

Digitalization of customs paperwork. Various research suggests that blockchain technology can improve the digitalization of operations and ease paperwork (Yang 2019). For instance, Belu (2019) argued that the uniqueness of the transactional mechanism in import–export activities lies in the crucial role of documents (Fig. 2) by which the transfer of ownership of products is conducted, and the payment for its value is made.

The digitization of customs declaration data facilitates the management of the documents required to obtain clearance. In the traditional system, customs clearance of goods requires a significant amount of documentary evidence, including the manifest, transport document (airway bill or bill of lading), packing list, certificate of origin, and invoice. (Fig. 2).

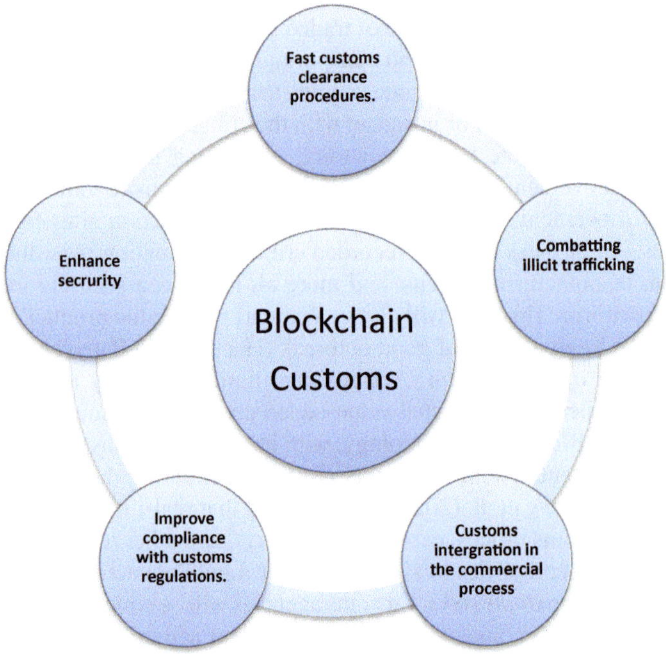

Fig. 2 Customs' place in a blockchain network (Medium 2018)

3.2 *Blockchain Technology in the Moroccan Context*

Focusing on technology can help explain the potential advancements in the commercial application of blockchain. According to Botton (2018), various shipment details, such as proof of purchase, customs forms, bills of lading, and insurance information, can become part of a transparent and accessible chain of control for suppliers, carriers, purchasers, regulators, and auditors. This comprehensive visibility enables customs to access essential and accurate data, including details of the seller, buyer, price, quantity, carrier, finance, insurance, and more, all associated with the commodities to be declared. Additionally, customs can track the real-time position and status of these goods. By integrating such total visibility into regulatory control, customs authorities can perform their daily tasks in a more informed and data-driven manner. Having access to this wealth of information empowers customs officials to make better-informed decisions and carry out their responsibilities more effectively, ultimately contributing to improved efficiency and compliance in international trade operations.

Utilizing blockchain technology, customs authorities and border services can significantly bolster their risk analysis and targeting capabilities, thereby facilitating international trade. While conventional blockchains like "permissionless" ones are open to anyone participating in the validation process, this openness may raise concerns about managing sensitive information and governing the network. In contrast, blockchains like PortNet with permission or private access offer a more controlled environment where specific administrators define roles, responsibilities, access levels, and validation rights for participants. This ensures impartiality while enabling seamless and reliable information exchange among unrelated parties, resulting in fully integrated end-to-end supply chain management.

For instance, in Morocco, PortNet S.A. integrates all foreign trade operators, and through its connection with the Moroccan customs' BADR[1] system, goods clearance is fully digitized based on controlled access to private blockchain networks of stakeholders. This enables customs to automatically extract information from primary sources for declaration purposes, ensuring enhanced accuracy, quality, and immutability of their data. By reducing the burden of manual verification and the resources required for declaration validation, customs can accelerate the processing of customs declarations, leading to a notable reduction in end-to-end processing time.

Initiated in 2008 by the National Ports Agency (Port Authority) in Morocco, PortNet is the result of a strategic alliance between port communities and the international trade sector. It is a government-owned tool aimed at implementing several sectoral strategies to improve the business climate, trade exchanges, port and logistics competitiveness, and widespread adoption of innovation in e-government services. In fact, PortNet is a successful model for the national partnership between the administration and the private sector, with the goal of enhancing the business climate. Therefore, the Moroccan international trade sphere is mainly represented by PortNet which is a robust tool in integrating all interested stakeholders such as the Moroccan

[1] BADR: Base Automatisé des Douanes en Réseau (Automated Customs Network Database).

Customs and Excise Administration, The National Office of Food Safety (ONSSA), The National Telecommunications Regulatory Agency (ANRT), The Ministry of Industry and Trade.

4 Research Methodology

4.1 Criteria of Inclusion

The interviewed individuals were chosen based on their work either within the Provincial Customs Directorate of Nouasser of Moroccan Customs and Excise Administration, category A freight forwarders operating in this Directorate, or companies in the aerospace sector located in Nouasser. Supply chain experts were also interviewed. The relevant individuals were personally invited to participate in an interview, and once they agreed, interview sessions were scheduled. The response to interview requests was generally very positive, and the interviewees were willing to share their expertise. The interviews were conducted through various means, such as individual discussions, phone calls, or, in one instance, via email. To ensure consistency, the interviews were structured around the eight principles outlined in McNamara's work (Chawinga and Majawa 2018).

The interviews were conducted using a semi-structured format, which is a common method of data collection in qualitative research (Myers and Newman 2007). Semi-structured interviews offer the advantage of not adhering strictly to a predetermined script, allowing for flexibility and development throughout the conversation, as highlighted by Myers and Newman (2007). This approach proves particularly beneficial when exploring relatively new technologies like Blockchain, as it allows for a more open and exploratory discussion, potentially leading to additional insights. The interviewees included individuals involved in facilitating import customs procedures, stakeholders from the aerospace sector, and categorized A freight forwarders. By engaging in semi-structured interviews with these relevant parties, the research aimed to gather valuable perspectives and knowledge about the potential applications of Blockchain in the context of customs and supply chain operations.

4.2 Criteria of Inclusion

The analysis process involved several key steps, including the use of tools such as word clouds, lexical analysis, and the application of a thematic analysis matrix using the Nvivo software. Firstly, we started our analysis by generating descriptive data and then creating a word cloud from the collected textual data. The word cloud will highlight the most frequently used terms, providing a visual overview of the key concepts related to how the PortNet blockchain facilitates import customs procedures.

This word cloud will help guide our analysis by identifying the most prominent aspects of the topic.

We then moved on to lexical analysis using NVivo software. This analysis will allow us to explore word associations and understand underlying meanings. We will examine term co-occurrences to identify specific relationships and contexts. Next, we will conduct a vertical analysis to study variations in vocabulary used in different import customs contexts and the overall international supply chain. This analysis will help us determine context-specific terms and assess their relative importance. Concurrently, we will also perform a condensed matrix analysis to compare trends and variations across different data sources, such as case studies and government reports. Finally, we will apply a specific model, perhaps an existing conceptual or theoretical framework, to analyze the results and interpret them in light of established conceptual frameworks.

By combining these different analyses—word clouds, lexical analysis, and the thematic analysis matrix—we will be able to generate a comprehensive understanding of the potential impact of blockchain on import customs procedures. These analyses will help us identify key themes, term relationships, contextual variations, and general trends while enabling us to draw solid conclusions and develop relevant recommendations in this field.

5 Results

The first bloc of analysis, word cloud, has highlighted key aspects of using blockchain to facilitate import customs procedures. Terms such as 'blockchain', 'PortNet', 'transparency', 'security', 'traceability', and 'automation' are observed. These words reflect the major advantages offered by blockchain in the customs domain. By utilizing blockchain, processes can become more transparent and secure, enabling customs authorities to authenticate and confirm the origin of imported goods. Furthermore, the traceability provided by blockchain allows for tracking the journey of goods at every stage, reducing the risks of fraud and counterfeiting. By automating certain tasks through blockchain, customs procedures can be expedited and simplified, thereby reducing the time and costs associated with importation. Overall, this word cloud illustrates how blockchain has the potential to revolutionize customs procedures and enhance the efficiency of international trade.

The proposed textual analysis comprehensively and accurately highlighted various aspects of facilitating import customs procedures through blockchain. It emphasizes the integration of different stakeholders, facilitation of cooperation with other services such as Ministry of Industrie and Trade, ONSSA, and ANRT streamlining customs clearance, digitizing commercial documents, and facilitating payment of financial charges. These key points perfectly illustrate the benefits and improvements brought about by the adoption of blockchain in customs clearance, international trade, and supply chain management. The integration of different stakeholders enables seamless collaboration and enhanced communication, facilitating the coordination of customs processes. Furthermore, facilitating cooperation with other services promotes better collaboration between customs authorities and other government agencies, thereby accelerating import procedures. Overall, this textual analysis provides a comprehensive view of the benefits offered by blockchain in facilitating import customs procedures. It highlights key points that enhance the efficiency, transparency, and security of customs processes, paving the way for smoother and more effective international trade.

The thematic analysis grid provides an in-depth understanding of the different themes and sub-themes related to this research topic. This method allows for breaking down the qualitative data collected into key elements, organizing them in a structured manner, and deriving meaningful insights. By using an inductive approach, we were able to identify relevant themes from the conducted interviews and initiate both horizontal and vertical analysis. The thematic analysis matrix facilitated the identification and categorization of emerging themes from a set of qualitative data. It is particularly relevant to this research topic as it allows for exploring the different dimensions of using blockchain in import customs procedures. For example, the horizontal analysis of the first interviewee allowed us to observe that they support our proposition that the PortNet blockchain facilitates customs procedures, streamlines the international supply chain, and digitizes international trade transactions. In fact, all the interviewees support the problem statement of this research.

Drawing from the semi-structured interviews and the coding of themes, this qualitative method demonstrated, through a vertical analysis of the theme "PortNet Blockchain and the facilitation of cooperation with other services," that interviewees support the idea that blockchain facilitates cooperation with other services by showing that almost all import authorization requests are exclusively made through the PortNet platform, thus making customs operations more flexible.

Indeed, our analysis revealed similarities in the opinions and perspectives of the interviewees. For example, the responses from the interviewees regarding the theme "PortNet Blockchain and the streamlining of customs clearance for imports" were almost identical, illustrating the extent to which people believe that the PortNet blockchain facilitates, streamlines, and simplifies customs clearance for all economic operators. Additionally, there is a unanimous consensus on the theme "Advantages of PortNet for managing import customs procedures" as all the respondents revealed that the most significant advantages of PortNet are the facilitation of customs procedures and the simplification of goods clearance.

To sum up, the thematic analysis matrix on blockchain and its facilitation of import customs procedures is a powerful method for exploring and understanding the various dimensions of this research topic. It allows for discovering emerging themes, analyzing the relationships between themes and sub-themes, and generating meaningful knowledge for the advancement of qualitative research. It can also serve as a basis for practical recommendations for policymakers or practitioners in the field.

6 Analyses and Discussions

Based on the results obtained, the analysis and discussion of the findings in this study highlight the positive aspects of using blockchain to facilitate and streamline import customs procedures in the Provincial Customs Directorate of Nouasser. This technology has shown positive implications, particularly in terms of simplifying international trade, managing the supply chain, and import customs clearance. The research focuses on three key areas: blockchain and international trade, blockchain and supply chain management, and the PortNet blockchain and its facilitation of import customs procedures. Drawing on the literature review and the empirical evidence, it appears that blockchain technology presents significant benefits, such as process automation, transaction transparency, goods traceability, and reduced timelines. These findings confirm that blockchain offers promising solutions to address challenges and inefficiencies associated with transactions in international trade.

Following the same reasoning, it is worth examining the potential benefits of using blockchain to reduce transaction costs, optimize document exchange processes, and strengthen security, transparency, and traceability among stakeholders. The respondents support the validity of a positive relationship between the PortNet blockchain and the international supply chain by providing valuable insights into potential improvements and advantages that blockchain can bring to this field. For example, one of the interviewees illustrated that a PortNet account would be sufficient to handle everything, starting from receiving import documents, making payments to suppliers through their bank, and concluding with the collection of goods. It seems that this technology has been able to confirm the extent to which the integration of blockchain technology into import customs procedures has reduced delays, simplified administrative formalities, and ensured transaction security. Theoretical findings have also shown how crucial blockchain technology is as a key performance indicator (KPI) for customs administrations. For instance, various research has emphasized that reducing customs clearance time is an indispensable KPI for this administration.

On the whole, based on the empirical evidence, blockchain technology offers many benefits and opportunities regarding import customs procedures by facilitating commerce operations, enhancing security and traceability, and integrating various operators in the same in field of international trade and supply chain management domain. However, it is important to critically examine certain aspects and challenges associated with this technology, including coordination among different stakeholders in the international logistics chain and the implementation of a powerful technological platform capable of addressing security challenges.

7 Limitations and Future Directions

Overall, the expected results have been achieved as the interviewed population positively acknowledges that the PortNet blockchain facilitates and streamlines import customs procedures. The study indicates that the expected outcomes have been achieved, as Customs inspectors, category A freight forwarders, and companies in the aerospace sector specifically recognize that the PortNet blockchain indeed facilitates and streamlines import customs procedures, thereby simplifying and improving the performance of the international supply chain.

However, since research on blockchain and import customs procedures is relatively new, there is a lack of available empirical and theoretical data to support the conclusions. Further research involving in-depth case studies, quantitative analyses, and real-world data can contribute to strengthening the empirical foundation of this research. Similarly, time and resource constraints can limit the scope of the research. Studies in this field often require extensive data collection, detailed analysis, and involvement of various stakeholders. The researchers were constrained by available time and resources, which can impact the overall theoretical findings. It is also important to note that the conclusions and recommendations made may not be directly applicable to other customs environments. Further research should be conducted in different regions and with diverse stakeholders to assess the extent and validity of this theoretical review. For example, including all freight forwarders from different categories in the research and exploring other industrial sectors such as the automotive industry.

Another major challenge relates to the dependence on existing technology. Although blockchain is designed to be interoperable with existing systems and infrastructure, technical limitations and integration challenges can hinder the practical implementation of blockchain. Further studies are needed to assess the practical implications of blockchain adoption and to propose solutions to potential integration issues. In fact, several public administrations are still far from being integrated into this technology, such as the Ministry of Agriculture, which continues to issue paper-based import permits as of the date of this study.

In summary, while this research has provided valuable insights into the use of blockchain in import customs procedures, it is important to acknowledge the study's limitations and constraints. Further research and close collaboration with stakeholders are necessary to deepen our understanding of this technology and its application in the customs domain. Indeed, future research can consider the following suggestions.

In order to enhance the scope of research on blockchain in supply chain management, particularly in customs procedures, an in-depth study on interoperability between different systems and platforms used in customs procedures is essential to ensure smooth functioning of blockchain technology. Future analysis could focus on the challenges related to interoperability between existing systems and blockchain solutions, and propose solutions to facilitate seamless integration. Several stakeholders in the supply chain are still less integrated into new technologies.

Additionally, exploring the relationship between blockchain implementation and change resistance is important, as understanding the factors influencing blockchain adoption in the customs domain is crucial for successful implementation. A future study could identify the factors that facilitate or hinder blockchain adoption by actors involved in customs procedures, as well as change management strategies to overcome potential resistance. We noticed that several interviewees, especially older ones, still rely on traditional methods such as physically verifying import proofs and do not solely rely on blockchain-based ones.

8 Conclusion

Exploring possibilities of artificial intelligence and machine learning would be a notable research direction, given recent advancements in this technology. Integrating AI and machine learning with blockchain could bring additional benefits to customs procedures. An interesting future direction would be to explore how these technologies can be used in combination with blockchain to enhance fraud detection, risk analysis, and decision-making in the customs domain.

References

Agarwal U, Rishiwal V, Tanwar S, Chaudhary R, Sharma G, Bokoro PN, Sharma R (2022) Blockchain technology for secure supply chain management: a comprehensive review. IEEE Access

Aitken J, Bozarth C, Garn W (2016) To eliminate or absorb supply chain complexity: a conceptual model and case study. Supply Chain Manag: Int J 21:759–774

Belu MG (2019) Application of blockchain in international trade: an overview. Roman Econ J 71:2–16

Benzidia S, Makaoui N, Bentahar O (2021) The impact of big data analytics and artificial intelligence on green supply chain process integration and hospital environmental performance. Technol Forecast Soc Chang 165:120557

Botton N (2018) Blockchain and trade: not a fix for brexit, but could revolutionise global value chains (if governments let it). ECIPE Policy Brief

Chang Y, Iakovou E, Shi W (2020) Blockchain in global supply chains and cross border trade: a critical synthesis of the state-of-the-art, challenges and opportunities. Int J Prod Res 58:2082–2099

Chawinga WD, Majawa F (2018) An assessment of Mzuzu University library after a fire disaster. Afr J Librar Arch Inf Sci 28

Chen MA, Wu Q, Yang B (2019) How valuable is FinTech innovation? Rev Financ Stud 32:2062–2106

Dargaoui S, Azrour M, El Allaoui A, Amounas F, Guezzaz A, Attou H, Hazman C, Benkirane S, Bouazza SH (2023) An overview of the security challenges in IoT environment. In: Mabrouki J, Mourade A, Irshad A, Chaudhry SA (eds) Advanced technology for smart environment and energy. Springer International Publishing, Cham, pp 151–160. https://doi.org/10.1007/978-3-031-25662-2_13.

Dobrovnik M, Herold DM, Fürst E, Kummer S (2018) Blockchain for and in logistics: what to adopt and where to start. Logistics 2:18

Douiba M, Benkirane S, Guezzaz A, Azrour M (2022) An improved anomaly detection model for IoT security using decision tree and gradient boosting. J Supercomput 1–20

Ganne E (2018) Can Blockchain revolutionize international trade? World Trade Organization Geneva

Gnata A. RELATORE: CH.MA PROF.SSA VALENTINA DE MARCHI

Guezzaz A, Asimi A, Mourade A, Tbatou Z, Asimi Y (2020) A multilayer perceptron classifier for monitoring network traffic. In: Big data and networks technologies, vol 3. Springer, pp 262–270

Guezzaz A, Benkirane S, Azrour M (2022) A novel anomaly network intrusion detection system for internet of things security. In: IoT and smart devices for sustainable environment. Springer, pp 129–138

Hackius N, Petersen M (2017) Blockchain in logistics and supply chain: trick or treat? In: Digitalization in supply chain management and logistics: smart and digital solutions for an industry 4.0 environment. In: Proceedings of the hamburg international conference of logistics (HICL), vol 23. epubli GmbH, Berlin, pp 3–18

Hastig GM, Sodhi MS (2020) Blockchain for supply chain traceability: business requirements and critical success factors. Prod Oper Manag 29:935–954

Hazman C, Guezzaz A, Benkirane S, Azrour M (2023a) Toward an intrusion detection model for IoT-based smart environments. Multimed Tools Appl. https://doi.org/10.1007/s11042-023-16436-0

Hazman C, Benkirane S, Guezzaz A, Azrour M, Abdedaime M (2023b) Intrusion detection framework for IoT-based smart environments security. In: Artificial intelligence and smart environment: ICAISE'2022. Springer, pp 546–552

Hazman C, Benkirane S, Guezzaz A, Azrour M, Abdedaime M (2023c) Building an intelligent anomaly detection model with ensemble learning for IoT-based smart cities. In: Advanced technology for smart environment and energy. Springer, pp 287–299

Hissou H, Benkirane S, Guezzaz A, Azrour M, Beni-Hssane A (2023) A novel machine learning approach for solar radiation estimation. Sustainability 15:10609. https://doi.org/10.3390/su151310609

Ioannou I, Demirel G (2022) Blockchain and supply chain finance: a critical literature review at the intersection of operations, finance and law. J Bank Financ Technol 6:83–107. https://doi.org/10.1007/s42786-022-00040-1

Khan S, Siddiqui T, Mourade A, Alabduallah BI, Alajlan SA, Almjally A, Albahlal BM, Alfaifi A (2023) Manufacturing industry based on dynamic soft sensors in integrated with feature representation and classification using fuzzy logic and deep learning architecture. Int J Adv Manuf Technol. https://doi.org/10.1007/s00170-023-11602-y

Kim H, Laskowski M (2017) A perspective on blockchain smart contracts: reducing uncertainty and complexity in value exchange. In: 2017 26th International conference on computer communication and networks (ICCCN). IEEE, pp 1–6

Kshetri N (2018) 1 Blockchain's roles in meeting key supply chain management objectives. Int J Inf Manag 39:80–89

Li X (2023) Inventory management and information sharing based on blockchain technology. Comput Ind Eng 179:109196

Mabrouki J, El Yadini A, Bencheikh I, Azoulay K, Moufti A, El Hajjaji S (2019) Hydrogeological and hydrochemical study of underground waters of the tablecloth in the vicinity of the controlled city dump mohammedia (Morocco). In: Advanced intelligent systems for sustainable development (AI2SD'2018), vol 3: advanced intelligent systems applied to environment. Springer, pp 22–33

Mabrouki MA (2021) Jamal: data analytics to assess the outbreak of the coronavirus epidemic: opportunities and challenges. In: Intelligent modeling, prediction, and diagnosis from epidemiological data. Chapman and Hall/CRC

Mabrouki J, Benbouzid M, Dhiba D, El Hajjaji S (2021) Internet of things for monitoring and detection of agricultural production. In: Intelligent systems in big data, semantic web and machine learning. Springer, pp 271–282
Mabrouki J, Fattah G, Kherraf S, Abrouki Y, Azrour M, El Hajjaji S (2022) Artificial intelligence system for intelligent monitoring and management of water treatment plants. In: Emerging real-world applications of internet of things. CRC Press, pp 69–87
Mohy-eddine M, Guezzaz A, Benkirane S, Azrour M (2023a) IoT-enabled smart agriculture: security issues and applications. In: Artificial intelligence and smart environment: ICAISE'2022. Springer, pp 566–571
Mohy-Eddine M, Azrour M, Mabrouki J, Amounas F, Guezzaz A, Benkirane S (2023b) Embedded web server implementation for real-time water monitoring. In: Mabrouki J, Mourade A, Irshad A, Chaudhry SA (eds) Advanced technology for smart environment and energy. Springer International Publishing, Cham, pp 301–311. https://doi.org/10.1007/978-3-031-25662-2_24
Myers MD, Newman M (2007) The qualitative interview in IS research: examining the craft. Inf Organ 17:2–26
Narayanan A, Bonneau J, Felten E, Miller A, Goldfeder S (2016) Bitcoin and cryptocurrency technologies: a comprehensive introduction. Princeton University Press
Okazaki Y (2018) Unveiling the potential of blockchain for customs. WCO Res Pap 45:1–24
Qu SQ, Dumay J (2011) The qualitative research interview. Qual Res Account Manag 8:238–264
Rahmanzadeh S, Pishvaee MS, Rasouli MR (2020) Integrated innovative product design and supply chain tactical planning within a blockchain platform. Int J Prod Res 58:2242–2262
Roeck D, Sternberg H, Hofmann E (2020) Distributed ledger technology in supply chains: a transaction cost perspective. Int J Prod Res 58:2124–2141
Saberi S, Kouhizadeh M, Sarkis J, Shen L (2019) Blockchain technology and its relationships to sustainable supply chain management. Int J Prod Res 57:2117–2135
Staples M, Chen S, Falamaki S, Ponomarev A, Rimba P, Tran AB, Weber I, Xu X, Zhu J (2017) Risks and opportunities for systems using blockchain and smart contracts. Data61. CSIRO, Sydney
Sunny J, Undralla N, Pillai VM (2020) Supply chain transparency through blockchain-based traceability: an overview with demonstration. Comput Ind Eng 150:106895
Suppatvech C, Godsell J, Day S (2019) The roles of internet of things technology in enabling servitized business models: a systematic literature review. Ind Mark Manag 82:70–86
Tapscott D, Kaplan A (2019) Blockchain revolution in education and lifelong learning. In: Blockchain research institute-IBM institute for business value
Tseng J-H, Liao Y-C, Chong B, Liao S (2018) Governance on the drug supply chain via gcoin blockchain. Int J Environ Res Public Health 15:1055
Tyagi NK, Goyal M (2021) Blockchain-based smart contract for issuance of country of origin certificate for Indian customs exports clearance. concurrency and computation: practice and experience e6249
Wang Y, Singgih M, Wang J, Rit M (2019a) Making sense of blockchain technology: how will it transform supply chains? Int J Prod Econ 211:221–236
Wang Y, Han JH, Beynon-Davies P (2019b) Understanding blockchain technology for future supply chains: a systematic literature review and research agenda. Supply Chain Manag: Int J 24:62–84
Wang Y, Jia F, Schoenherr T, Gong Y, Chen L (2020) Cross-border e-commerce firms as supply chain integrators: the management of three flows. Ind Mark Manage 89:72–88
Wouda HP, Opdenakker R (2019) Blockchain technology in commercial real estate transactions. J Proper Inves Financ 37:570–579
Yang C-S (2019) Maritime shipping digitalization: blockchain-based technology applications, future improvements, and intention to use. Transp Res Part E: Logist Transp Rev 131:108–117

Mapping of Low Flow Rates in the Srou River Basin, Upstream of Oum Er Rbia River, Morocco

Omar Ghadbane, Mohamed Chakir, Hassan Ouakhir, and Mohamed El Ghachi

Abstract Effective and sustainable water resource management necessitates precise and specific understanding of these resources. Surface flows arise from intricate interactions among subsurface hydrogeological processes, spatial and temporal fluctuations in meteorological inputs, and the unique characteristics of river systems in basins. However, current water resource management heavily relies on data from hydrometric and rainfall stations, which frequently exhibit gaps and inaccuracies. Furthermore, most of these stations are positioned at the outlets of basins, thereby monitoring extensive areas. It's crucial to note that basins might not always be the optimal choice for an in-depth water resource analysis, as sub-basins can exhibit significant geological, lithological, and topographical differences. Consequently, the cumulative flow rates recorded at these stations amalgamate contributions from various aquifers, each responding differently, particularly during periods of low flow. These distinctions become complex to redistribute spatially once consolidated. This study introduces a methodology aimed at generating hydrological data on a more refined scale, shifting the focus from the entire basin to more standardized units termed as unitary basins. This approach has been implemented in the Srou basin, a key tributary of the Oum Er Rbia river. The basin was divided into smaller catchments, with flow measurements conducted between 2016 and 2019, excluding periods of direct rainfall. These assessments underscore the diverse flow characteristics within the Srou river system, highlighting the necessity of considering these variations for effective water resource management.

Keywords Mapping · Low flow · Srou river basin · Hydrological profile · Water yield · Recession curve

---

O. Ghadbane (✉) · M. Chakir · H. Ouakhir · M. E. Ghachi
Laboratory of Landscape Dynamics, Risks, and Heritage, Sultan Moulay Slimane University, Beni Mellal, Morocco
e-mail: om.ghadbane@usms.ma

M. Azrour et al. (eds.), *Sustainable and Green Technologies for Water and Environmental Management*, World Sustainability Series,
https://doi.org/10.1007/978-3-031-52419-6_7

1 Introduction

Contemporary worries of scientists and policymakers primarily revolve around matters pertaining to climate change and environmental variability. These concerns stem from the combined impact of these phenomena on both society and the ecosystem, encompassing both immediate and long-term effects (Edenhofer 2014; Jouilil et al. 2013).

The ramifications of climate change extend across diverse sectors, encompassing agriculture, food security, water availability, biodiversity, public health, and the economy. Occurrences like prolonged droughts, heightened flooding, escalating sea levels, and alterations in marine and terrestrial ecosystems carry significant repercussions for both human communities and ecosystems (Meybeck et al. 2017). Simultaneously, climate shifts exert a notable influence on global water resources. Alterations in climatic patterns, encompassing increases in average temperatures, fluctuations in rainfall regimes, and glacier melting, directly impact water availability, distribution, and quality (Attaoui et al. 2023). In the face of these challenges, it becomes imperative to adopt climate change adaptation strategies for sustainable water resource management (Nascimento 2017). This could involve various measures such as water conservation, more efficient use of water resources, construction of innovative storage infrastructures, adoption of climate-resilient agricultural practices, and implementation of water management policies based on current climatic data (Brunel 2010).

Morocco, a country facing water resource constraints, holds about 22.5 billion cubic meters of surface water, of which only 16 billion are usable under current technical and economic conditions (4ème Rapport 2022). These resources are primarily concentrated in the North and Northwest regions, creating a significant deficit in the rest of the country. The increasing frequency of droughts in recent years exacerbates this situation, affecting almost the entire territory (Abdellaoui et al. 2002). Mountainous regions, often overlooked in hydro-climatic studies, also face drought-related constraints. The droughts observed in Morocco since 1980 highlight that these regions aren't spared from water shortages, especially during low flow periods.

In this backdrop, understanding low flow rates becomes pivotal for effective and sustainable water resource management, particularly in water-scarce areas like Morocco. The growing interest within the scientific community regarding these flows underscores their importance in water management (Devos et al. 1999; Lang Delus 2011). An in-depth knowledge of these rates offers significant advantages. It not only allows for anticipation of drought periods and the implementation of suitable adaptive measures but also optimizes water allocation among different uses, preserves fragile aquatic ecosystems, and strategically plans hydraulic infrastructures considering specific constraints (Smakhtin 2001).

However, defining and describing low flow periods remains intricate due to their extensive nature, which makes their identification challenging. The characterization of these periods can rely on statistical criteria such as daily, monthly flow rates, or rolling averages over several days, or even on flow thresholds (Lang Delus 2007; Ghachi 2007; Catalogne 2012).

Viewed from this perspective, the mapping of flow rates during periods of low flow holds immense importance for evaluating the geographic distribution of these rates. This mapping offers indispensable insights for sustainable water resource management, identifying regions where water shortages could be particularly acute. It aids in predicting and strategically addressing shortages, pinpointing susceptible water sources, guiding hydraulic infrastructure development by assessing needed storage capacities and requisite flows, supporting the safeguarding of aquatic ecosystems by evaluating minimum flow requirements, and informing planning within the agricultural and industrial sectors through crucial data for optimal water utilization. The aim of this study is to address the precise quantification of water resources in the Srou basin using an approach focused on measuring and monitoring flow rates during low flow periods, excluding the effect of rainfall at a finer scale, called a unitary basin.

To tackle this concern, the adopted methodology hinges on gathering streamflow data within the basin during low flow periods, meticulously excluding the influence of rainfall. This approach generates data tailored to distinct hydrological conditions.

The guiding hypotheses for this study are formulated to shape the approach for quantifying water resources in the Srou basin. The initial hypothesis posits that low flow periods within this basin, devoid of direct rainfall impact, present a chance to gauge aquifer flow rates. A separate hypothesis addressed pertains to the spatial dispersion of aquifer yields.

2 Methodology

2.1 Presentation of Study Area

The study is particularly focused on the Srou basin, a tributary of the Oued Oum-Er-Rbia, situated in the heart of the Central Middle Atlas. This basin encompasses an area of 1374 km^2 at the hydrometric station of Chacha N'mallah. Geographically, it stretches across the northwest region of the Middle Atlas, with its boundaries defined by longitudinal coordinates ranging from 5°6' to 6°50' west and latitudinal coordinates from 38°4' to 39°1' north (Fig. 1).

Administratively, the Oued Srou basin spans across the province of Khénifra and encompasses a small section of the Ifrane province. Its geographical boundaries are defined by the Hercynian massif of the central plateau to the west, the limestone plateau of the Middle Atlas to the north, and the Maroco-Oranian plateau to the southeast. The terrain of the basin is rugged, with elevations ranging from 700 to 2400 m. The landscape exhibits remarkable diversity, showcasing distinctive structural formations. The main river primarily flows from northeast to southwest.

The basin encompasses two distinct geological regions: the Mesetean domain to the west and the Atlas domain to the east. It is characterized by a range of geological

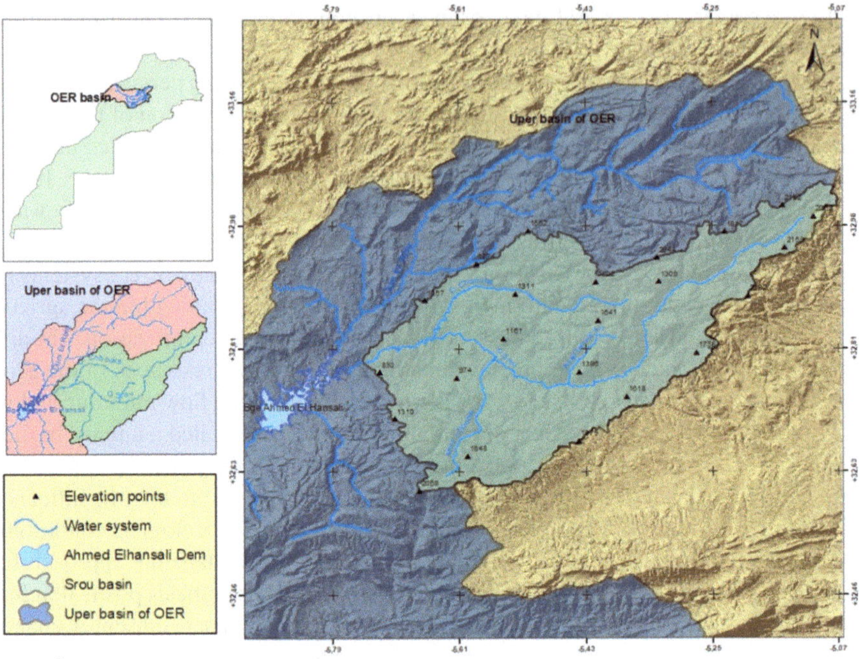

Fig. 1 Localization of the study area at different scales

formations, including early Ordovician sandstone-shale deposits in the west, layers of evaporitic clay intercalated with basaltic lava from the Triassic in the central region, and various limestone and marl-limestone formations from the Jurassic and Cretaceous in different parts of the basin. Quaternary deposits, such as limestone crusts and silts, are situated in the lower areas. This geological diversity significantly impacts the hydrological attributes and the distribution of water resources within the basin (Ghadbane 2022).

The diversity of rocks in the basin leads to different permeabilities. About 42% of the area is composed of impermeable materials such as clay, marl, and basalts. The main aquifer formations, covering about 43% of the area, are made up of Lias limestone and dolomite-limestone. This promotes surface flows and water retention. In the Chbouka river sub-basin, permeable limestone formations dominate, covering about 65% of its area, allowing significant flows despite its reduced size (Ghadbane et al. 2023).

The studies conducted by Ghadbane et al. (2021), Qadem (2015a), Lahlou (2021), Orfi et al. (2020), Ghachi and Morchid (2015), Serbout (2001) all converge towards the observation of marked interannual variability in precipitation. The rainfall fluctuations recorded at the Elhri (downstream) and Chacha N'mallah (upstream) stations reflect this trend. The results indicate a prolonged period of drought in the region, mainly between 1981 and 1994. Although interspersed with a few episodes of higher

Table 1 Hydrological characteristics of the Srou river basin

Basin	S in (Km^2)	Permeability (%)	Average annual flow (m/s^3)	Specific Q ($l/s/Km^2$)
Srou	1377	52.75	7.7	5.5
Chbouka	276	68.8	2.4	8.6

precipitation, these moments were sporadic and short-lived, extending from 1 to 4 years depending on the stations (Table 1).

This temporal sequence of drought is not unique to the region, as similar observations have been made in other parts of Morocco and Africa. References such as Assani (2000), Soro et al. (2011), Kouassi et al. (2018), Brou et al. (2005) support these conclusions. These observations underscore the persistence of the water availability problem in these regions and highlight the importance of understanding local climate trends for adaptive and sustainable water resource management.

2.2 *Field Work and Data Creation*

While we possess a comprehensive comprehension of lunar soil composition, our understanding of the Seine's flow in Paris remains relatively constrained (Qadem 2015b; Aldegheri 1979).

Recent hydrological studies aim to leverage the data supplied by hydrological services. Given the advent of climate change and the escalating scarcity of water resources, particularly in arid and semi-arid regions, attaining a profound understanding of this vital resource has become imperative. Our endeavor centers on generating hydrological data on a significant scale, achieved through the subdivision of watersheds into more compact and homogenous units termed as unitary basins. Within the Oued Srou basin, we embraced this subdividing approach, creating sub-basins assessed at the limnometric scale. Gauging campaigns were conducted in these designated areas, referred to as stations, from 2016 to 2019.

This methodology, already employed by the Rhine-Meuse Basin Agency in France to establish a catalogue of low flows (François et al. 1993), was taken up by Lebaut (2000) in the late 90s to study underground flow in sandy-gravelly formations. A similar approach was adopted in Morocco, particularly in the upper Sbou region (Devos et al. 1999), as well as by Qadem (2015b).

Monitoring of hydrometric Data. Measurements will be carried out when the hydrological conditions are stabilized at the exit points of the unit basins. The data collected during the gauging campaigns must be organized chronologically to provide an instantaneous overview of the overall flow throughout the basin. The flow rates will be calculated from the water speed measurements using a micro-propeller.

Selection of Measurement Locations. The measurement locations are carefully chosen considering the homogeneity of the sub-basins and the consistency of the flows, especially during low flow periods. Following the field surveys between February and June 2016 in the Oued Srou basin, three sub-basins were characterized for their permanent flow. These sub-basins are as follows: the Chbouka river sub-basin, the Assif Tamharrart sub-basin, and the Assif N'douma sub-basin. Despite their varying areas, these sub-basins exhibit surprisingly homogeneous lithology (Table 2) (Figs. 2 and 3).

Along the main watercourse, we have identified four strategically placed measurement points. The initial point is positioned upstream, chosen for its optimal accessibility, facilitating seamless monitoring. This precise location marks the commencement of the Oued Srou's distinctive nomenclature. The second point lies downstream from the confluence with the Assif Tamharrart tributary (upstream of the Lkbab bridge), while the third point is situated further downstream from this confluence (downstream of the Lkbab bridge). The final point is situated in the downstream

Table 2 Geological characteristics of Sub-basins in the River Srou Basin with perennial flow

Unitary basin	Area (km^2)	Dominant lithology	Permeability
Assif Tamharrart	41	Red clay	Impermeable
Assif Ndouma	85	Limestone	Permeable
Chbouka	331	Limestone and dolomite	Permeable

Fig. 2 Photo of Assif Tamharrart and Arrogo rivers dry in January 2016

Fig. 3 Flow of Assif Tamharrart and Assif N'douma in June 2016

area, away from urbanized zones and upstream of quarries that perpetually modify the riverbed, attributed to the construction of small dams for water retention. Additionally, this point is positioned downstream of the most densely populated sector within the Oued Srou watershed. In this region, two urban agglomerations discharge noteworthy volumes of wastewater into the Oued Srou, thereby disrupting its natural flow. The location of these points is represented on the map (Fig. 4).

To simplify the graphical representation of the measurement points, we have used the following coding: the measurement points are identified by uppercase letters derived from the initials of the Oued Srou and its tributaries (Table 3).

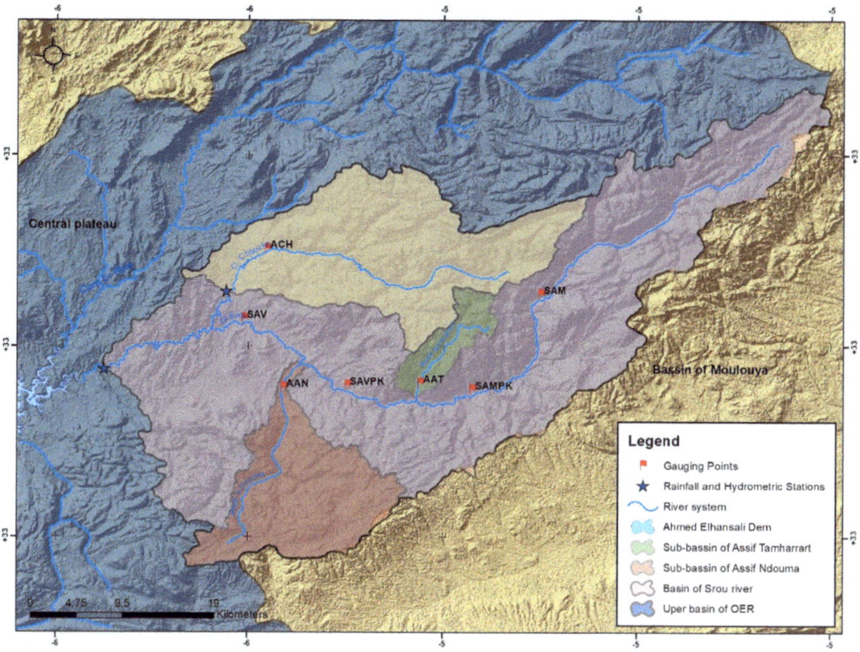

Fig. 4 Location of gauging points (DEM 2016)

Table 3 Coordinates and coding of gauging points

Gauging point	Code	X (m)	Y (m)	Z (m)
Assif Tamharrart Tributary	AAT	494,967	241,691	1188
Assif Ndouma Tributary	AAN	483,233	241,375	965
Chbouka Tributary	ACH	478,655	251,486	932
Upstream River Srou	SAM	505,615	250,393	1507
Upstream Srou at Lakbab Bridge	SAMPK	499,688	240,418	1373
Downstream Srou at Lakbab Bridge	SAVPK	488,814	241,000	1180
Downstream River Srou	SAV	480,003	242,284	923

Fig. 5 Limnimetric scale installed on the Srou river and the Assif N'douma tributary

Installation of Limnimetric Stations. A gauging station represents a location where various operations will be carried out to measure flow rates based on water heights, considering that the relationship between height and flow remains constant over time (Laborde 1645). Water heights will be measured directly using a graduated scale, fixed vertically at the edge of the watercourse. This scale, known as the limnimetric scale, is constructed of metal sheet (Fig. 5).

The installation of this scale is essential for monitoring flow variations throughout the year. The choice and installation of a station must meet the following criteria, in accordance with (Hydrologie 2008):

- The stretch of the watercourse must be straight.
- The cross-section must remain stable, with low sedimentation.

We positioned six limnimetric scales at the outlet of the sub-basins, as well as along the main watercourse of the Srou river.

Complete gauging method with micro-current meter. We chose a gauging method considering two criteria: the availability of suitable equipment, given our background in humanities and social sciences, and the fact that our study does not require sophisticated devices. We plan to carry out gauging during low flow periods, with low flow rates. We collaborated with a technician from ABHOER who has the necessary equipment.

We opted for the complete gauging method with a micro-current meter. This approach involves measuring the flow depth using a rod, simultaneously measuring speeds with the micro-current meter, and determining the widths of the cross-section. By combining this data, we can calculate partial flows for each segment and determine the total flow.

Fig. 6 Preparation and cleaning of the gauging section (Assif Tamharrart)

Measurement of the cross-section. The exact location of the gauging is not necessarily linked to the limnimetric scale as the flow remains relatively constant near it. The criteria for choosing the gauging site include a flow speed greater than 0.15 m/s, a stable riverbed, and a depth of at least 0.03 m. Width measurements and intervals between points are taken from the banks. The measuring points are arranged to reflect speed variations and ensure the accuracy of the results (Hydrologie 2008) (Fig. 6).

The vertical locations vary depending on the topography of the bed. The interval between two points generally does not exceed 1/20 of the total width of the cross-section. The depth is measured with the graduated rod of the micro-current meter, usually between 2/10 and 8/10 of the total depth (Hydrologie 2008).

Full Gauging Method with Micro-Current Meter. We chose a gauging method considering two criteria: the availability of suitable equipment, given our humanities and social sciences profile, and the fact that our study does not require sophisticated devices. We plan to conduct gauging during low flow periods with low discharge rates. We collaborated with a technician from ABHOER who has the necessary equipment (Fig. 7).

We opted for the full gauging method with a micro-current meter. This approach involves measuring the flow depth with a rod, simultaneously measuring speeds with the micro-current meter, and determining the widths of the cross-section. By combining these data, we can calculate the partial discharges for each segment and determine the total discharge. Next, the water speed at that specific point is calculated using the propeller's calibration equation.

$$v = k \cdot n + C$$

With:

- **k**: hydraulic pitch of the impeller of the flowmeter [m], determined through calibration tests conducted in a calibration channel.

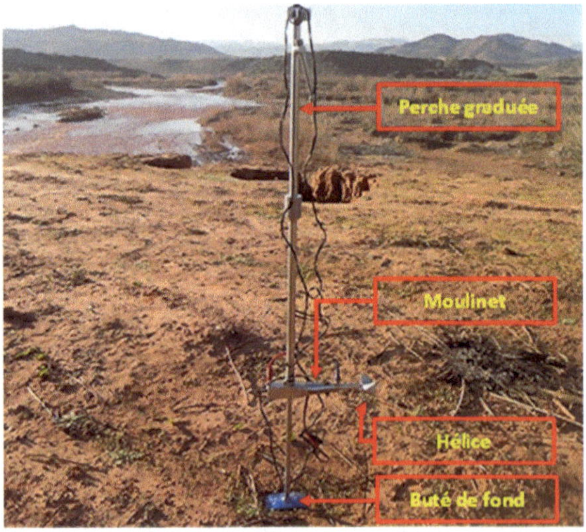

Fig. 7 OTT-type micro-current meter with main constituent elements

- **n**: number of revolutions of the impeller per second [1/s].
- **C**: impeller constant [m/s], determined through calibration tests conducted in a calibration channel.

Recording of Results and Gauging Analysis. During the gauging operations, flow velocity measurements are taken using the micro-current meter. These data, combined with the width and depth of the cross-section, are recorded on special sheets called "gauging sheets." Details such as the date, time, name of the gauger, and equipment information are noted. After data collection in the field, the processing, or analysis, is performed in the office. The raw data are transformed into average velocities and flows in cubic meters using the "DEPJAU" software, developed by the French Geological Survey (BRGM) of France. This software generates graphs from the raw data, using interpolation methods to obtain precise and visually informative results (Fig. 8).

Planning of Gauging Campaigns. Gaugin campaigns were organized between 2016 and 2019 to complement the data after 2016, focusing on periods of low flows. The months chosen for these campaigns correspond to the lowest flow periods of the year, based on statistical analyses of low flows, generally observed in July and August. These measurements were taken in stable hydrological conditions throughout the Srou watershed. In planning, precautions were taken to avoid disturbances due to precipitation, by allowing a five-day delay after rains. Particularly, the 2017–2018 hydrological year was dedicated to monthly campaigns to better understand flows during high water periods and to calibrate measurement equipment by establishing rating curves.

Hydrological Profiles. The hydrological profile, as described by Arts and François (1998) depicts the progression of river discharge from the upstream to downstream

Fig. 8 Measurement of width and depth of the stream section

points. This profile is crucial for understanding how a watercourse receives its supply, particularly during low-water periods, whether based on its length or area, as mentioned by Obda et al. (2001). These profiles, reflecting the accumulation of discharge along a watercourse, convey the geographical peculiarities of the watershed. According to Sary (1993), they highlight the numerous natural parameters that characterize the basin.

To establish a hydrological profile, having discharge data from a minimum of three points along the river is essential. Lejeune et al. (1989) indicate that these profiles are often graphically represented by curves illustrating the evolution of either raw or specific discharges, relative to the watershed area or distance along the river.

Regarding analysis, the slope of each segment on the graph provides a precise understanding of the specific discharge's evolution for a given section of the basin:

- A slope equal to 1 signifies that the specific discharge remains uniform throughout that section.
- If the slope exceeds 1, it suggests an exponential increase in discharge relative to the area.
- Conversely, a slope below 1 indicates that the area expands more rapidly than the growth of discharge.

Following the analysis of hydrological profiles, two major observations emerge:

- Discharges tend to align along nearly straight segments, and these patterns often parallel from one measurement session to another.
- Noticeable differences exist between different gauging sessions, thus illustrating variations in hydrological conditions specific to each session.

Water yield. Gauging campaigns provide insights into the structuring of flows outside the influence of precipitation. Discharges during low-water periods are converted into specific discharges and visualized through hydrological yield mapping, following (Zumstein 1969). This approach aims to globalize the analysis of hydrological profiles for all the unit basins in the relevant area.

The mapping illustrates the specific discharge of each elementary basin, thus highlighting variations in water between two measurement points while disregarding lateral contributions (Arts and François 1998; Arts and Sary 2000). It reveals the spatial distribution of aquifer drainage, largely influenced by the hydrogeological properties of the terrain. Through this mapping, it is possible to identify areas with low yields, and even areas with losses, as well as areas with average or high yields. This approach differs from traditional hydrological studies as it emphasizes the spatial heterogeneity of flows in limestone regions, considering spatial variabilities and geographical specificities (Sary 1993).

Recession Coefficients. Unlike large hydrological basins, unit basins are distinguished by their lithological homogeneity. This specificity allows for more stable recession coefficients that are unaffected by the hydrological phase (Lebaut 2000).

The analysis of this coefficient focuses on the relationship $Q = Q(t)$ without direct influence from precipitation (Hall 1968). By utilizing theoretical equations governing groundwater flows, it is possible to model the reduction of aquifer discharges based on the specific attributes of each aquifer. In essence, the recession coefficient contains data about storage capacity and aquifer characteristics (Tallaksen 1995).

By studying various recession formulas in a homogeneous environment, Berkaloff (1966) found that the Maillet formula was predominant, with 59% of the samples following this law.

The mathematical formulation to describe this decrease in discharge from an initial discharge (Q0) as a function of time (t) is:

$$Q_0 = Q_t.e^{-\alpha t}$$

3 Results and Discussion

The outcomes derived from the diverse gauging campaigns are gathered and subjected to an all-encompassing analytical procedure. The central aim of this analysis is to comprehend the flow patterns of the rivers within the Oued Srou watershed during periods of low flow.

3.1 Hydrological Profiles of the Srou River (2016–2019)

The hydrological profile of the Srou River basin was formulated using measurements obtained from four distinct points along the primary river course. These measurements were carried out during periods of low flow, with precautions taken to ensure that precipitation did not impact the regime during these intervals (Fig. 9).

The discontinuities evident in the hydrological profiles denote abrupt shifts or interruptions in water flow characteristics along a river or within a watershed. These disparities manifest across an area of approximately 619.04 km^2, specifically downstream of the Lkbab bridge. This might indicate a substantial alteration in the water flow dynamics at this specific point.

In regions spanning about 438.58 km^2, where the slopes are more gradual, lighter variations in slope are observable within the hydrological profiles. This phenomenon can be attributed to the presence of a smaller tributary known as Assif Tamharrart. The influx of water from this tributary trigger a localized surge in discharge, leading to the recorded slope fluctuations in the hydrological profiles. Tributaries often wield a significant influence on flow dynamics, occasionally inducing temporary alterations in flow attributes along the primary river course.

An intriguing finding is the existence of slopes below 1 observed over an expanse of 619.04 km^2 during the gauging campaigns on 05/08/2017 and 02/08/2019. This occurrence can be linked to the diversion and confinement of water bodies caused by quarries positioned upstream. Quarries can serve as artificial obstacles impeding

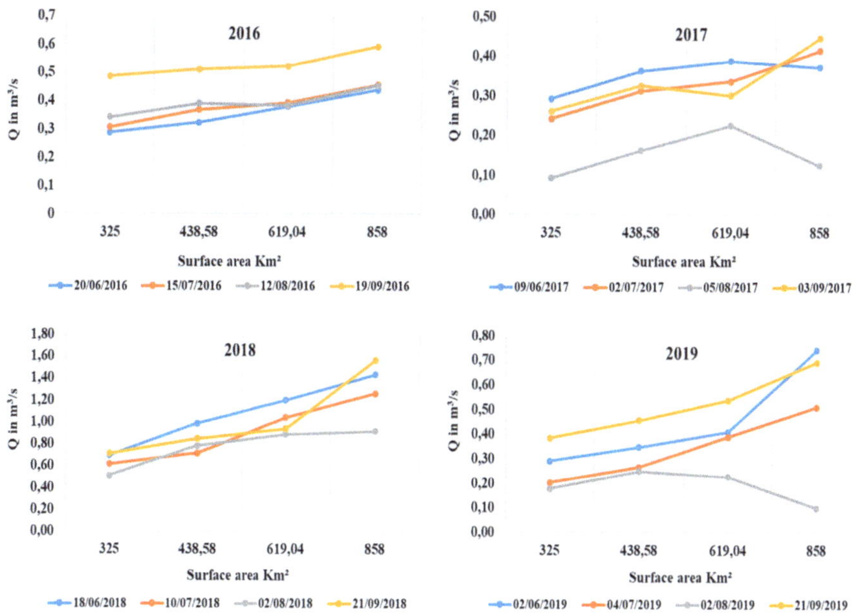

Fig. 9 Hydrological profiles of the main channel of the Srou River

the natural water flow, resulting in temporary redirection and a consequent decrease in discharge. This scenario could lead to a phase of low flow, thereby elucidating the slopes below 1 noted in the hydrological profiles during those specific periods.

3.2 The Water Yield of the Srou River Basin and Its Tributaries

The cartographic depiction of average yields per unit basin within the Oued Srou watershed (Fig. 10) underscores notable discrepancies in the allocation of water resources over a span of three years, spanning from 2016 to 2019. This examination unveils substantial divergences in water availability across various sectors of the basin.

Stations positioned in the upper portion of the basin display greater water yields in comparison to their downstream counterparts. This suggests a link between the volume of accessible water and the specific location within the basin. Upstream regions, notably the Chbouka river basin, exhibit elevated yields, while downstream

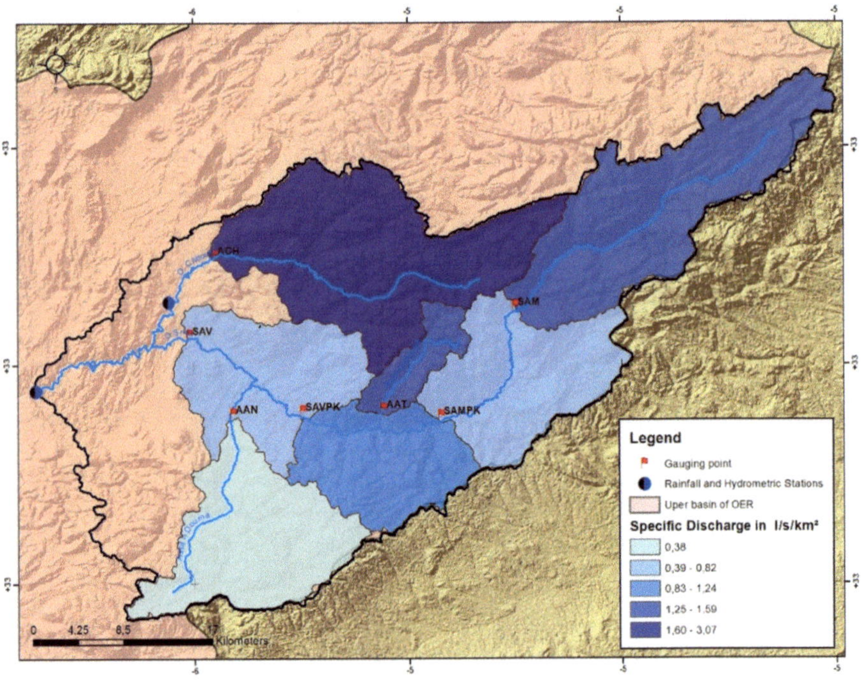

Fig. 10 Map of the average yield of unitary basins in the Srou River basin (based on 27 flow values) (DEM 2016)

zones, particularly those characterized by impermeable tertiary formations, exhibit lower yields.

The Chbouka river basin, situated in the northeastern area, stands out with yields reaching approximately 3.07 $l/s/km^2$. The prevalence of limestone formations, significant elevations, and substantial precipitation in the form of both rain and snow contribute to these heightened yields.

Furthermore, the region boasts abundant vegetation, including forests, which play a pivotal role in water regulation. Karstic terrains, renowned for their high absorption and retention capacities, along with sources like Arougou and lakes such as Tigalmamine, Agoulmane Miami, and Agoulmane Oumlil, serve as testaments to the water potential inherent in this area.

The upstream basin of the Srou river, also located in the northeastern region, enjoys a yield of approximately 1.47 $l/s/km^2$. Similar to Chbouka, the Srou region is characterized by limestone formations, notable elevations, and dense vegetation coverage. These attributes, coupled with ample precipitation, create a conducive environment for high yields, albeit slightly lower than those of Chbouka.

The Assif Tamharrart basin exhibits a yield of 1.59 $l/s/km^2$. Geological formations, in tandem with contributions from the Assif Tamharrat tributary, serve as pivotal factors in determining this yield.

Conversely, the Assif N'douma basin, despite favorable terrain, possesses a yield of approximately 0.365 $l/s/km^2$. Overexploitation of water resources, particularly for irrigation purposes, has detrimentally impacted its availability.

The significance of geology and river interactions in yield regulation is underscored by the contributions of the Assif Tamharrat tributary and areas characterized by impermeable tertiary terrains.

In conclusion, the examination of yields underscores the impact of geology, climate, vegetation, and human activities on water availability within the Oued Srou river basin, particularly within the Srou and Chbouka regions.

3.3 Recession Coefficients for the Srou River Basin and Its Unitary Basins

The determination of recession coefficient values is grounded in four measurement campaigns carried out in June and August 2016, along with June and September 2017. Each campaign is separated by intervals ranging from 77 to 108 days. The initial discharge, denoted as Q0, corresponds to the peak flow recorded during the inaugural June 2016 campaign, marked as t0. Subsequently, the subsequent campaign occurred in August of the same year (t1), with a 77-day interval between t0 and t1. These timeframes were deliberately selected to minimize the influence of rainfall on flow measurements. A gradual reduction in flows was observed throughout this phase of the year.

Subsequently, for the first measurement campaign in June 2017, the flow value is positioned along the recession curve. By identifying the days' interval corresponding to this flow value (noted as t3), it was determined that this span is 22 days for station ACH. Finally, the temporal difference between the two 2017 measurement campaigns is calculated by adding 22 days to the 86-day interval between these measurements. This sum designates the flow value for the concluding measurement campaign along the curve (Table 4) (Fig. 11).

A thorough examination of these recession coefficients provides us with a precise insight into the hydrogeological properties and water usage patterns within each basin. Let's delve into these observations.

Table 4 The monitored periods for Identifying recession coefficients in the Srou River basin and its Unitary Basins

Sub-basin	1st campaign	2nd campaign	Number of days	α
ACH	03/06/2016	19/08/2016	77	0,007
	09/06/2017	03/09/2017	108	
SAV	06/05/2019	02/08/2019	88	0,026
	02/07/2017	05/08/2017	64	
AAT	09/06/2017	05/08/2017	57	0,017
	02/06/2019	02/08/2019	64	
AAN	18/06/2018	02/08/2018	45	0,009
	09/06/2017	02/07/2017	52	

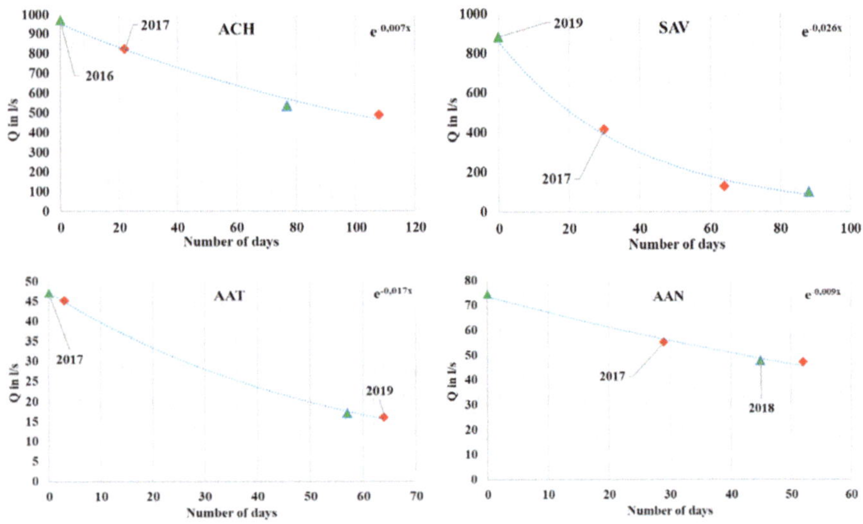

Fig. 11 Construction of recession curves for the Srou River basin and its Unitary Basins

The Oued de Srou basin, characterized by a coefficient of 0.026, signifies a swift drainage of its aquifer. A notable portion of this basin, particularly the middle and downstream regions, comprises impermeable formations that impede groundwater recharge. However, its flow primarily originates from the permeable upstream section. This geological setup, in combination with factors like intensive groundwater extraction and limited precipitation, could exacerbate the depletion of water reserves. This configuration renders the aquifer particularly susceptible to unregulated exploitation.

In contrast, the Assif Tamharrart basin, with a coefficient of 0.017, displays analogous dynamics of rapid drainage as observed in Oued de Srou. Soil composition, extraction rates, and rainfall might serve as decisive factors. A meticulous examination of this area is crucial to discern the underlying causes of this phenomenon.

Oued de Chbouka, with a modest coefficient of 0.007, indicates a more gradual drainage. The majority of this basin is covered with permeable formations, facilitating regular recharge and enhanced water retention. These circumstances, coupled with more abundant precipitation, could account for the resilience of its aquifer.

The Assif N'douma basin, featuring a coefficient of 0.009, demonstrates a similar propensity to that of Oued de Chbouka, although the specific determinants could vary.

In summary, these coefficients elucidate discrepancies among basins concerning aquifer recharge and drainage. They underscore the imperative of accounting for local geological particulars in judicious water resource management, ensuring the enduring sustainability of these invaluable reserves.

4 Conclusion

In summary, our meticulous exploration of the flow and hydrological characteristics of Oued de Srou has illuminated fundamental facets of water dynamics within this basin. The employed methodologies, ranging from gauging to hydrological profiles, and encompassing analyses of yields and recession coefficients, have collectively crafted a comprehensive depiction of how water circulates and is preserved within this region.

Hydrological profiles have emerged as indispensable tools for comprehending the fluctuations in flow along the river and its tributaries. These variations, often influenced by geographical and hydrogeological nuances, have unveiled how distinct sections of the basin respond to external triggers. Moreover, the scrutiny of hydrological yields has enabled the identification of regions within the basin possessing an elevated potential for water generation, offering insights into areas most suited for water supply.

Conversely, recession coefficients have played a pivotal role in determining the aquifers' capability to maintain stable flows. These coefficients not only mirror the underlying geology but also the permeability and retention capacity of formations.

Beyond comprehension alone, these findings carry significant ramifications for future water resource management in the region. With seasonal shifts in flow, the adjustment of irrigation strategies becomes imperative to ensure optimal utilization of available water. These data, when coupled with our comprehension of the basin's attributes, can steer the formulation of integrated water resource management plans, guaranteeing judicious conservation and utilization to meet both current and future demands of the region.

References

Abdellaoui B, Merzouk A, Aberkan M, Albergel J (2002) Bilan hydrologique et envasement du barrage Saboun (Maroc). Rev Sci Eau J Water Sci 15:737–748. https://doi.org/10.7202/705478ar

Aldegheri M (1979) Mesure des débits à partir des vitesses. Paris

Arts I, François D (1998) Profils hydrologiques du bassin de la Meurthe : Proposition d'une typologie. Profils Hydrol. Bassin Meurthe Propos. Une Typol 23:169–182

Arts I, Sary M (2000) Cartographie des débits d'étiage : application au bassin de la Meurthe. Rev Géographique Est 40. https://doi.org/10.4000/rge.4175

Assani AA (2000) Analyse de la variabilité temporelle des précipitations (1916–1996) à Lubumbashi (Congo-Kinshasa) en relation avec certains indicateurs de la circulation atmosphérique (oscillation australe) et océanique (El Niño/La Niña). Sci Chang Planétairessécheresse 10:245–252

Attaoui ZE, Sossi FZA, Khatori YE (2023) La gestion des risques menaçant la qualité de l'eau: application du PGSSE et la méthode HACCP. SHS Web Conf 175:01037. https://doi.org/10.1051/shsconf/202317501037

Berkaloff E (1966) Limite de validite des formules courantes de tarissement du debit. Bureau de Recherches Geologiques et Minières : Direction Scientifique, Service d'Hydrogéologie, Paris

Brou YT, Akindès F, Bigot S (2005) La variabilité climatique en Côte d'Ivoire : entre perceptions sociales et réponses agricoles. Cah Agric 14:533–540 (1)

Brunel S (2010) La biosécurité dans le bassin méditerranéen : les risques associés au commerce de plantes et aux produits végétaux

Catalogne C (2012) Amélioration des méthodes de prédétermination des débits de référence d'étiage en sites peu ou pas jaugés. https://www.theses.fr/2012GRENU030

Devos A, Nejjari A, Gille E, Akdim B, Benjelloul M (1999) Ressources et usages de l'eau dans le bassin supérieur de l'oued Guigou (Moyen-Atlas - Maroc). In: Etudes de géographie physique, pp 191–195

Edenhofer O (2014) Technical summary in: climate change 2014: Mitigation of climate change. contribution of working group III to the fifth assessment report of the intergovernmental panel on climate change. Technical report

El Ghachi M, Morchid FZ (2015) Analyse des tendances pluviométriques dans la ville de Khénifra dans un contexte de variabilité climatique (Rainfall trend analysis in Khenifra city in a context of climate variability). J Mater Environ Sci 6:3346–3358

El Orfi T, Ghadbane O, El Ghachi M, Lebaut S (2020) Variabilité des précipitations dans le bassin versant de l'oued Oum ErRbia en amont du barrage Ahmed El Hansali. In: Renforcement de la résilience des hydrosystèmes face aux changements globaux : de la mesure hydrologique aux modèles de gestion. Work bureau, FLSH Béni Mellal—Maroc

François D, Gille E, Zurnstein JF (1993) Analyse statistique des séries chronologiques et applications aux données hydroclimatiques

Ghachi M (2007) La Seille : un système fluvial anthropisé (Lorraine, France). https://hal.univ-lorraine.fr/tel-01748994
Ghadbane O, El Ghachi M, Chakir M (2021) Étude de la variabilité climatique et de ses impacts sur le régime hydrologique de l'Oued Srou, affluent de rive gauche de l'Oued Oum Er Rbia. Afr Sci Rev Int Sci Technol 18:69–80
Ghadbane O, Chakir M, Ouakhir H, Elhawari J, Achehboune A, Rahimi O, Bouhafa Y, El Ghachi M (2023) Variabilité des débits d'étiages « VCNd » dans un contexte climatique semi-aride: Cas du bassin versant de l'oued Srou, Haut Oum Er Rbi, Maroc [Variability of low-flow rates « MAMd » in a semi-arid climatic context: Case of the watershed of the river Srou, Oum Er Rbia upstream, Morocco. Int J Innov Sci Res 66:171–184
Ghadbane O (2022) Étiage et tarissement dans le bassin versant de l'oued de Srou (amont Oum-Er-Rbia-Maroc) (1976–2019) : identification, analyse et impact. https://doi.org/10.13140/RG.2.2.17090.35524.
Hall FR (1968) Base-flow recessions—a review. Water Resour Res 4:973–983. https://doi.org/10.1029/WR004i005p00973
Hydrologie—De la mesure à l'information hydrologique. Mondiale, Organisation Météorologique., 7 bis, avenue de la Paix, Case postale 2300, CH-1211 Genève 2, Suisse (2008)
Jouilil I, Bitar K, Salama H, Amraoui A, Mokssit A, Tahiri M (2013) Sècheresse météorologique au bassin Hydraulique Oum Er Rbia durant les dernières décennies. LARHYSS J P-ISSN 1112-3680 E-ISSN 2521-9782
Kouassi AM, Nassa RAK, Bi TMN, Kouame KF, Biemi J (2018) Caracterisation des debits d'etiage dans un contexte de changements climatiques: Cas du bassin versant du N'zi (Cote d'Ivoire). Agron Afr 30:215–223. https://doi.org/10.4314/aga.v30i3
Laborde J-P (2000) Eléments d'hydrologie de surface. Université de Nice—Sophia Antipolis, 1645, Route des Lucioles—06410 BIOT—France
Lahlou N (2021) Les crues dans le haut bassin de l'Oum Er-Rbia (Amont station Machraa Edahk) entre 1934–2018 : identification, mesure, suivie et extraction (Maroc). https://hal.science/tel-03563681
Lang Delus C (2007) Étiages et tarissements : vers quelles modélisations ? L'approche conceptuelle et l'analyse statistique en réponse à la diversité spatiale des écoulements en étiage des cours d'eau de l'Est Français. https://theses.hal.science/tel-00534656
Lang Delus C (2011) Les étiages : définitions hydrologique, statistique et seuils réglementaires. Cybergeo Eur J Geogr. https://doi.org/10.4000/cybergeo.24827
Lebaut S (2000) L'apport de l'analyse et de la modélisation hydrologiques de bassins versants dans la connaissance du fonctionnement d'un aquifère : les grès d'Ardenne-Luxembourg. https://hal.univ-lorraine.fr/tel-01748991
Lejeune O, Devos A, Marre A (1989) Spatialisation des débits d'étiage de la Blaise (Marne-France)
4ème Rapport sur l'État de l'Environnement du Maroc- version intégrale. Ministère de l'Energie, des Mines et de l'Environnement. Département de l'Environnement. Observatoire National de l'Environnement et du Développement Durable, Maroc (2022)
Meybeck A, Laval E, Lévesque R, Parent G (2017) Sécurité alimentaire et nutrition à l'heure des changements climatiques. In: Communication présentée au Actes du Colloque international organisé par le gouvernement du Québec en collaboration avec la FAO, Québec, pp 24–27
Nascimento J da R (2017) Gestion intégrée de l'eau et développement durable : le cas du Cap-Vert. https://theses.hal.science/tel-01814328
Obda K, Gille E, Akdim B (2001) Application de la méthodologie des profils hydrologiques dans la gestion des ressources en eau : cas des oueds Zloul (Moyen Atlas septentrional) et Nekor (Rif central). In: Mosella, pp 109–120
Qadem A (2015a) Quantification, modélisation et gestion de la ressource en eau dans le bassin versant du Haut Sebou (Maroc)
Qadem A (2015b) Quantification, modélisation et gestion de la ressource en eau dans le bassin versant du haut Sebou (Maroc). Géographie Univ, Lorraine

Sary M (1993) Signification du profil hydrologique de basses eaux. Cas de la Plaine dans le Massif Vosgien. In: L'eau, la terre et les hommes. Au fil de l'eau, pp 129–136

Serbout R (2001) Le bassin versant de l'oued Oum er Rébia en amont de Déchra el Oued (moyen-Atlas) : contribution à l'étude des phénomènes hydro-climatiques : thèse.... https://www.theses.fr/2001NICE2043

Smakhtin VU (2001) Low flow hydrology: a review. J Hydrol 240:147–186. https://doi.org/10.1016/S0022-1694(00)00340-1

Soro TD, Soro N, Oga YM-S, Lasm T, Soro G, Ahoussi KE, Biémi J (2011) La variabilité climatique et son impact sur les ressources en eau dans le degré carré de Grand-Lahou (Sud-Ouest de la Côte d'Ivoire). Physio-Géo Géographie Phys Environ 55–73. https://doi.org/10.4000/physio-geo.1581

Tallaksen LM (1995) A review of baseflow recession analysis. J Hydrol 165:349–370. https://doi.org/10.1016/0022-1694(94)02540-R

Zumstein J-F (1976) Contribution à l'élaboration de cartes hydrologiques. Cartes des étiages mensuels des bassins de la Meuse et de la Moselle de 1969 à 1973. Houille Blanche 569–581

The Temporal and Spatial Dynamic of Soil Erosion and Suspended Sediment Concentration in the Upstream Part of Oum Err Bia River (Middle Atlas/ Morocco)

Halouan Said, Ennaji Nadia, Ouakhir Hasan, and Abahrour Mohamed

Abstract Soil erosion is one of a major environmental issue in many regions of the world, especially in the Mediterranean areas. Due to their climate aggressivity and soil fragility, majority of Mediterranean countries are affected by water erosion. With its localization in the Middle Atlas of Morocco, the upstream part of Oum Err Bia basin is one of the influenced areas by soil erosion. This part of Morocco characterized by steep slopes and weak rocks which make soils vulnerable to water erosion. The purpose of this study is to quantify soil erodibility by applying Revised Universal Soil Loss Equation (RUSLE(, and validate it in the four selected catchments (Tamadout, Faraoun, Almou and Ait Addi catchments). Besides, 32 bottles of samples were collected in the outlet of representative catchments so as to estimate the concentration of sediment. As a result, the obtained annual soil loss average was 145 75 t ha^{-1} yr^{-1}. However, the concentration of sediment ranged from 0.2 to 234.5 g/l^{-1}. The results confirmed that different factors affect and control the soil loss like intense rainfall events, land uses type, lithology and human activities.

Keywords Upstream part of Oum Err Bia Basin · RUSLE model · Soil loss · Concentration of sediment transport

H. Said (✉) · E. Nadia · A. Mohamed
Sidi Mohamed Ben Abdellah University, Taza, Morocco
e-mail: said.halouan@usmba.ac.ma

O. Hasan
Sultan Moulay Slimane University, Beni Mellal, Morocco

M. Azrour et al. (eds.), *Sustainable and Green Technologies for Water and Environmental Management*, World Sustainability Series,
https://doi.org/10.1007/978-3-031-52419-6_8

1 Introduction

Soil erosion emerges as a natural process, constituting a customary geological occurrence intertwined with the intricacies of the hydrologic cycle (Ouakhir et al. 2022). It arises through the detachment caused by the kinetic energy of raindrops and the subsequent movement of soil particles by water from their original location. This process leads to the deterioration of water quality, soil fertility, and a decrease in dam capacity (Benchettouh et al. 2017). Unconsolidated soils have the potential to be carried from level or undulating landscapes through strong winds (Kouli et al. 2009). Simultaneously, the erosion caused by water predominantly involves the degradation of earth materials through processes of rill and interrill erosion. Rill erosion encompasses the detachment and conveyance of soil particles due to the focused movement of surface water, whereas interrill erosion transpires as raindrops combine with shallow overland flow (Beskow et al. 2009). Having just 19% of its land suitable for farming, Morocco confronts significant environmental hurdles and is under substantial stress concerning its soil and water reserves (Benmansour et al. 2013). According to the findings of Namr and Mrabet (2004) erosion exerts its primary on-site influence on the agricultural scenery of Morocco by diminishing soil fertility, while its main off-site consequence involves heightened siltation in water reservoirs. A study revealed that among the 22.7 million hectares of land with potential use in Northern Morocco, as highlighted by Sadiki et al. (2007), a substantial 77% are susceptible to exceedingly severe erosion hazards.

Indeed, approximately 100 million t/yr^{-1} of soil are eroded, leading to an estimated reduction in reservoir capacity due to sedimentation amounting to 75 million m^{-3}/yr^{-1}, as stated by the Secretariat of water, environment and forest in Morocco during 2001. Additionally, within the Middle Atlas region, Mosaid et al. (2022) demonstrated significant soil erodibility using the EPM model. The recorded maximum annual soil loss rates ranged from 5 to 652 m^3/km^2/yr^{-1}, with an average of 49 m^3/km^2/yr^{-1}. Furthermore, Ennaji et al. (2022a) found in two catchments within middle atlas, a high soil loss ranged from 39,007 to 164,420 m^3/km^2/yr^{-1}. In this specified area of Morocco, uncontrolled soil erosion and the resulting land degradation have rendered extensive catchments economically nonproductive (Ouakhir and Ghachi 2023). Hence, the absence of sufficient information prevents the determination of the vulnerability of agricultural sectors, which would necessitate specific water and soil management preservation techniques. The aim of this present paper is to apply a model for evaluating water erosion in the upper part of the Oum Err Bia River located in the Middle Atlas region of Morocco. The Revised Universal Soil Loss Equation (Renard et al. 2000) constitutes an empirical framework tailored for application in scenarios encompassing runoff plots or individual hillslopes (Fu et al. 2000). Leveraging the capabilities of GIS and remote sensing technologies, the RUSLE model has been harnessed as a distributed framework for assessing soil loss. Its utility extends beyond regions characterized by moderate slopes, encompassing topographically intricate landscapes as well (Sun et al. 2014).

Nevertheless, the predictive capability of RUSLE can extend to ungauged catchments, relying on insights into catchment characteristics and localized hydroclimatic conditions (Angima et al. 2003). However, the estimation of sediment yield remains beyond its reach (Renard et al. 2000). Within this investigation, the utilization of RUSLE has extended to encompass expansive watershed scales, and various approaches, including the upscaling technique, have been explored to calculate the constituent factors of RUSLE.

2 Presentation of Study Area

The research was conducted in the upper portion of the Oum Err Rbia River basin (OERB) (6273.33 km^2) located in the Middle Atlas region of Morocco (Fig. 1a, b). OERB is among the largest basins in Morocco, covering nearly 35,000 km^2, which is about 7% of Morocco's total (Ouatiki et al. 2017). The study area, situated in the northeast part of OERB (Fig. 1b), has a rugged topography with elevations ranging from 405 m a.s.l at the outlet to 2400 m a.s.l (Boudhar et al. 2020). The annual precipitation in the study area varied from 1100 mm in the Middle-Atlas Mountains to 300 mm in the river outlet region, with a mean of 550 m (ABHOER 2022). In terms of lithology, the most dominant lithology rocks are limestone formation sub- tabular, Liasic dolomitic cretaceous, doleritic basalt of Trias, quartzites and red clay (Fig. 1b) (Mosaid et al. 2022). This rainfall is concentrated from October to April, and above 800 m, it snows for 20 days on average each year. The temperatures in the study area ranged from 5 to 50 °C, with mean minimum and maximum values of 3.5 °C (January) and 38 °C (August) (USAID, 2010). The climate in OERB varies significantly, ranging from humid in coastal and mountainous regions to semi-arid in the plains, with cold winters and dry summers (Jazouli et al. 2019). This climatic diversity creates favorable conditions and potential for agriculture and water resources. Additionally, numerous reservoirs and dams, such as the Ahmed El Hanssali dam, have been constructed on OER to produce hydroelectric power, reserve water for irrigation, and supply water for domestic and industrial use. The OER and its tributaries pass through regions with significant agricultural activities, including large irrigated areas and livestock, as well as industrial activities such as oil mills, phosphate extraction, and sugar beet processing (Barakat et al. 2019).

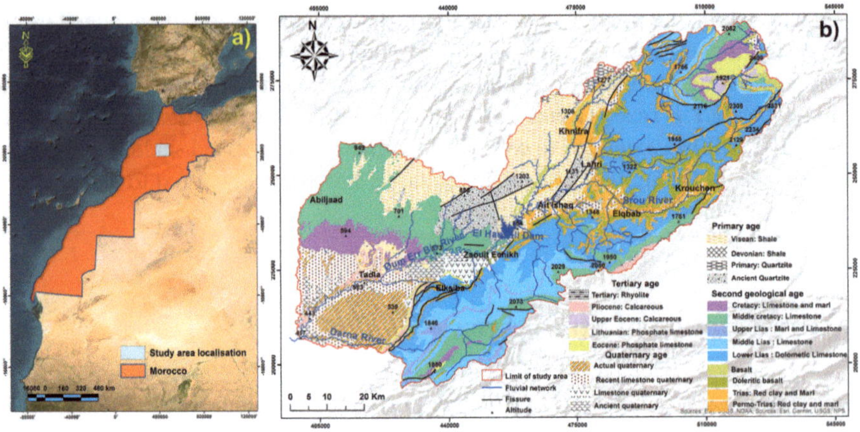

Fig. 1 **a** Localisation of study area in Morocco and **b** the lithology map of the upper part of Oum Err Bia River basin (OERB)

3 Methods and Laboratory Work

3.1 Methodology

The Revised Universal Soil Loss Equation (RUSLE) is a widely used model for predicting soil erosion (Ebrahimzadeh et al. 2018). It takes into account several factors that contribute to soil erosion: R (rainfall erosivity), K (soil erodibility), LS (slope length and steepness), C (cover management), and P (erosion control practices).

- *R (Rainfall Erosivity)*

Rainfall erosivity represents the energy of rainfall as it impacts the soil surface and causes erosion. It takes into account the intensity, duration, and frequency of rainfall events. High-intensity rainfall events are more erosive than light rain. R is usually quantified using indices like the R-factor, which combines the rainfall characteristics into a single value. A high R-factor indicates a higher potential for soil erosion. Regions with high R-values are more prone to erosion, which means they need greater soil conservation measures (Ouakhir et al. 2022).

- *K (Soil Erodibility)*

Soil erodibility refers to the susceptibility of soil to erosion. It considers factors like soil texture, structure, organic matter content, permeability, and cohesion. Soils with higher erodibility are more prone to erosion. For instance, sandy soils are generally more erodible than clay soils due to their loose structure. The K-factor is used to quantify erodibility and is incorporated into the RUSLE equation. Management practices that improve soil structure and organic matter content can reduce the K-value and decrease erosion potential (Xu et al. 2008).

- *LS (Slope Length and Steepness)*

The LS factor accounts for the combined effect of slope length and slope steepness on erosion. Longer slopes and steeper gradients increase the risk of soil erosion because they enhance the speed and volume of water flow across the surface. The LS factor is calculated based on a mathematical relationship between slope length and steepness. Areas with high LS values are more susceptible to erosion and require specific erosion control strategies, such as terracing, contour farming, or planting cover crops (Foster et al. 2003).

- *C (Cover Management)*

The C-factor represents the land cover and land management practices that influence soil erosion. It considers how much of the soil surface is covered and protected by vegetation or residue. A higher C-factor indicates less protective cover, leading to higher erosion potential. Cropland without cover or with minimal vegetation has a higher C-factor compared to areas with dense vegetation cover or residue from previous crops. Implementing cover crops, conservation tillage, or maintaining natural vegetation can reduce the C-value and minimize erosion (Demirci and Karaburun 2012).

- *P (Erosion Control Practices)*

The P-factor represents the effectiveness of erosion control practices that are applied to the land. It reflects the impact of measures taken to reduce erosion, such as terracing, contour ploughing, and grassed waterways. A higher P-factor signifies better erosion control, resulting in reduced soil loss. Properly designed and maintained erosion control practices can significantly lower erosion rates and protect the soil resource (Brahim et al. 2020).

In summary, the RUSLE model factors (R, K, LS, C, and P) collectively help to quantify and predict soil erosion risk. By understanding and managing these factors, land managers and conservationists can implement strategies to reduce erosion, protect soil fertility, and ensure sustainable land use practices (Fig. 1). the equation of RUSLE model can be expressed as follows:

$$\mathbf{A} = \mathbf{R} \times \mathbf{K} \times \mathbf{LS} \times \mathbf{C} \times \mathbf{P} \tag{1}$$

A stand for the mean soil loss per unit area due to erosion (in t ha^{-1} $year^{-1}$). R represents the factor of rainfall erosivity (in MJ mm ha^{-1} h^{-1} $year^{-1}$), K symbolizes the soil erodibility factor (in t ha h MJ^{-1} ha^{-1} mm^{-1}), while L denotes the factor of slope length, S stands for the steepness factor, C represents the factor related to cover and management practices, and P signifies the factor associated with conservation support practices. The values attributed to L, S, C, and P are dimensionless (Xu et al. 2008) (Fig. 2).

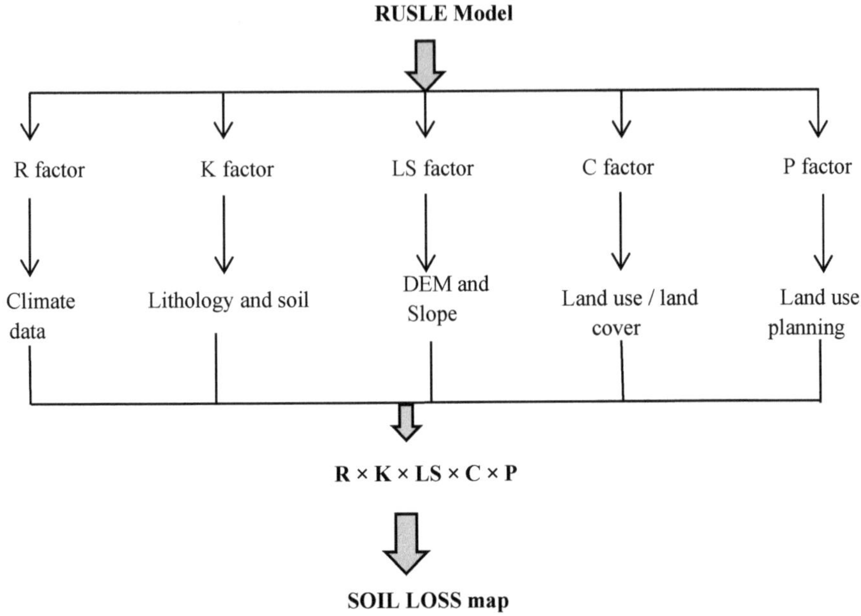

Fig. 2 Diagram explained the factors of the RUSLE model

3.2 *Laboratory Work and Validation of Model*

The RUSLE model was validated by measuring the suspended sediment concentration (SSC) in the outlet of four selected catchments within the study area. Extracting the SSC (g/l^{-1}) in laboratory work involves several steps. The first one was collecting 32 representative samples of the sediment from the water source in the outlet, usually by filtering 250 ml volume of water through a pre-weighed filter. The filter and sediment are then dried in an oven to remove any remaining water. Next, the dry filter and sediment are weighed again to determine the weight of the sediment. This weight is then divided by the volume of water filtered to obtain the concentration of suspended sediment in the water. Different parameters were measured in the laboratory (Turbidity (N.T.U), salinity (g/l), Conductivity (μ S/cm d'), PH and SSC (g/l^{-1}). Additionally, laboratory protocols and quality control measures followed to ensure reproducibility and accuracy of results.

4 Results

4.1 Presentation of RSULE Results

- *Rainfall erosivity (R)*

Erosivity factor (R)is a measure of the potential for rainfall to cause erosion, and the factor R quantifies this potential based on the intensity and amount of precipitation (Ennaji et al. 2022a). The values range from 50.07 to 81.56 n/h, indicating variations in erosivity potential within the studied area. Figure 3 represents the map of R factor in the studied area, the highest erosivity factor observed is 81.56, suggesting that during that specific period, there was a relatively intense and substantial amount of rainfall that could lead to increased erosion. The lowest erosivity factor recorded is 50.07n/h, indicating a period with milder rainfall conditions in terms of erosivity potential. The values in between exhibit moderate to relatively high erosivity potential, showcasing fluctuations in precipitation patterns that could impact erosion levels. There are instances of repeating values, such as 53.74 and 51.15 n/h suggesting similar erosivity conditions during those periods. The overall pattern indicates a mix of moderate to high erosivity periods, with some fluctuations in between. Consequently, it's important to consider these values alongside other relevant data, such as land cover, soil types, and topography (Renard et al. 2000). This will help provide a clearer understanding of how erosivity may interact with the landscape and potentially influence erosion processes in the upstream part of the Oum Err Bia River basin.

- *Soil Erodibility factor (K)*

The Fig. 4 represent the soil erodibility factor (k) values in the upstream part of the Oum Err Bia River basin, which is a measure of how susceptible the soil is to erosion (Jazouli et al. 2017). The range of values is from 0.2 to 0.5.

The range from 0.2 to 0.5 indicates that the upstream part of the basin consists of different types of soils with varying erodibility characteristics. A higher k value suggests that the soil is more susceptible to erosion, while a lower value indicates that the soil is relatively resistant to erosion.

The fact that the k values fall within this range suggests that the soils in the area generally possess moderate erodibility. This means that they might be prone to erosion under certain conditions but might also have some inherent resistance against erosion. Furthermore, these values emphasize the importance of considering soil erodibility when planning land use and management practices in the region. Areas with higher k values may require more erosion control measures and careful land management to prevent soil loss.

Areas with higher k values may benefit from adopting soil conservation measures like terracing, cover cropping, and maintaining vegetative cover to minimize erosion.

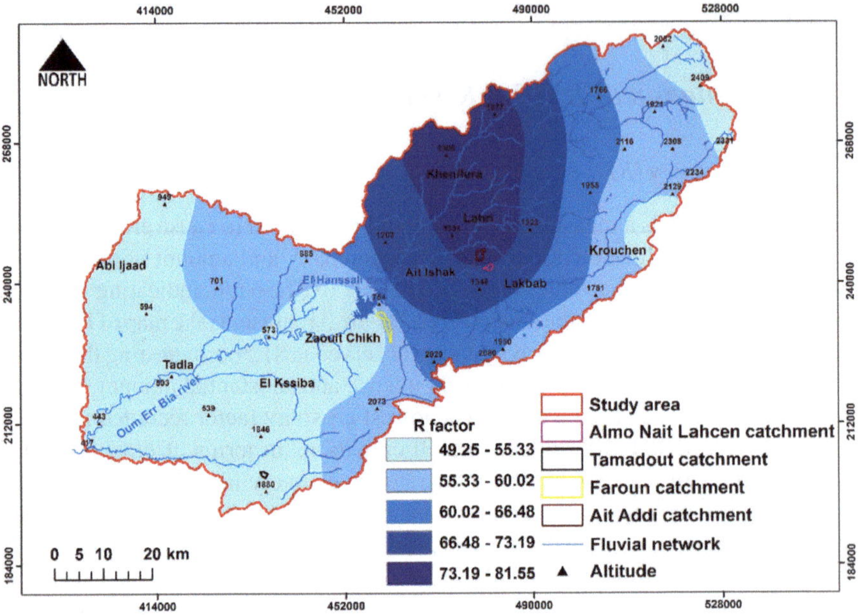

Fig. 3 Result of R factor

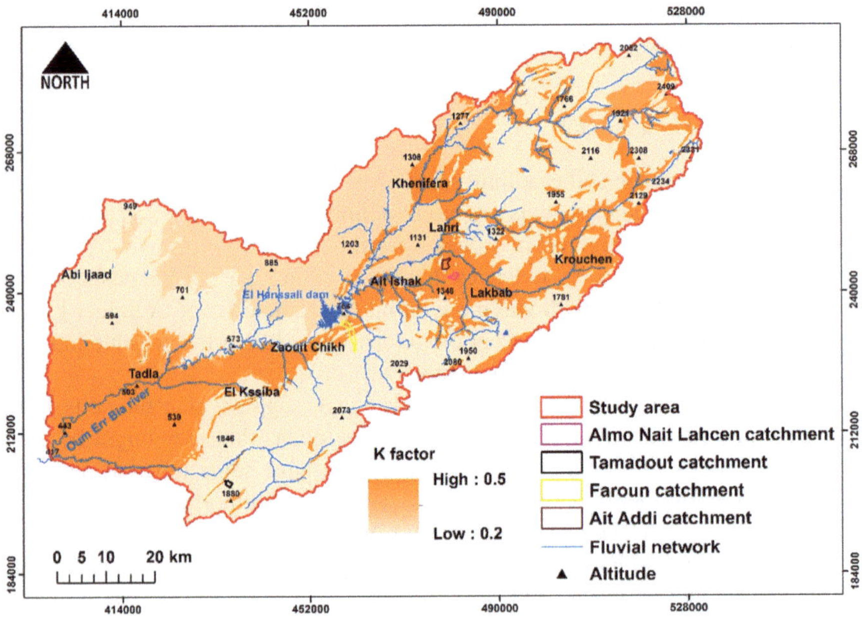

Fig. 4 Map of soil erodibility factor

In summary, the range of soil erodibility factor (k) values from 0.2 to 0.5 in the upstream part of the Oum Err Bia River basin indicates a moderate susceptibility of the soils to erosion. Proper land management practices should be considered to mitigate potential erosion risks and maintain soil health in the region (Fig. 4).

- *Slope and length factor (LS)*

The slope factor refers to the inclination or steepness of the land surface (Mati et al. 2000). In the context of a river basin, slope plays a crucial role in determining the speed at which water flows downstream (Rapp et al. 1972). Steeper slopes generally lead to faster water flow, which can impact erosion, sediment transport, and flooding potential. A lower slope might indicate flatter terrain with slower water movement (Ouakhir et al. 2016). Basing on the Fig. 5, The characteristics of the studied area can influence the entire river's behaviour downstream. LS values ranged from 0 to 211.20, which indicated a high inclination of the studied area.

- *Land cover management (C)*

Certainly, it appears basing on the Fig. 6 that a provided list of values, representing different levels of protections of land cover management in the studied area. In general, the trend of increasing values suggests an improvement in land cover management over time with intervention. This might involve activities such as reforestation, erosion control, sustainable agricultural practices, and more.

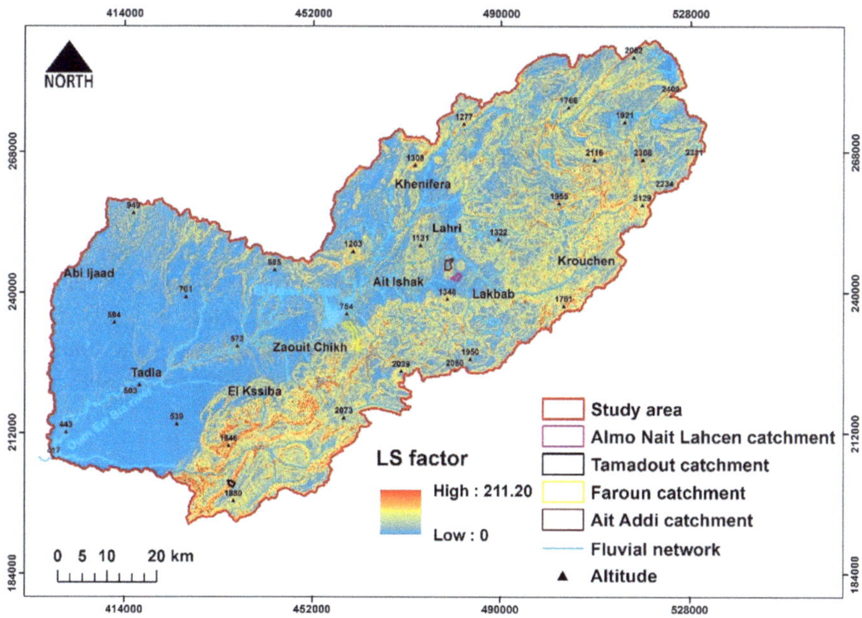

Fig. 5 Map of LS factor

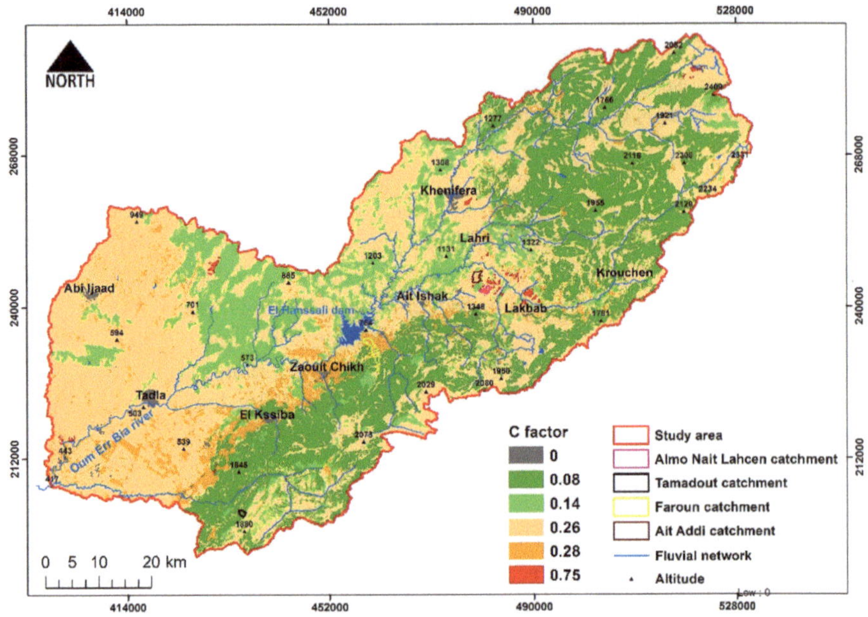

Fig. 6 The map of land management factor (C)

- *Annual soil loss (A)*

The table 1 and figure present the results of soil loss in the studied basin. A very low soil loss was registered in the downstream part of the basin, where soil less than 10.44 t/ha/yr^{-1}, covering approximately 3703.28 km^2, which accounts for about 59.10% of the total area. This category encompasses areas with slightly more erosion than. The soil loss falls within the range of 10.44–32.63 t/ha/yr^{-1}. These areas cover around 1897.44 km^2 and about 30.28% of the total area. Besides Moderate erosion covers areas with moderate soil loss. The soil loss range is 32.63–65.91 t/ha/yr^{-1}. These areas occupy 523.46 km^2 and approximately 8.35% of the total area. This category includes areas with higher levels of soil loss. The soil loss range is 65.91–154.67 t/ha/yr^{-1}. These areas account for 134.67 km^2 and about 2.15% of the total area. Furthermore, very high erosion was shown in the Srou basin, where soil loss is greater than 154.67 t/ha/yr^{-1} (Fig. 7 and Table 1).

These areas cover a relatively small area of 7.04 km^2 and about 0.11% of the total area. Overall, this table presents a comprehensive breakdown of soil erosion levels, corresponding soil loss ranges, and their distribution over the specified area. It's a valuable tool for understanding the extent and severity of soil erosion in different regions within the area of interest.

Table 1 Distribution of areas of erosion degree in the studied basin

Degree of erosion	Soil loss in t/ha^{-1}/yr^{-1}	Area (km)	Area (%)
Very low	<10.44	3703.28	59.10
Low	10.44–32.63	1897.44	30.28
Moderate	32.63–65.91	523.46	8.35
High	65.91–154.67	134.67	2.15
Very high	>154.67	7.04	0.11
Total		6265.89	100

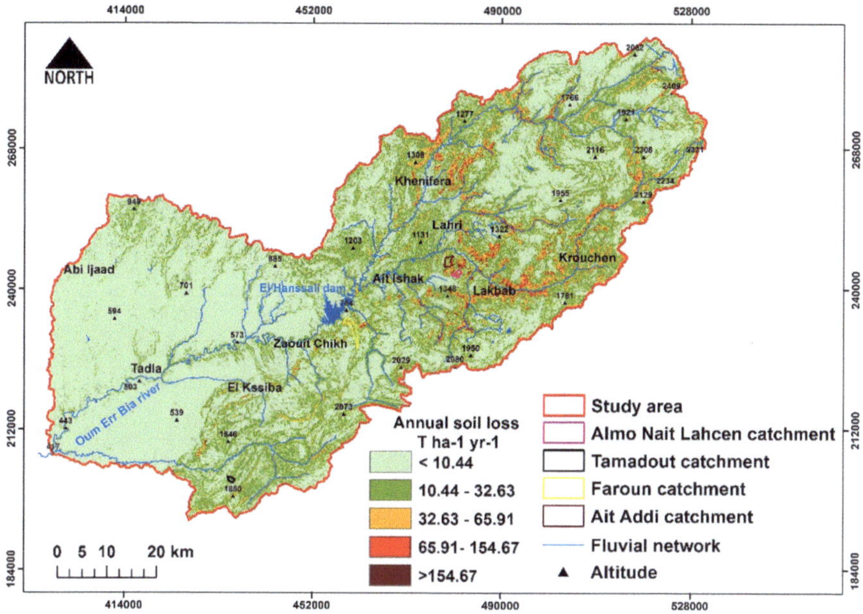

Fig. 7 Annual soil loss map (A)

4.2 *Dynamic of Suspended Sediment Transport in the Validate Areas*

The followed table provided a dataset with various parameters measured in different catchments within laboratory. These parameters are PH, conductivity (Cond), resistivity (Resis), salinity (Sal), turbidity (NTU), and suspended sediment concentration (SSC) (Table 2).

Table 2 Results of laboratory work of the studied catchments

Catchment name	Date	SP	PH	Cond	Resis	Sal	NTU	(SSC) g/l
Ait Addi	27/11/2021	A1	3.7	731.5	7.3	0.5	110	114.94
	27/11/2021	A2	3.5	683	3	0.5	4000	110
	3/12/2021	A3	3.3	711	2.8	0.4	4000	13.28
	26/12/2021	A4	4	696.7	2.3	0.5	4000	178.88
	26/12/2021	A5	3.7	342.8	1.5	1.3	4000	133.5
	6/3/2022	A6	2.6	228	1.3	1	4000	14.24
	6/3/2022	A7	2.4	196.7	1.5	1	4000	12.08
	30/3/2022	A8	2.3	171.8	1.3	1.5	4000	10.53
	26/4/2022	A9	2	168.2	2	1	4000	10.64
Almou Nait Lahcen	27/11/2021	B1	1	230.5	2.6	2	4000	11.58
	27/11/2021	B2	2	511	1.5	1	4000	181.06
	3/12/2021	B3	1.7	370.5	2.4	1	4000	2.85
	26/12/2021	B4	1.3	258	2.5	1.5	4000	64.76
	26/12/2021	B5	2.4	814	1.5	1.3	4000	46.5
	6/3/2022	B6	1.8	731.7	1.3	1.4	4000	1.064
	6/3/2022	B7	2.4	833.5	1.5	2	4000	84.4
	30/3/2022	B8	1.7	671.4	1.7	2	4000	96.7
	6/5/2022	B9	1.3	533.2	1.6	2.3	4000	88.3
Faraoun	27/11/2021	C1	7.6	445	2.24	0.2	0.9	0
	27/11/2021	C2	6.4	337.4	2.3	0.1	4000	0
	26/12/2021	C3	9.88	257	3.89	0.01	15	15
	6/3/2022	C4	8.94	2.77	442	0.02	0.6	58.16
	6/3/2022	C5	7	3.05	333.9	0.1	0.72	43
	30/3/2022	C6	7.06	3.5	275	0.02	0.83	44.1
	6/5/2022	C7	6.33	2.8	286	0.01	0.82	43.5
	2/9/2022	C8	8.43	755	1327	0.3	190	78
	17/9/2022	C9	8.6	662	1511	0.3	33	0.2
Tamadout	3/12/2021	D1	2.2	370	1086	0.2	378.05	182.45
	3/12/2022	D2	3.4	300	1217.3	0.1	258.7	210.03
	5/4/2022	D3	2.7	270.3	1533	0.2	178	234.5
	2/9/2022	D4	5	188.4	1603	0.2	135.8	117.44
	13/9/2022	D5	2.8	141.3	1177	0.2	0.93	64.44

Basing on the obtained results, at Ait Addi catchment SSC ranges from 10.53 to 178.88 g/l^{-1}. The catchment experiences notable variations in sediment concentration, likely influenced by events affecting sediment transport, such as rainfall or runoff. SSC varies from 1.064 g/l to 181.06 g/l. Besides, at Almou Nait Lahcen catchment, SSC varies from 1.064 to 181.06 g/l, this result shows significant variability in SSC, suggesting changes in sediment sources and transport processes. While at Faroun catchment, SSC was very low, reached 190 g/l as maximum value. However, this catchment exhibits a low generation of sediment due to the effect of land cover. At Tamadout catchment SSC varies from 64.44 g/l to 234.5 g/l, which confirm a significant fluctuation in SSC. The higher SSC values might be influenced by local erosion and sediment transport. Consequently, catchments with higher SSC values (e.g., Ait aadi, Almou N Lahcen Oaali, Tamadout) might have more pronounced sediment sources, due to higher erosion rates, land use practices, and runoff events. The Faraoun catchment stands out with instances of very low sediment. This could indicate effective sediment control practices, a lack of sediment sources, and sediment deposition within the catchment.

The variability in SSC within each catchment suggests the influence of different factors, including meteorological events, land use changes, and sediment transport dynamics (Goumih et al. 1090).

5 Discussion

The presented findings highlight the substantial variability in suspended sediment concentration (SSC) across different catchments, each characterized by unique environmental conditions and land use patterns. These results prompt a broader discussion on the factors influencing soil erosion within Mediterranean regions. Collaborating with fellow researchers, we can explore various aspects contributing to sediment dynamics in these areas.

- *Hydrological events and sediment transport*

Meteorological events, such as rainfall and runoff, play a pivotal role in sediment transport (Estrany and Grimalt 2014). In Mediterranean climates, characterized by sporadic but intense rainfall, soil surfaces become susceptible to erosion during these events (Wiley et al. 2009). Collaborative research could delve into the frequency and intensity of precipitation, their relationship with sediment transport, and the potential for extreme events to trigger substantial erosion (Fortesa et al. Nov. 2020).

- *Land use practices and vegetation cover*

The varying levels of SSC across catchments suggest a direct correlation with land use practices. Agricultural activities, deforestation, and urbanization can expose soil surfaces to erosion (Peter et al. 2014). Collaborators could investigate the influence of different land covers on sediment generation, considering the effectiveness of vegetation in stabilizing soils and reducing erosion rates (Nadal-Romero et al. 2013).

(a) (b)

Fig. 8 Dynamic of gully erosion in **a** Tamadout and **b** Ait Addi catchments

- *Human interventions and sediment control practices*

The low SSC observed in the Faraoun catchment raises questions about effective sediment control practices. Collaborative efforts could explore the measures undertaken to mitigate erosion, including terracing, reforestation, and soil conservation techniques (Borges et al. 2008). Understanding the success of these practices could offer insights into sustainable erosion management strategies applicable to other regions (Briak et al. 2016).

- *Geomorphology and topography*

Local geomorphological features and topography play a significant role in sediment transport dynamics. Collaborative research could analyse how catchment characteristics, such as slope gradients, soil types, and drainage patterns, interact with erosion processes (Ouallali et al. 2020). This could shed light on the susceptibility of specific areas to erosion and guide land management decisions (Ouakhir and Ghachi 2023) (Fig. 8).

- *Climate change implication*

Given the changing climate, collaborative investigations might explore how shifts in precipitation patterns, temperature, and extreme weather events impact erosion dynamics in Mediterranean regions. Researchers could assess the potential intensification of erosion due to altered hydrological regimes and plan adaptation strategies accordingly.

- *Socioeconomic factors and land use planning*

Collaboration between researchers and social scientists could provide insights into the socioeconomic drivers of land use changes. Exploring the balance between economic development, population growth, and environmental conservation could reveal the underlying causes of erosion and guide policy decisions.

- *Long-term monitoring and data sharing*

Collaborative efforts in establishing long-term monitoring networks across catchments could provide valuable data on sediment dynamics. Sharing and comparing data between catchments could help identify regional trends and develop a comprehensive understanding of erosion processes.

In summary, the wide range of SSC values observed across the Mediterranean catchments underscores the intricate interplay of multiple factors in soil erosion processes. Collaborative research endeavours can contribute to a holistic understanding of these factors, allowing for the development of targeted erosion control and mitigation strategies. Such collaborations will be essential in addressing the challenges posed by soil erosion in Mediterranean regions, ensuring the sustainable management of natural resources and ecosystem health.

6 Conclusion

In conclusion, the comprehensive analysis of soil erosion and suspended sediment concentration (SSC) within different catchments of the Oum Err Bia basin has revealed a significant variability across the Mediterranean regions. The observed suspended sediment concentration ranges in Ait Addi, Almou Nait Lahcen, Faroun, and Tamadout catchments reflect distinct environmental conditions and land use patterns. These findings have paved the way for a deeper discussion on the intricate factors that govern soil erosion processes in these regions, setting the stage for collaborative research aimed at addressing erosion challenges and promoting sustainable resource management.

The collaborative exploration of various influencing factors holds the potential to enhance our understanding of erosion dynamics within the Mediterranean context. Meteorological events, particularly rainfall and runoff patterns, are crucial drivers of sediment transport, especially in Mediterranean climates marked by intense but sporadic rainfall. Investigating the relationship between precipitation patterns and erosion potential could shed light on extreme event-induced erosion.

Land use practices have been identified as a primary contributor to varying soil erosion and sediment transport levels. Collaborative efforts can delve into the impact of diverse land covers, emphasizing the role of vegetation in stabilizing soils and mitigating erosion risks. Additionally, the successful implementation of sediment control practices, evident in the Faraoun catchment, presents an opportunity to dissect effective strategies and replicate their success elsewhere.

The geomorphological characteristics of catchments, including topography and soil properties, interact intricately with erosion processes. Collaborative research can uncover these relationships, aiding in pinpointing erosion-prone areas and informing land management decisions. Furthermore, considering the implications of climate change on erosion dynamics and assessing the socioeconomic drivers of land use changes through interdisciplinary collaboration could offer invaluable insights.

Long-term monitoring networks across catchments, coupled with data sharing initiatives, form the basis for identifying regional trends and comprehending erosion processes comprehensively. Such collaborative endeavors will be instrumental in shaping erosion control and mitigation strategies tailored to the unique challenges of Mediterranean regions.

In summary, the complex interplay of multiple factors in soil erosion processes across Mediterranean catchments underscores the importance of collaborative research. By jointly addressing these factors, researchers and stakeholders can develop targeted strategies to tackle soil erosion challenges, ensuring the sustainable stewardship of natural resources and the preservation of ecosystem health in these regions.

References

Angima SD, Stott DE, O'Neill MK, Ong CK, Weesies GA (2003) Soil erosion prediction using RUSLE for central Kenyan highland conditions. Agric Ecosyst Environ 97(1–3):295–308. https://doi.org/10.1016/S0167-8809(03)00011-2

Barakat A, Ouargaf Z, Khellouk R, El Jazouli A, Touhami F (2019) Land use/land cover change and environmental impact assessment in Béni-Mellal district (Morocco) using remote sensing and GIS. Earth Syst Environ 3(1):113–125. https://doi.org/10.1007/s41748-019-00088-y

Benchettouh A, Kouri L, Jebari S (2017) Spatial estimation of soil erosion risk using RUSLE/GIS techniques and practices conservation suggested for reducing soil erosion in Wadi Mina watershed (northwest, Algeria). Arab J Geosci 10(4). https://doi.org/10.1007/s12517-017-2875-6

Benmansour M et al (2013) Assessment of soil erosion and deposition rates in a Moroccan agricultural field using fallout 137Cs and 210Pbex. J Environ Radioact 115:97–106. https://doi.org/10.1016/j.jenvrad.2012.07.013

Beskow S, Mello CR, Norton LD, Curi N, Viola MR, Avanzi JC (2009) Soil erosion prediction in the Grande River Basin, Brazil using distributed modeling. CATENA 79(1):49–59. https://doi.org/10.1016/j.catena.2009.05.010

Borges N, Estrany J, Garcia C, Batalla RJ (2008) Groundwater control on the suspended sediment load in the Na Borges River, Mallorca, Spain. Geomorphology 106:292–303. https://doi.org/10.1016/j.geomorph.2008.11.008

Boudhar A, Ouatiki H, Bouamri H, Lebrini Y, Karaoui I, Hssaisoune M, Arioua A, Benabdelouahab T (2020) Hydrological response to snow cover changes using remote sensing over the oum er rbia upstream basin, morocco. In: Advances in science, technology and innovation. Springer International Publishing. https://doi.org/10.1007/978-3-030-21166-0_9

Brahim B, Meshram SG, Abdallah D, Larbi B, Drisss S, Khalid M, Khedher KM (2020) Mapping of soil sensitivity to water erosion by RUSLE model: case of the Inaouene watershed (Northeast Morocco). Arab J Geosci 13:1–15

Briak H, Moussadek R, Aboumaria K, Mrabet R (2016) Assessing sediment yield in Kalaya gauged watershed (Northern Morocco) using GIS and SWAT model. Int Soil Water Conserv Res 4(3):177–185. https://doi.org/10.1016/j.iswcr.2016.08.002

Demirci A, Karaburun A (2012) Estimation of soil erosion using RUSLE in a GIS framework: a case study in the Buyukcekmece Lake watershed, northwest Turkey. Environ Earth Sci 66(3):903–913. https://doi.org/10.1007/s12665-011-1300-9

Ebrahimzadeh S, Motagh M, Mahboub V, Mirdar Harijani F (2018) An improved RUSLE/SDR model for the evaluation of soil erosion. Environ Earth Sci 77(12). https://doi.org/10.1007/s12665-018-7635-8

El Jazouli A, Barakat A, Khellouk R, Rais J, El Baghdadi M (2019) Remote sensing and GIS techniques for prediction of land use land cover change effects on soil erosion in the high basin of the Oum Er Rbia River (Morocco). Remote Sens Appl Soc Environ 13:361–374. https://doi.org/10.1016/j.rsase.2018.12.004

Ennaji N, Ouakhir H, Halouan S, Abahrour M (2022a) Assessment of soil erosion rate using the EPM model: case of Ouaoumana basin, Middle Atlas, Morocco. IOP Conf Ser Earth Environ Sci 1090(1). https://doi.org/10.1088/1755-1315/1090/1/012004

Ennaji E, Ouakhir H, Halouan H, Abahrour M (2022b) sediment transport index modelling using the GIS at small agricultural catchment, pp 16–20

Estrany J, Grimalt M (2014) Catchment controls and human disturbances on the geomorphology of small Mediterranean estuarine systems. Estuar Coast Shelf Sci 150(PB):230–241. https://doi.org/10.1016/j.ecss.2014.03.021

Fortesa J, Latron J, García-Comendador J, Company J, Estrany J (2020) Runoff and soil moisture as driving factors in suspended sediment transport of a small mid-mountain Mediterranean catchment. Geomorphology 368:107349. https://doi.org/10.1016/J.GEOMORPH.2020.107349

Foster GR, Toy TE, Renard KG (2003) Comparison of the USLE, RUSLE1.06c, and RUSLE2 for application to highly disturbed lands. First Interag Conf Res Watersheds 27(30):154–160. http://citeseerx.ist.psu.edu/viewdoc/download?doi=10.1.1.462.6311&rep=rep1&type=pdf

Fu BJ, Zhao WW, Chen LD, Zhang QJ, Lü YH, Gulinck H, Poesen J (2000) Assessment of soil erosion at large watershed scale using RUSLE and GIS: a case study in the Loess Plateau of China'. Land Degrad Dev 16(1):73–85. https://doi.org/10.1002/ldr.646

Goumih M, El Ghachi M, Ouakhir H, Ennaji N, Midaoui A (2022) Hydrological response of the Assif Melloul basin in the Central High Atlas (Morocco): monitoring and creating a rating curve. IOP Conf Ser Earth Environ Sci 1090(1). https://doi.org/10.1088/1755-1315/1090/1/012003

Ibno Namr K, Mrabet R (2004) Influence of agricultural management on chemical quality of a clay soil of semi-arid Morocco. J Afr Earth Sci 39(3–5):485–489. https://doi.org/10.1016/j.jafrearsci.2004.07.016

El Jazouli A, Barakat A, Ghafiri A, El Moutaki S, Ettaqy A, Khellouk R (2019) Soil erosion modeled with USLE, GIS, and remote sensing: a case study of Ikkour watershed in Middle Atlas (Morocco). Geosci Lett 4(1). https://doi.org/10.1186/s40562-017-0091-6

Kouli M, Soupios P, Vallianatos F (2009) Soil erosion prediction using the Revised Universal Soil Loss Equation (RUSLE) in a GIS framework, Chania, Northwestern Crete, Greece. Environ Geol 57(3):483–497. https://doi.org/10.1007/s00254-008-1318-9

Mati BM, Morgan RPC, Gichuki FN, Quinton JN, Brewer TR, Liniger HP (2000) Assessment of erosion hazard with the USLE and GIS: a case study of the Upper Ewaso Ng'iro North basin of Kenya. ITC J 2(2):78–86. https://doi.org/10.1016/S0303-2434(00)85002-3

Mosaid H, Barakat A, Bustillo V, Rais J (2022) Modeling and mapping of soil water erosion risks in the Srou Basin (Middle Atlas, Morocco) using the EPM model, GIS and magnetic susceptibility. J Landsc Ecol Republic 15(1):126–147. https://doi.org/10.2478/jlecol-2022-0007

Nadal-Romero E, Lasanta T, García-Ruiz JM (2013) Runoff and sediment yield from land under various uses in a Mediterranean mountain area: Long-term results from an experimental station. Earth Surf Process Landforms 38(4):346–355. https://doi.org/10.1002/esp.3281

Ouakhir H, El Ghachi M (2023) Temporal dynamic of soil erosion and rainfall erosivity within Srou River Basin (Middle Atlas/Morocco). Am J Mech Appl 11(1):1–5. https://doi.org/10.11648/j.ajma.20231101.11

Ouakhir H, El Ghachi M, Goumih M, Ennaji N (2020) Fluvial dynamic in Oued El Abid Basin: monitoring and quantification at an upstream river section in Bin El Ouidane Dam—2016/2017—(Central High Atlas/Morocco). https://doi.org/10.11648/j.ajma.20200804.11

Ouakhir H, Ennaji N, Goumih M, Chakir M, Ghadbane O, Lahlou N, Nafia K, Ouzanni H, El Ghachi M (2022) Estimation of soil erosion rate applying the RUSLE model within the El Abid river basin. Central High Atlas , Morocco

Ouakhir H, El Ghachi M, Ennaji N, Goumih M (2023) River response to fluvial dynamic in a meandering within the downstream part of el abid river basin (Central High Atlas/, no January, 2023)

Ouallali A, Briak H, Aassoumi H, Beroho M, Bouhsane N, Moukhchane M (2020) Hydrological foretelling uncertainty evaluation of water balance components and sediments yield using a multi-variable optimization approach in an external Rif's catchment. Morocco. Alexandria Eng J 59(2):775–789. https://doi.org/10.1016/j.aej.2020.02.017

Ouatiki H, Boudhar A, Tramblay Y, Jarlan L, Benabdelouhab T, Hanich L, El Meslouhi M, Rachid Chehbouni A (2017) Evaluation of TRMM 3B42 V7 rainfall product over the Oum Er Rbia watershed in Morocco. Climate 5(1):1–17. https://doi.org/10.3390/cli5010001

Peter KD, d'Oleire-Oltmanns S, Ries JB, Marzolff I, Ait Hssaine A (2014) Soil erosion in gully catchments affected by land-levelling measures in the Souss Basin, Morocco, analysed by rainfall simulation and UAV remote sensing data. Catena 113:24–40. https://doi.org/10.1016/j.catena.2013.09.004

Rapp A, Axelsson V, Berry L, Murray-Rust DH (1972) Soil erosion and sediment transport in the Morogoro river catchment, Tanzania. Geogr Ann Ser A Phys Geogr 54(3–4):125–155. https://doi.org/10.1080/04353676.1972.11879863

Renard KG, Foster GR, Weesies GA, Mccool DK, Yoder DC (2000) <CAT10827029PDF.pdf>

Sadiki A, Faleh A, Navas A, Bouhlassa S (2007) Assessing soil erosion and control factors by the radiometric technique in the Boussouab catchment, Eastern Rif, Morocco. CATENA 71(1):13–20. https://doi.org/10.1016/j.catena.2006.10.003

Sun W, Shao Q, Liu J, Zhai J (2014) Assessing the effects of land use and topography on soil erosion on the Loess Plateau in China. CATENA 121:151–163. https://doi.org/10.1016/j.catena.2014.05.009

Wiley J, Estrany J, Garcia C, Batalla RJ (2009) Suspended sediment transport in a small Mediterranean agricultural catchment Suspended sediment transport. Earth Surf Process Landforms 34:197–9337. https://doi.org/10.1002/esp.1777

Xu YQ, Shao XM, Bin Kong X, Jian P, Cai YL (2008) Adapting the RUSLE and GIS to model soil erosion risk in a mountains karst watershed, Guizhou Province, China. Environ Monit Assess 141(1–3):275–286. https://doi.org/10.1007/s10661-007-9894-9

Improving the Capabilities of Medical Imaging Scanners by Incorporating Backstepping Control

Said Ziani, Essahel Said, and Amine Elammari

Abstract In this paper, we address the aforementioned limitations of conventional scanners in cardiac imaging and propose a solution to optimize their functionality. Specifically, we explore the integration of Backstepping control as a means to overcome these challenges. By incorporating Backstepping control algorithms into the imaging process of a conventional scanner, we aim to enhance image quality and accuracy in capturing cardiac images. Backstepping control is a control theory approach known for its ability to compensate for the motion of dynamic systems. By precisely tracking the movement of the heart during image acquisition, we can mitigate motion artifacts and improve the clarity of cardiac images. Through experimental investigations and simulations, we evaluate the effectiveness of the Backstepping control integration in a conventional scanner for cardiac imaging. We compare the results with traditional imaging techniques and specialized cardiac scanners, such as Cardiac CT or Coronary CT, to demonstrate the potential of our approach.

Keywords Scanners · Backstepping · Control

1 Introduction

In the medical field, medical imaging plays a crucial role in the diagnosis and monitoring of pathologies. Among the most commonly used imaging modalities, the scanner is a versatile diagnostic tool that provides detailed images of the inside of the human body. Over the years, scanners have seen significant advancements, offering increasingly advanced features and improved image quality. Motorization is one of the key aspects of a scanner's operation, enabling the rotation of the gantry

S. Ziani (✉) · A. Elammari
ENSAM, Mohammed V University, Rabat, Morocco
e-mail: s.ziani@um5r.ac.ma

E. Said
Sciences and Engineering Laboratory Polydisciplinary, Sidi Mohammed Ben Abdellah University, PB 1223 Taza-Gare, Taza, Morocco

M. Azrour et al. (eds.), *Sustainable and Green Technologies for Water and Environmental Management*, World Sustainability Series,
https://doi.org/10.1007/978-3-031-52419-6_9

and rapid image acquisition (Jiang et al. 2020; Ito et al. 2017). An efficient and precise motorization is essential to ensure image quality and reduce examination time, thereby improving patient care and optimizing hospital resources.

In this context, this thesis proposes an innovative approach to optimize scanner functionality by integrating Backstepping control. Backstepping control is an advanced control technique, known for its robustness and ability to meet the demands of complex systems.

The main objective of this work is to demonstrate how Backstepping control can be applied to the motorization of a scanner, particularly for conventional and cardiac scanners. We aim to show that this control approach allows for precise and efficient responses to the specific needs of each type of scanner, resulting in a significant improvement in image quality and a reduction in examination time.

The paper is structured as follows: firstly, a thorough study of scanner operation is presented, highlighting the significance of motorization in the image acquisition process. Next, a model of a permanent magnet synchronous machine is developed for integration into the scanner system. We then proceed with a comparison of different control techniques to determine the suitability of Backstepping control for our application.

Subsequently, we propose a practical solution by simulating the Backstepping control with specific speeds of conventional and cardiac scanners. The results demonstrate the ability of Backstepping control to precisely follow desired speed profiles while ensuring a rapid transition between different speeds.

Finally, we put forth a proposal for a scanner that combines the features of both a conventional scanner and a cardiac scanner, using Backstepping control with a Permanent Magnet Synchronous Machine (PMSM). This proposal aims to improve image quality, reduce examination time, and optimize hospital resources.

2 Proposed Method

2.1 *Performance of Electric Motors for a Scanner*

The electric motor is crucial for the operation of the scanner as it enables the rotation of heavy components to produce precise images. It needs to deliver high torque and speed accuracy to allow precise positioning of elements during scanning. Motors must be designed to provide high precision, low speed variation, and high positioning resolution to meet the scanner's requirements (Ghedamsi et al. 2008).

The electric motors commonly used in scanners include stepper motors, direct current (DC) motors, and asynchronous motors. Stepper motors are often employed for precise movements, such as axial displacements of the table, while DC and asynchronous motors are used for rotations and faster movements. The performance of an electric motor used in a scanner can vary depending on several factors such as the brand, model, size, and weight of the scanned object, as well as the required rotational speed. Typical performance characteristics of an electric motor for a scanner include maximum torque, maximum rotational speed, positioning accuracy, and maximum acceleration. These performance attributes are usually optimized to meet the specific requirements of each scanner application, hence they may vary from one model to another.

2.2 *Motorization of a Cardiac Scanner*

Within a medical scanner, motorization plays a particularly crucial role in the component responsible for rotating the gantry. The gantry is an essential part of the scanner that supports the X-ray tube and detector. It is designed to perform rotational movements around the patient to acquire the necessary images. The motorization of the gantry enables precise and real-time control of these rotational movements. With advanced motors and positioning devices, the gantry can perform smooth and accurate rotations, resulting in high-quality images. The gantry's rotation is pivotal in the image acquisition process, as it allows for volumetric scans by performing a series of rotations around the patient. This technique, known as helical computed tomography (CT), provides high-resolution cross-sectional images in various orientations. The gantry's motorization must deliver precise and synchronized movements with other scanner components, such as the X-ray tube and detector, to ensure consistent and high-quality image acquisition. It also needs to be robust enough to handle significant loads and operate reliably under intensive usage conditions (Chen et al. 2021).

By utilizing advanced control techniques like Backstepping control, the gantry's motorization can be optimized to achieve high positioning accuracy and suitable rotation speed. This improvement enhances the quality of acquired images, reduces examination times, and offers a more comfortable experience for patients. The cardiac scanner is designed to track cardiac activity and capture detailed images of the heart at specific moments in the cardiac cycle. To achieve this, it is often necessary to connect the cardiac scanner to an electrocardiogram (ECG) that records the heart's electrical activity. The fundamental principle of the cardiac scanner is to synchronize image acquisition with different phases of the cardiac cycle, including systole (heart contraction) and diastole (heart relaxation). This ensures sharp and precise images of the heart, even while it is in motion (Ay et al. 2009).

Ensuring synchronized image acquisition relies on the crucial role of the motorization component of the cardiac scanner. Important motorization parameters include rotation speed and positioning accuracy. The rotation speed must be fast enough to track the rapid movements of the heart and capture images at specific moments in the cardiac cycle (Ben Achour 2023; Achour et al. 2022; Chaou et al. 2022; Youssef et al. 2022; Ziani et al. 2023b). In practice, modern cardiac scanners can achieve high rotation speeds, typically exceeding 0.3 s per rotation, equivalent to 200 revolutions per minute (Ziani et al. 2023a). This rapid speed helps reduce motion artifacts and obtain high-quality images even during the contraction and relaxation phases of the heart. Positioning accuracy is also vital to ensure that images are acquired at the right moment in the cardiac cycle. Advanced positioning control mechanisms are used to ensure precise scanner movement and optimal synchronization with cardiac activity, thereby obtaining consistent and reliable images.

2.3 Application of Backstepping to the Control of PMSM

In this section, we explore the application of the Backstepping control technique to achieve precise and efficient control of Permanent Magnet Synchronous Machines (PMSM). By integrating Backstepping into the control system, we aim to enhance the motor performance and address the specific challenges associated with PMSM control. The Backstepping method, known for its robustness and ability to handle complex systems, offers a promising approach to optimize the functionality of PMSMs in various applications, including medical imaging scanners. We will delve into the theoretical framework and practical implementation of Backstepping for PMSM control, analyzing its performance and effectiveness in achieving desired speed profiles and accurate positioning during scanning processes (Hafid 2023; Laabab et al. 2023a; Laabab et al. 2022; Youssef et al. 2023; Echoukairi et al. 2023). The equations of the (d-q) model of the Permanent Magnet Synchronous Machine can be written in the following form:

$$
\begin{aligned}
\frac{dI_d}{dt} &= -\frac{R_s}{L_d}I_d + \mathrm{p}\frac{L_q}{L_d}\Omega I_q + \frac{V_d}{L_d} \\
\frac{dI_q}{dt} &= -\frac{R_s}{L_q}I_q - \mathrm{p}\frac{L_d}{L_q}\Omega I_d - \mathrm{p}\frac{\Omega\varphi_m}{L_q} + \frac{V_q}{L_q} \\
\frac{d\Omega}{dt} &= \frac{p}{J}(L_d - L_q)I_d I_q - \frac{f}{J}\Omega + \frac{\mathrm{p}}{J}\varphi_m I_q - \frac{C_r}{J}
\end{aligned}
\tag{1}
$$

where:

Rs	Stator resistance (Ω)
L_d	Self-inductances of d and q axes (H)
φ_m	Mutual flux due to the permanent magnet (Wb)
I_d, I_q	Currents of d and q axes (A)
Ω	Angular velocity (rad/s)
J	Moment of inertia (Kg·m^2)
F	Damping constant (N/rad/s)
P	Number of pole pairs
Cr	Load torque (N·m)y

The first virtual control of our system for I_{qref} can be expressed as follows:

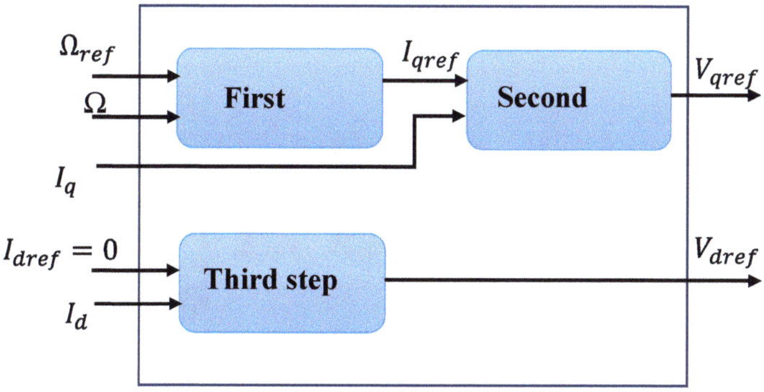

Fig. 1 Internal structure of the Backstepping controller block

$$I_{qref} = \frac{J}{\mathrm{p}\left[(L_d - L_q)I_d + \varphi_m\right]}\left[\frac{d\Omega^*}{dt} + \frac{f}{J}\Omega + \frac{C_r}{J} + K_\Omega Z_\Omega + K'_\Omega(\Omega^* - \Omega)\right] \quad (2)$$

The second virtual control of our system for V_{qref} can be expressed as follows:

$$V_{qref} = L_q[K_q Z_q + \dot{I}_{qref} + \frac{R_s}{L_q}I_q + \frac{\mathrm{p}\Omega}{L_q}(L_d I_d + \varphi_m)] \quad (3)$$

The control law for our system V_{dref} is of the following form:

$$V_{dref} = L_d[K_d Z_d + \frac{R_s}{L_d}I_d - \mathrm{p}\frac{L_q}{L_d}\Omega I_q] \quad (4)$$

Figure 1 shows the internal structure of the Backstepping controller block.

2.4 *Backstepping Control Simulation*

In this section, we conduct a simulation to demonstrate the effectiveness and performance of the Backstepping control technique in our system, utilizing MATLAB Simulink. The simulation is performed using a model of the Permanent Magnet Synchronous Machine (PMSM) integrated with the Backstepping controller. Figure 2 presents the schematic of the Backstepping control simulation in MATLAB Simulink. We set up the control parameters and input references within the MATLAB Simulink environment to execute the Backstepping control algorithm. This enables us to generate the necessary control actions to accurately regulate the currents and voltages of the PMSM. Throughout the simulation, we analyze the system's response under various operating conditions, such as changes in load torque, angular velocity, and reference currents. By observing the system's behavior, we can evaluate how

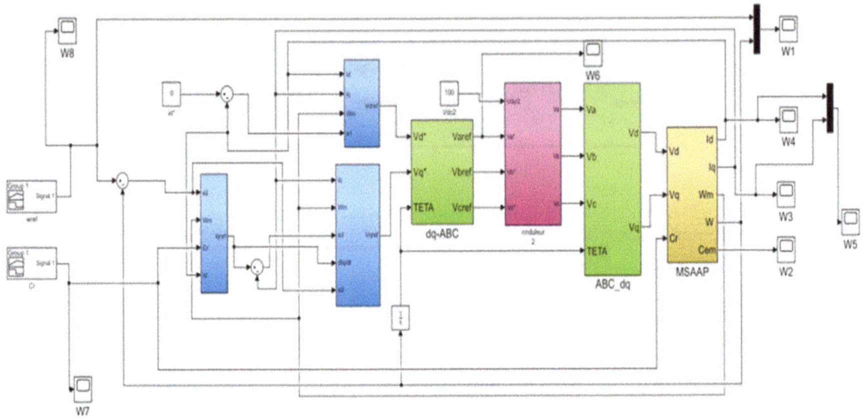

Fig. 2 Backstepping control simulation diagram

well the Backstepping control approach tracks the desired trajectories, maintains stability, and achieves precise control of the PMSM. Additionally, we compare the performance of the Backstepping controller with other commonly used control techniques for PMSM applications, all within the MATLAB Simulink environment. The results obtained from this simulation provide valuable insights into the advantages and limitations of the Backstepping control strategy, demonstrating its suitability for real-world applications. The use of MATLAB Simulink for the simulation allows us to perform a comprehensive and dynamic analysis of the Backstepping control, making it a powerful tool to validate the effectiveness and robustness of this control approach for our specific PMSM system. This simulation study sets the foundation for potential real-world implementation, including applications such as medical imaging scanners.

3 Results and Discussion

3.1 Simulation Results

To ensure the functionalities of both conventional and cardiac scanners, it is possible to integrate the Backstepping control applied to the MSAP machine into the scanner's operation. Backstepping control is a robust and effective control technique, known for its ability to manage complex systems with high precision and stability. By employing this control approach, the scanner system can accurately track desired speed profiles and quickly respond to changes in operating conditions (Laabab et al. 2023b). The Backstepping control algorithm utilizes a recursive design methodology, explicitly considering the system dynamics and uncertainties. This results in improved tracking performance, error reduction, and enhanced system robustness (Ouhadou 2019).

Moreover, Backstepping control can effectively meet the stringent requirements of cardiac scanners, which demand rapid speed transitions to capture dynamic heart movements.

To grasp the significance of high rotation speed in a cardiac scanner, it is crucial to consider the phases of the heart's contraction and relaxation, known as systole and diastole, respectively (Fig. 3). During systole, the heart contracts to pump blood from the ventricles to the arteries, whereas during diastole, the heart relaxes to refill with blood (Ouhadou, et al. 2018). In cardiac imaging, it is critical to capture these dynamic heart movements to obtain accurate information about its structure and function. A high rotation speed of the scanner is necessary to minimize motion artifacts and obtain clear images during both systolic and diastolic phases. We conducted a simulation of the Backstepping control to assess its ability to track speed setpoints for both a conventional scanner and a cardiac scanner. For the conventional scanner, we set a rotational speed setpoint of 8 rad/s, while for the cardiac scanner, the setpoint was 20 rad/s. The results are presented in Fig. 3.

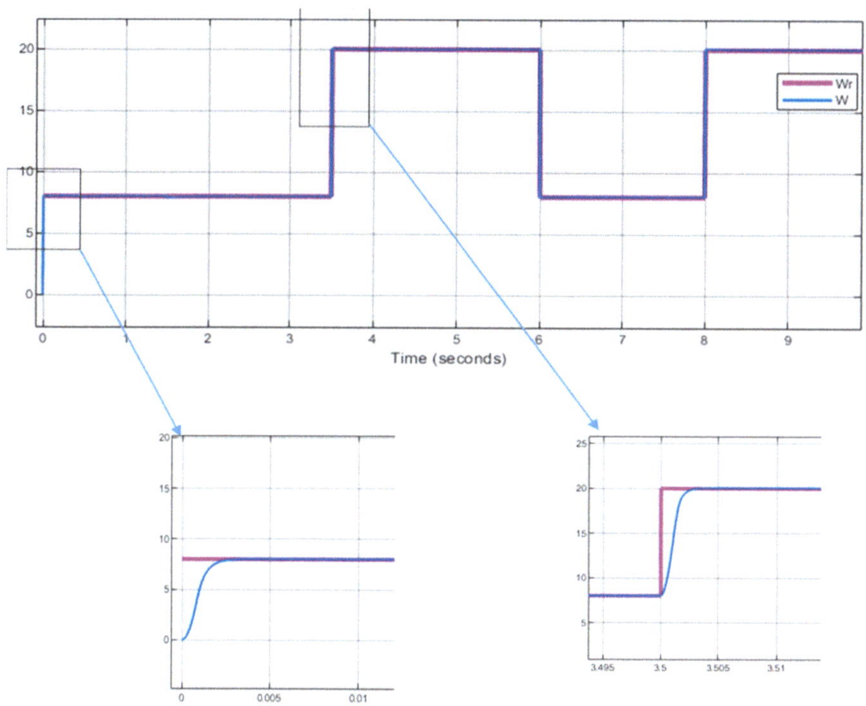

Fig. 3 Backstepping control simulation with the speed of a conventional scanner and a Cardiac scanner

3.2 Discussion

The obtained results are highly promising. The time taken to transition from 0 to 8 rad/s is less than 0.005 s, demonstrating the responsiveness and speed of the Backstepping control in tracking the speed setpoint. Additionally, the transition from 8 rad/s to 20 rad/s also occurs in less than 0.005 s, highlighting the control's ability to quickly adapt to setpoint variations. These results confirm the effectiveness and precision of Backstepping control in the context of scanner motorization, for both conventional and cardiac scanners. The control's precise tracking of speed setpoints ensures optimal scanner operation, guaranteeing high-quality image acquisitions and reduced examination times. These observations reinforce the value of Backstepping control as a reliable and high-performance solution for medical scanner motorization. It offers an advanced approach that meets the specific requirements of various medical imaging applications, ensuring precise control of the scanner's rotational speed. The proposed motorization model in this paper can also be utilized for diagnosing cardiac malformations in fetuses, and mathematical control methods based on wavelet theory (Said and Hassouani 2020a; Said et al. 2019; Said et al. 2017) and artificial intelligence algorithms such as Deep Learning can even be contemplated (Said et al. 2023). Indeed, real-time control based on wavelets in scanners is a highly promising technique in the field of biomedical engineering. This approach holds significant potential for enhancing the accuracy and effectiveness of diagnostic processes, particularly in the context of fetal cardiac anomaly detection. By harnessing the capabilities of wavelet analysis (Said et al. 2018a,2018b; Said and El Hassouani 2019; Jana et al. 2021; Said 2022), it becomes feasible to extract nuanced patterns and subtle anomalies from medical imaging data, enabling early and accurate identification of cardiac malformations. This advancement aligns with the broader trend of incorporating advanced mathematical methods and artificial intelligence algorithms, such as Deep Learning, to revolutionize medical diagnostics and treatment strategies. As a result, the fusion of wavelet-based real-time control with cutting-edge technologies has the potential to usher in a new era of precision medicine and improved patient outcomes (Sornsen et al. 2021; Said 2023; Said and Hassouani 2020b; Chak et al. 2022).

4 Conclusion

In summary, this research paper has focused on optimizing scanner functionalities by incorporating Backstepping control. The study highlighted the benefits of this approach for both conventional and cardiac scanners. Simulation results confirmed that Backstepping control enables precise and efficient speed response, leading to improved image quality and reduced examination time. The study also underscored the significance of motorization in scanner operation, emphasizing the need for robust and responsive control. By integrating Backstepping control with an MSAP machine, the researchers demonstrated its ability to meet the requirements of both scanner types, offering a versatile and high-performance solution. Moving forward, future research could explore enhancing Backstepping control without dedicated sensors by utilizing advanced observation techniques and state estimation. This could simplify system design and reduce costs associated with additional sensors. Furthermore, investigating novel control and optimization techniques for scanner motorization, such as machine learning methods, could further improve the system's adaptability and performance by optimizing parameters in real-time based on individual patient and scanner characteristics.

References

Achour HB et al (2022) Permanent magnet synchronous motor PMSM control by combining vector and PI controller. WSEAS Trans Syst Control 17:244–249

Ay MR, Ahmadian A, Ghadiri H, Maleki A, Ghafarian P, Zaidi H (2009) The influence of X-ray spectra filtration on image quality and patient dose in the GE VCT 64-slice cardiac CT scanner. In: 2009 3rd international conference on bioinformatics and biomedical engineering, Beijing, China, pp 1–4. https://doi.org/10.1109/ICBBE.2009.5163626

Ben Achour H et al (2023) PI controller and quadratic feedback of synchronous machine. In Farhaoui Y, Rocha A, Brahmia Z, Bhushab B (eds) Artificial intelligence and smart environment. ICAISE 2022. Lecture notes in networks and systems, vol 635. Springer, Cham

Chak WH, Saito N, Weber D (2022) The scattering transform network with generalized morse wavelets and its application to music genre classification. In: 2022 international conference on wavelet analysis and pattern recognition (ICWAPR), Toyama, Japan, pp 25–30. https://doi.org/10.1109/ICWAPR56446.2022.9947091

Chaou Y et al (2022) Nonlinear control of the permanent magnet synchronous motor PMSM using backstepping method. WSEAS Trans Syst Control 17:56–61

Chen X, Zhou B, Xie H, Shi L, Liu H, Liu C (2021) Investigation of direct and indirect approaches of deep-learning-based attenuation correction for general purpose and dedicated cardiac SPECT scanners. In: IEEE nuclear science symposium and medical imaging conference (NSS/MIC). Piscataway, NJ, USA 2021:1–2. https://doi.org/10.1109/NSS/MIC44867.2021.9875517

Ghedamsi K, Aouzellag D, Berkouk E (2008) Control of wind generator associated to a flywheel energy storage system. Renew Energy 33:2145–2156

Hafid BA et al (2023) A quadratic observer for sensorless drive system controller. In: Farhaoui Y, Rocha A, Brahmia Z, Bhushab B (eds) Artificial intelligence and smart environment. In: ICAISE 2022. Lecture notes in networks and systems, vol 635. Springer, Cham

Hutabarat YA, Rizal A, Mukthar H, Ziani S (2023) Individual identification using lower body skeleton joint data from kinect V2 sensors, 2023 3rd International Conference on Intelligent Cybernetics Technology & Applications (ICICyTA), Denpasar, Bali, Indonesia, pp 12–16, https://doi.org/10.1109/ICICyTA60173.2023.10428906.

Ito S, Yoo HW, Schitter G (2017) Comparison of modeling-free learning control algorithms for galvanometer scanner's periodic motion. In: 2017 IEEE international conference on advanced intelligent mechatronics (AIM), Munich, Germany, pp 1357–1362. https://doi.org/10.1109/AIM.2017.8014207

Jana J, Tripathi S, Chowdhury RS, Bhattacharya A, Bhaumik J (2021) An area efficient VLSI architecture for 1-D and 2-D discrete wavelet transform (DWT) and inverse discrete wavelet transform (IDWT). In: Devices for Integrated Circuit (DevIC). Kalyani, India 2021:378–382. https://doi.org/10.1109/DevIC50843.2021.9455902

Jiang J et al (2020) Augmented whole-body scanning via magnifying PET. IEEE Trans Med Imaging 39(11):3268–3277. https://doi.org/10.1109/TMI.2019.2962623
Laabab I et al (2023a) A review of the application of artificial intelligence for weather prediction in solar energy: using artificial neural networks. In: Farhaoui Y, Rocha A, Brahmia Z, Bhushab B (eds) Artificial intelligence and smart environment. ICAISE 2022. Lecture notes in networks and systems, vol 635. Springer, Cham
Laabab I et al (2023b) A solar panels overheating protection: a review. Indones J Electr Eng Comput Sci 29(1):49–55
Ouhadou M et al (2018) Experimental modeling of the thermal resistance of the heat sink dedicated to SMD LEDs passive cooling. In: Proceedings of the 3rd international conference on smart city applications
Ouhadou M et al (2019) Experimental investigation on thermal performances of SMD LEDs light bar: junction-to-case thermal resistance and junction temperature estimation. Optik 182
Rizal A, Hadiyoso S, Aulia S, Wijayanto I, Triwiyanto, Said Z (2023) Multiscale hjorth descriptor on epileptic EEG classification. J. Electr. Comput. 4961637:11 https://doi.org/10.1155/2023/4961637
Said Z (2022) Contribution to single-channel fetal electrocardiogram identification. Traitement Du Signal 39(6):2055–2060
Said Z, Farhaoui Y, Moutaib M (2023) Extraction of fetal electrocardiogram by combining deep learning and SVD-ICA-NMF methods. Big Data Min Anal 6(3):301–310. https://doi.org/10.26599/BDMA.2022.9020035
Said Z (2023) Fetal electrocardiogram identification using statistical analysis. In: Farhaoui Y, Rocha A, Brahmia Z, Bhushab B (eds) Artificial intelligence and smart environment. ICAISE 2022. Lecture notes in networks and systems, vol 635. Springer, Cham
Said Z, El Hassouani Y (2019) Fetal-maternal electrocardiograms mixtures characterization based on time analysis. In: 2019 5th international conference on optimization and applications (ICOA), Kenitra, Morocco, pp 1–5. https://doi.org/10.1109/ICOA.2019.8727619
Said Z, El Hassouani Y (2020a) A new approach for extracting and characterizingfetal electrocardiogram. Traitement du Signal 37(3):379–386. https://doi.org/10.18280/ts.370304
Said Z, El Hassouani Y (2020b) Fetal electrocardiogram analysis based on LMS adaptive filtering and complex continuous wavelet 1-D. In: Farhaoui Y (eds) Big data and networks technologies. BDNT 2019. Lecture notes in networks and systems, vol 81. Springer, Cham. https://doi.org/10.1007/978-3-030-23672-4_26
Said Z, Jbari A, Belarbi L (2017) Fetal electrocardiogram characterization by using only the continuous wavelet transform CWT. IN: 2017 international conference on electrical and information technologies (ICEIT), Rabat, Morocco, pp 1–6. https://doi.org/10.1109/EITech.2017.8255310
Said Z, Jbari A, Bellarbi L (2018a) QRS complex characterization based on non-negative matrix factorization NMF. In: 2018 4th international conference on optimization and applications (ICOA), Mohammedia, Morocco, pp 1–5. https://doi.org/10.1109/ICOA.2018.8370548
Said Z, Jbari A, Bellarbi L, Farhaoui Y (2018b) Blind maternal-fetal ECG separation based on the time-scale image TSI and SVD—ICA methods. Proc Comput Sci 134:322–327. ISSN 1877-0509
Said Z, El Hassouani Y, Farhaoui Y (2019) An NMF based method for detecting RR interval. In: International conference on big data and smart digital environment 2019. Springer
Sornsen I, Suppitaksakul C, Kitpaiboontawee R (2021) Partial discharge signal detection in generators using wavelet transforms. In: 2021 international conference on power, energy and innovations (ICPEI), Nakhon Ratchasima, Thailand, pp 195–198. https://doi.org/10.1109/ICPEI52436.2021.9690682
Youssef C et al (2022) Electric vehicle backstepping controller using synchronous machine. In: The international conference on artificial intelligence and smart environment, pp 367–373
Youssef C et al (2023) Backstepping control of the permanent magnet synchronous generator (PMSG) used in a wind power system. In: Farhaoui Y, Rocha A, Brahmia Z, Bhushab B (eds)

Artificial intelligence and smart environment. ICAISE 2022. Lecture notes in networks and systems, vol 635. Springer, Cham

Ziani S (2023) Enhancing fetal electrocardiogram classification: A hybrid approach incorporating multimodal data fusion and advanced deep learning models. Multimed Tools Appl

Ziani S, Rizal MS (2023) A time-scale image analysis for detection of fetal electrocardiogram. Multimed Tools Appl. https://doi.org/10.1007/s11042-023-17165-0

Characterization of Fetal Electrocardiogram Using Short Time Fourier Transform

Said Ziani and Essahel Said

Abstract This paper presents a novel method for blind source extraction of the fetal electrocardiogram (ECG) using the short-time Fourier transform. Unlike previous studies that primarily focused on time–frequency methods for speech signals, this research addresses the more challenging task of analyzing non-stationary biomedical signals, such as the fetal and maternal electrocardiograms. One significant advantage of these biomedical signals is their remarkable energy distribution variation over time and frequencies, which is evident in their spectrograms. The proposed approach aims to accurately separate these mixed signals without prior knowledge of their sources. Through comprehensive simulations and rigorous experimental evaluations, the results demonstrate the high performance and effectiveness of the method in successfully extracting fetal and maternal ECG signals. This advancement has promising implications for enhancing fetal monitoring and maternal health assessment during pregnancy.

Keywords Blind separation BSS · Fetal electrocardiogram · Short fourier transform · Spectrogram

1 Introduction

The monitoring and assessment of the fetal electrocardiogram (ECG) play a critical role in ensuring the well-being of both the unborn child and the expectant mother during pregnancy. Extracting the fetal ECG from the composite signals is a challenging task due to the presence of various interfering components. In recent years, the application of advanced signal processing techniques has shown promising results

S. Ziani (✉)
ENSAM, Mohammed V University in Rabat, Rabat, Morocco
e-mail: s.ziani@um5r.ac.ma

E. Said
Sciences and Engineering Laboratory Polydisciplinary, Faculty of Taza, Sidi Mohammed Ben Abdellah University, Taza-Gare, 1223 Taza, PB, Morocco

M. Azrour et al. (eds.), *Sustainable and Green Technologies for Water and Environmental Management*, World Sustainability Series,
https://doi.org/10.1007/978-3-031-52419-6_10

in tackling this issue. In this research paper, we focus on characterizing the fetal ECG using the Short Time Fourier Transform (STFT)—a powerful signal processing tool that facilitates the analysis of non-stationary signals in both the time and frequency domains. The dynamic nature of the fetal ECG signal poses a significant challenge in its extraction (Mondal et al. 2022; Singh and Hota 2022; Ghedamsi et al. 2008; Ziani and El Hassouani 2020), making the application of advanced signal processing techniques essential for accurate results. While previous works (Ziani et al. 2017, 2018a, b, 2019; Ziani and El Hassouani 2019; Ziani 2022). have made progress in this area, most studies have primarily concentrated on time–frequency methods applied to speech signals and have not fully addressed the complexity of non-stationary biomedical signals, such as the fetal and maternal ECGs. These biomedical signals exhibit distinct energy distribution patterns over time and frequencies, which are clearly evident in their spectrograms. The primary objective of this study is to explore the potential of the STFT in extracting the fetal ECG from the mixture of signals, thereby enabling accurate and reliable fetal monitoring. By characterizing the fetal ECG through the STFT and analyzing its spectrogram, the aim is to achieve improved separation performance and provide valuable insights into the health status of the fetus. In the following sections, we will present the methodology, experimental setup, and results of the proposed approach. The simulation and experimental evaluations will demonstrate the effectiveness and high performance of our method in characterizing and extracting the fetal ECG. The findings of this study hold significant promise in advancing the field of fetal monitoring and enhancing maternal healthcare during pregnancy.

2 Proposed Method

2.1 *Model*

The purpose of blind source separation is to cancel source signal contributions S from a set of linear instantaneous mixtures of these sources A, without knowledge of A and using only the observations X.

We can write:

$$X = AS \tag{1}$$

with

- The M observations $X \in IRM \times T$
- The mixture matrix $A \in IRM \times N$
- The sources $S \in IRN \times T$
- $M \geq N$.

Our investigation includes the determined system, i.e. the configuration involving the same number of sources and observations (N = M = 2).

So Eq. (1) can be written as:

$$\begin{cases} x_1(t) = a_{11}s_F(t) + a_{12}s_M(t) \\ x_2(t) = a_{21}s_F(t) + a_{22}s_M(t) \end{cases} \tag{2}$$

where

$$x = \begin{pmatrix} x_1 \\ x_2 \end{pmatrix},\ A = \begin{pmatrix} a_{12} & a_{12} \\ a_{21} & a_{22} \end{pmatrix} \text{and } s = \begin{pmatrix} s_F \\ s_M \end{pmatrix}$$

The coefficients a_{ij} of the mixing matrix A are real, constant and different from zero. In our study S_F presents the fetal heart activity FECG and S_M presents the maternal heart activity MECG. The goal of this paper is to estimate S_F by using Short Time Fourier Transform STFT.

2.2 *Time-Frequency Analysis*

The STFT of $x_i(\tau)$ is then given by:

$$X_i(t, \omega) = \frac{1}{\sqrt{2\pi}} \int_{-\infty}^{+\infty} x_i(\tau)h(\tau - t)e^{-j\omega\tau}d\tau \tag{3}$$

where

- $h(\tau - t)$ is a shifted real-valued window function.
- $X_i(t, \omega)$ is the contribution of signal x_i in the short time and frequency windows, respectively, centered on t and ω.

A spectrogram is a graphical representation that illustrates the pointwise magnitude of the Fourier transform applied to a segment of an ECG signal. In our study, we calculate these spectrograms using a total of 2048 samples. It is important to note that the window size, which determines the number of samples used for each spectrogram, plays a critical role in this representation. If the window size is set too short, the spectrogram may not capture all the relevant information present in the ECG signal. On the other hand, if the window size is excessively long, the temporal resolution of the spectrogram decreases, potentially leading to the loss of important details.

To strike a balance between capturing essential information and preserving temporal resolution, we compute the spectrogram feature representation with a stride of 512 samples. This means that each consecutive spectrogram has a 75% overlap with the previous one in the time series. By adopting this approach, we ensure that significant signal variations are adequately represented. However, it is worth noting that selecting shorter strides will result in a higher-resolution representation of the signal. Nevertheless, this comes at the cost of increased computational demands since a greater number of spectrograms need to be computed. Hence, the choice of the stride parameter involves a trade-off between computational efficiency and obtaining a more detailed representation of the ECG signal.

Our approach hinges on utilizing the complex ratio of mixtures as a fundamental principle. This involves carefully analyzing the interplay between various mixed signals and computing their complex ratios to reveal crucial insights and patterns. By leveraging the complex ratio of mixtures, we can gain a deeper understanding of the underlying relationships among the signals and effectively extract the relevant information required for our objectives. This powerful technique forms the cornerstone of our methodology and allows us to achieve accurate and reliable results in the context of our research. The complex ratio of mixtures is mathematically expressed by Eq. (4). This equation encapsulates the essential relationship between the mixed signals and enables us to precisely quantify their interdependencies. By applying Eq. (4), we can effectively analyze the complex interactions among the signals and leverage this information to extract the desired components with a high level of accuracy. This equation serves as a fundamental tool in our approach, providing a robust mathematical framework to tackle the challenges posed by signal mixing and separation. Through the application of Eq. (4), we aim to unravel the intricate relationships within the mixtures and ultimately achieve our research objectives.

$$\alpha(t_j, \omega_k) = \frac{X_1(t_j, \omega_k)}{X_2(t_j, \omega_k)} \tag{4}$$

where

$$X_1(t, \omega) = \frac{1}{\sqrt{2\pi}} \int_{-\infty}^{+\infty} x_1(\tau) h(\tau - t) e^{-j\omega\tau} d\tau \tag{5}$$

and

$$X_2(t, \omega) = \frac{1}{\sqrt{2\pi}} \int_{-\infty}^{+\infty} x_2(\tau) h(\tau - t) e^{-j\omega\tau} d\tau \tag{6}$$

thus

$$\alpha(t_j, \omega_k) = \frac{a_{11}\mathbf{S}_1(t_j, \omega_k) + a_{12}\mathbf{S}_2(t_j, \omega_k)}{a_{21}\mathbf{S}_1(t_j, \omega_k) + a_{22}\mathbf{S}_2(t_j, \omega_k)} \tag{7}$$

where

$$S_1(t, \omega) = \frac{1}{\sqrt{2\pi}} \int_{-\infty}^{+\infty} s_F(\tau)h(\tau - t)e^{-j\omega\tau} d\tau \tag{8}$$

and

$$S_2(t, \omega) = \frac{1}{\sqrt{2\pi}} \int_{-\infty}^{+\infty} s_M(\tau)h(\tau - t)e^{-j\omega\tau} d\tau \tag{9}$$

Figure 1 illustrates the overlap in the time–frequency domain. The x-axis represents time, while the y-axis represents frequency. The figure shows how the spectrogram of a signal is computed using a specific window size and stride. The window size determines the number of samples used to compute each segment of the spectrogram. If the window size is too short, the spectrogram may not capture all the relevant information in the signal, resulting in incomplete representations. On the other hand, if the window size is too long, the spectrogram loses temporal resolution, making it challenging to discern fast changes in the signal.

Hence, our objective is to find the values of coefficients a_{11}, a_{12}, a_{21}, and a_{22}, which will enable us to solve Eq. (1). Additionally, it is essential to note that the fetal heart rate is significantly higher than that of the mother. Consequently, we can identify at least two pairs (ti, ωi) and (tj, ωj) in the time–frequency domain where s_F or s_M is equal to zero.

If: $S_F \neq 0$ and $S_M = 0$ So,

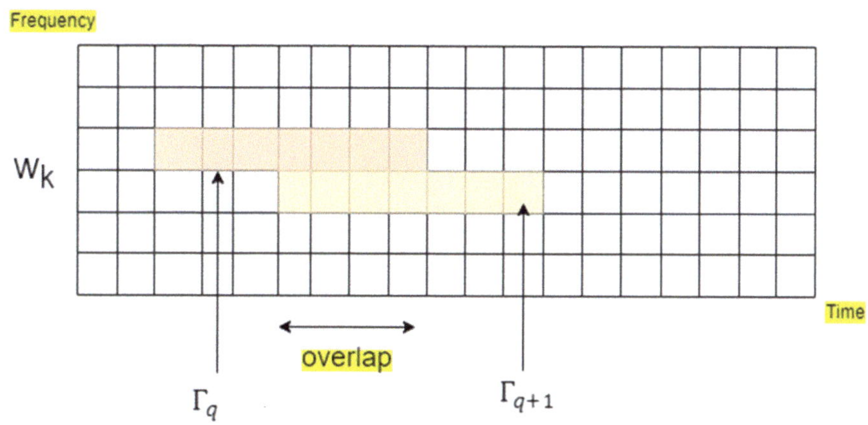

Fig. 1 Overlap in the time–frequency domain

$$\frac{a_{11}}{a_{21}} = \Gamma_1 \tag{10}$$

If: $S_F = 0$ and $S_M \neq 0$ So,

$$\frac{a_{12}}{a_{22}} = \Gamma_2 \tag{11}$$

So by replacing Eqs. 10 and 11 in Eq. 1 we obtain the expression of sF as depicted in Fig. 2:

$$\begin{cases} x_1 = a_{11}s_F + a_{12}s_M \\ x_2 = a_{21}s_F + a_{22}s_M \end{cases} \Rightarrow$$

$$\begin{cases} x_1 = \Gamma_1.a_{21}s_F + \Gamma_2.a_{22}s_M & \text{(a)} \\ x_2 = a_{21}s_F + a_{22}s_M & \text{(b)} \end{cases}$$

(a) Γ_2.(b) implies:

$$s_F = \frac{x_1 - \Gamma_2.x_2}{a_{21}(\Gamma_1 - \Gamma_2)} \tag{12}$$

s_F presents the Fetal ECG: FECG.

The process of calculating Γ_1 and Γ_2 reveals a fascinating ability to effectively identify the target signal. This research showcases the notable simplicity in determining the ratios Γ_1 and Γ_2. Employing the Short Fourier Transform (STFT) serves

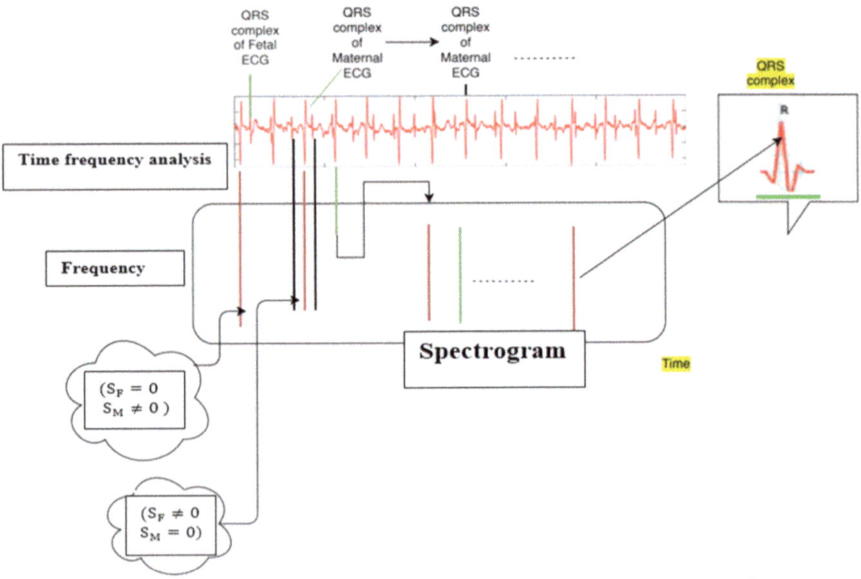

Fig. 2 The process of identifying the coordinates (ti, wi)

as a mathematical microscope, facilitating the detection of a signal's frequency content, especially when analyzing irregularities within electrocardiograms. The QRS complex in electrocardiograms plays a crucial role, holding vital medical parameters associated with the heart's electrical activity. An outstanding feature of this method is its capability to capture both the fetal and maternal QRS complexes in the signal recorded by the electrode placed on the mother's abdomen. Despite the fetal electrocardiogram having a significantly lower amplitude, its heart rate surpasses that of the mother. This characteristic allows for the clear identification of the QRS complexes in the time-scale domain or scalogram, as vividly demonstrated in Fig. 3. As a result, the algorithm can easily pinpoint all points (ti, wi) where either the fetal ($SF = 0$ and $SM \neq 0$) or maternal ($SF \neq 0$ and $SM = 0$) sources occur independently. By calculating the specified ratio (Eq. 7) at each of these points and identifying the repetition period, the algorithm efficiently computes the average of all the values obtained (Eqs. 11 and 12). This process further enhances the precision and accuracy of the extraction.

The presented Fig. 3 illustrates the step-by-step process of extracting the fetal electrocardiogram from the raw composite signal acquired through electrodes placed on

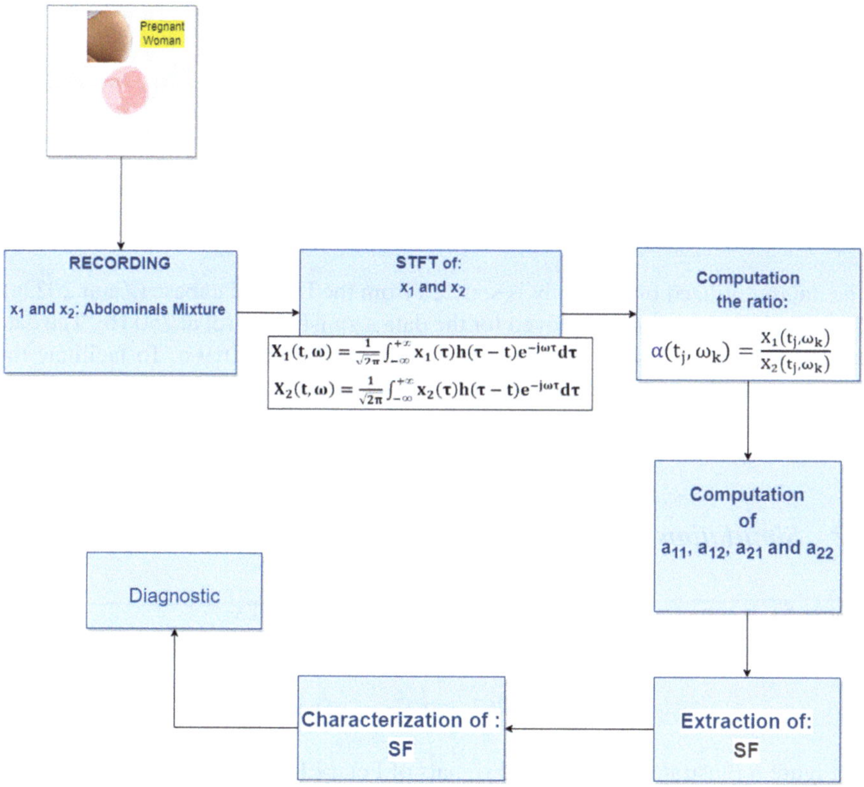

Fig. 3 Global algorithm

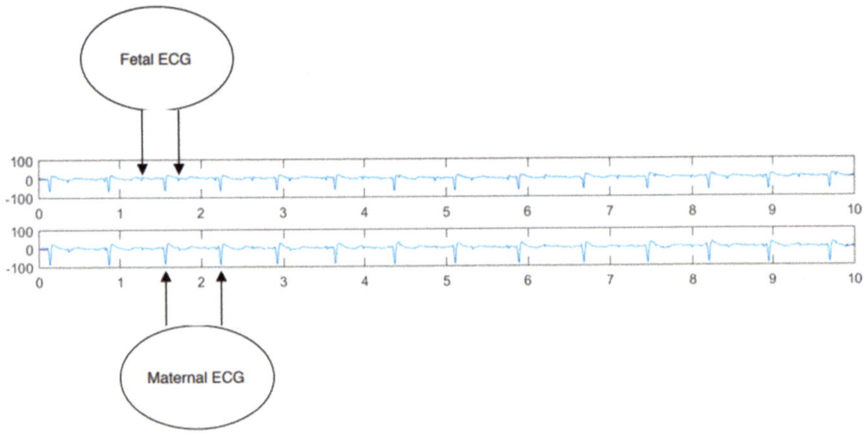

Fig. 4 Dataset input

the mother's abdomen. This algorithm employs sophisticated techniques to disentangle the maternal and fetal components, enabling a precise assessment of the fetal heart's electrical activity.

3 Results and Discussion

3.1 Data Recording

The dataset utilized in this study is sourced from the DaISy database (Ziani 2023a). The sampling frequency employed for the data acquisition is set at 250 Hz. The data processing and analysis were conducted using MATLAB 2019 a. To facilitate the investigation, the mixed signals x_1 and x_2 are parameterized with a time duration of $T = 2500$ s, as depicted in Fig. 4.

3.2 Simulation Results

Figure 5a, b shows respectively the spectrogram of: x_1 and x_2.

The simulation of the proposed algorithm yielded the following results:

$$\Gamma_1 = 0.7863 \text{ and } \Gamma_2 = -0.8864$$

Figure 6 illustrates the extracted signals of Fetal Electrocardiogram (FECG) and Maternal Electrocardiogram (MECG). These signals have been obtained using the

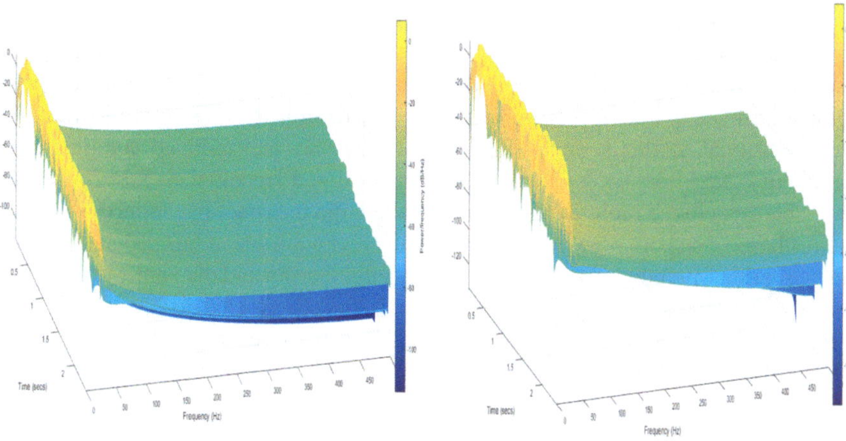

(a)
Spectrogram of: x_1

(b)
Spectrogram of: x_2

Fig. 5 Spectrogram of: x_1 and x_2 on 3D visualization

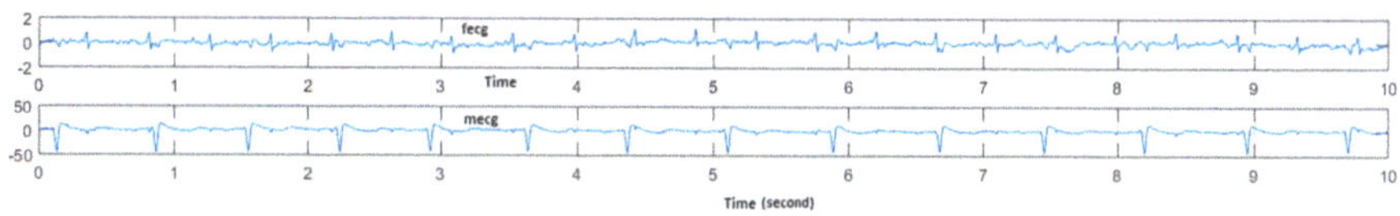

Fig. 6 Fetal and maternal signal

proposed algorithm and are presented for visual analysis and further interpretation. The FECG signal represents the electrical activity of the fetal heart, while the MECG signal corresponds to the maternal heart's electrical activity. These extracted signals are crucial in providing valuable insights into the health and well-being of both the fetus and the mother during pregnancy. The clear separation and accurate extraction of these signals are indicative of the effectiveness and success of the algorithm in isolating the distinct electrocardiogram components from the composite signal acquired through the electrodes placed on the mother's abdomen.

3.3 Discussion

The simulation of the proposed algorithm has successfully extracted the signals of Fetal Electrocardiogram (FECG) and Maternal Electrocardiogram (MECG) from the composite data acquired from the DaISy database. The clear visualization of these extracted signals (as shown in Fig. 5) confirms the efficacy of the algorithm in accurately isolating the fetal and maternal heart activities. This achievement is crucial for providing valuable insights into the health and well-being of both the fetus and the mother during pregnancy. The results of our study are consistent with previous research efforts that have also aimed to extract fetal and maternal electrocardiogram signals (Ziani et al. 2023a). However, the proposed algorithm has certain advantages over traditional methods, such as improved accuracy and reduced computational complexity. Additionally, by parameterizing the mixed signals with $T = 2500$ s, we were able to achieve more detailed and reliable signal extraction. Indeed, real-time control based on short-time Fourier transforms and wavelets in biomedical domains can be seamlessly integrated into both energy engineering and the control systems of electrical machines (Ziani and El Hassouani 2020; Achour et al. 2022, 2023; Chaou et al. 2022; Youssef et al. 2022), specifically in the context of electric scanners that constitute a pivotal component of medical imaging. Moreover, in the realm of electrical machine control, particularly within the domain of electric scanners utilized in medical imaging, the integration of these techniques can lead to breakthroughs in accuracy and efficiency. Employing wavelet-based methods for real-time control can enable the extraction of crucial information from electrical signals, allowing for precise adjustments and fine-tuning of the machines' performance. This not only contributes to the improvement of medical imaging processes but also extends its impact to broader applications in industry and technology (Ziani 2023b, Ziani et al. 2023b; Hafid et al. 2023; Laabab et al. 2022, 2023b).

Ultimately, the convergence of real-time control techniques rooted in short-time Fourier transforms and wavelets bridges the gap between biomedical engineering, energy systems, and electrical machinery. This interdisciplinary approach opens doors to innovative solutions that have the potential to revolutionize various sectors, showcasing the remarkable synergy between mathematical theory and practical applications (Youssef et al. 2023; Echoukairi et al. 2023; Laabab et al. 2023a; Ouhadou et al. 2018, 2019).

4 Conclusion

In conclusion, the proposed algorithm has demonstrated its efficacy in accurately extracting FECG and MECG signals from composite data. These extracted signals hold significant potential for various clinical applications, including fetal heart rate monitoring and maternal health assessment during pregnancy. However, further research and refinement are warranted to address potential limitations and enhance the algorithm's performance. Overall, the findings of this study contribute to the growing body of knowledge in fetal electrocardiogram analysis and pave the way for improved prenatal care and monitoring.

References

Achour HB et al (2022) Permanent magnet synchronous motor PMSM control by combining vector and PI controller. WSEAS Trans Syst Contr 17:244–249

Achour HB et al (2023) PI controller and quadratic feedback of synchronous machine. In: Farhaoui Y, Rocha A, Brahmia Z, Bhushab B (eds) Artificial intelligence and smart environment. ICAISE 2022. Lecture Notes in Networks and Systems, vol 635. Springer, Cham

Chaou Y et al (2022) Nonlinear control of the permanent magnet synchronous motor PMSM using backstepping method. WSEAS Trans Syst Contr 17:56–61

Echoukairi A et al (2023) Improved methods for automatic facial expression recognition. Int J Interact Mobile Technol (iJIM) 17(06):33–44. https://doi.org/10.3991/ijim.v17i06.37031

Ghedamsi K, Aouzellag D, Berkouk E (2008) Control of wind generator associated to a flywheel energy storage system. Renew Energy 33:2145–2156

Hafid BA et al (2023) A quadratic observer for sensorless drive system controller. In: Farhaoui Y, Rocha A, Brahmia Z, Bhushab B (eds) Artificial intelligence and smart environment. ICAISE 2022. Lecture Notes in Networks and Systems, vol 635. Springer, Cham

Laabab I et al (2022) A literature review of solar cell overheating control. In: Farhaoui Y, Rocha A, Brahmia Z, Bhushab B (eds) Artificial Intelligence and Smart Environment. ICAISE 2022. Lecture Notes in Networks and Systems, vol 635. Springer, Cham

Laabab I et al (2023a) A Solar panels overheating protection: a review. Indonesian J Electr Eng Comput Sci 29(1):49–55

Laabab I et al (2023b) A review of the application of artificial intelligence for weather prediction in solar energy: using artificial neural networks. In: Farhaoui Y, Rocha A, Brahmia Z, Bhushab B (eds) Artificial Intelligence and Smart Environment. ICAISE 2022. Lecture Notes in Networks and Systems, vol 635. Springer, Cham

Mondal A, Manikandan MS, Pachori RB (2022)Convolutional neural network based ECG quality assessment using derivative signal. In: 2022 fourth international conference on cognitive computing and information processing (CCIP), Bengaluru, India, pp 1–5. https://doi.org/10.1109/CCIP57447.2022.10058688

Ouhadou M et al (2018) Experimental modeling of the thermal resistance of the heat sink dedicated to SMD LEDs passive cooling. In: Proceedings of the 3rd international conference on smart city applications

Ouhadou M et al (2019) Experimental investigation on thermal performances of SMD LEDs light bar: Junction-to-case thermal resistance and junction temperature estimation. Optik 182

Singh K, Hota MK (2022) Design and development of DSP enabled low-cost ECG machine. In: 2022 first international conference on electrical, electronics, information and communication technologies (ICEEICT), Trichy, India, pp 1–5. https://doi.org/10.1109/ICEEICT53079.2022.9768423

Youssef C et al (2022) Electric vehicle backstepping controller using synchronous machine. In: International conference on artificial intelligence and smart environment, pp 367–373

Youssef C et al (2023) Backstepping control of the permanent magnet synchronous generator (PMSG) used in a wind power system. In: Farhaoui Y, Rocha A, Brahmia Z, Bhushab B (eds) Artificial intelligence and smart environment. ICAISE 2022. Lecture Notes in Networks and Systems, vol 635. Springer, Cham

Ziani S (2022) Contribution to single-channel fetal electrocardiogram identification. Traitement Du Signal 39(6):2055–2060

Ziani S (2023a) Fetal electrocardiogram identification using statistical analysis. In: Farhaoui Y, Rocha A, Brahmia Z, Bhushab B (eds) Artificial intelligence and smart environment. ICAISE 2022. Lecture Notes in Networks and Systems, vol 635. Springer, Cham

Ziani S (2023b) Enhancing fetal electrocardiogram classification: A hybrid approach incorporating multimodal data fusion and advanced deep learning models. Multimed Tools Appl. https://doi.org/10.1007/s11042-023-17305-6

Ziani S, El Hassouani Y (2019) Fetal-maternal electrocardiograms mixtures characterization based on time analysis. In: 2019 5th international conference on optimization and applications (ICOA), Kenitra, Morocco, pp 1–5. https://doi.org/10.1109/ICOA.2019.8727619
Ziani S, El Hassouani Y (2020a) A new approach for extracting and characterizingfetal electrocardiogram. Traitement du Signal 37(3):379–386. https://doi.org/10.18280/ts.370304
Ziani S, El Hassouani Y (2020b) Fetal electrocardiogram analysis based on LMS adaptive filtering and complex continuous wavelet 1-D. In: Farhaoui Y (eds) Big data and networks technologies. BDNT 2019. Lecture Notes in Networks and Systems, vol 81. Springer, Cham. https://doi.org/10.1007/978-3-030-23672-4_26
Ziani S, Jbari A, Belarbi L (2017) Fetal electrocardiogram characterization by using only the continuous wavelet transform CWT. In: 2017 international conference on electrical and information technologies (ICEIT), Rabat, Morocco, pp 1–6. https://doi.org/10.1109/EITech.2017.8255310
Ziani S, Jbari A, Bellarbi L (2018a) QRS complex characterization based on non-negative matrix factorization NMF. In: 2018 4th international conference on optimization and applications (ICOA), Mohammedia, Morocco, pp 1–5. https://doi.org/10.1109/ICOA.2018.8370548
Ziani S, Jbari A, Bellarbi L, Farhaoui Y (2018b) Blind maternal-fetal ECG separation based on the time-scale image TSI and SVD—ICA methods. Procedia Comput Sci 134:322–327. ISSN 1877-0509
Ziani S, El Hassouani Y, Farhaoui Y (2019) An NMF based method for detecting RR interval. In: International conference on big data and smart digital environment 2019. Springer
Ziani S, Farhaoui Y, Moutaib M (2023a) Extraction of fetal electrocardiogram by combining deep learning and SVD-ICA-NMF methods. Big Data Mining Analyt 6(3):301–310. https://doi.org/10.26599/BDMA.2022.9020035
Ziani S, Suchetha M, Rizal A (2023b) Time-scale image analysis for detection of fetal electrocardiogram. Multimed Tools Appl. https://doi.org/10.1007/s11042-023-17165-0

Flow Duration Curve a Tool for Assessing the Intensity and Seasonal Patterns of Low-Flow Periods: Case Study of Tassaout River Basin Within the Upstream of Moulay Youssef Dam (1978–2016) (Morocco)

Mohamed Chakir, Omar Ghadbane, Hassan Ouakhir, and Mohamed El Ghachi

Abstract Flow duration curve combines in one curve the flow characteristics of a stream throughout the range of discharge. It can also facilitate the assessment of both the severity and seasonal fluctuations of low-flow periods. This curve offers a valuable tool for comprehensively understanding the dynamics of water scarcity over time. The purpose of this study is to evaluate the intensity and seasonal patterns of low flow in the Tassaout River basin, upstream of the Moulay Youssef dam. This basin experiences a semi-arid climate characterized by limited rainfall. Furthermore, alongside the constrained distribution of water resources, the region faces growing demands due to ongoing climate change. Low-flow events are recurrent on a seasonal basis, and their severity can be influenced by annual variations in rainfall. This article aims to contribute to the examination of low-flow period severity and seasonality through the analysis of recorded water deficits during these periods. For this purpose, the Sequent Peak Algorithm (SPA) method was employed to extract water deficit volumes. The low-flow threshold was determined using the Q75 indicator derived from the flow duration curve, which was constructed from daily discharge data collected at the Ait Tamlil and Tamsemat gauging stations, spanning the period from 1978 to 2016.

Keywords Tassaout river basin · Low-flow · Flow duration curve · Threshold · Seasonality · Water deficit

---

M. Chakir (✉) · O. Ghadbane · H. Ouakhir · M. E. Ghachi
Dynamics of Landscapes, Risks and Heritage Laboratory, Faculty of Letters and Human Sciences, Sultan Moulay Slimane University, Beni Mellal, Morocco
e-mail: chakirmohamed.doc@hotmail.com

M. Azrour et al. (eds.), *Sustainable and Green Technologies for Water and Environmental Management*, World Sustainability Series,
https://doi.org/10.1007/978-3-031-52419-6_11

1 Introduction

Hydrological studies are of great importance today, especially in view of the dramatic increase in population densities around the world, and researchers are interested in the reasons behind the exacerbation and optimization of drought, as well as the effects of droughts. The process of holistic assessment of hydrological risks is very difficult (Raml and Arzek 2018). Drought is one of the long-term hydrological disasters, because it is a natural imbalance that man suffers or contributes to worsening (Larue and Giret 1989). The role of human activity and pressure in increasing the sensitivity of the phenomenon cannot be ignored (Corbonnois and Humbert 2000). Consequently, the analysis of the low-flow event is directly linked to the problem of water scarcity, because the magnitude is serious when the scarcity of water resources affects the ability to meet human needs. Given the scarcity of water resources in areas with semi-arid climates, we wish to use this article to understand the mechanisms and processes that are the first link in a river basin management system, belonging to the Central High Atlas, which is a model of water management.

The Tassaouat basin (upstream of the M. Youssef dam) is characterised by significant anthropic dynamism, which has led to the development of agricultural activities, a large area of irrigated land and a major hydro-agricultural network through which irrigation water is conveyed and distributed to agricultural areas. Considering that the study of runoff is a problem mainly linked to the climatic and human factors of the river basin, which lead to complex variations in time and space, we have tried to approach the problem in its geographical context, to better contextualise the work. Atmospheric or meteorological drought is characterised by the evaporative capacity of the atmosphere being such that it outweighs precipitation; the water balance is therefore in deficit during the period ($P < E$) (Lambert 1975). This article proposes a quantification of the water deficit during low-flow periods in the Tassaout basin (upstream of the Moulay Youssef dam), in terms of surface water resources mobilised for the benefit of several irrigated agricultural areas.

The study basin offers us a good environment for understanding hydrological behaviour, because the known risk that we can expect is a more controlled risk that can be contained and better managed (Chakir 2022). Consequently, before tackling the imperatives of rational management of water resources, an in-depth and specific study of the various natural characteristics and a meticulous analysis of hydro-climatic data are essential.

2 General Context of the Study

2.1 *Presentation of the Study Area*

The Tassaout river basin (upstream of the Moulay Youssef dam) is part of the central High Atlas Mountains and lies to the south-east of the Oum El Rabia basin. It covers an area of 1307 km^2. It rises at an altitude of 3978 m and flows from north-east to north-west. The right bank is characterised by relatively short and few, often temporary, tributaries. The largest tributaries are on the left bank.

The geological bedrock is mainly composed of permeable sandstone, dolomite and limestone belonging to the Triassic and Lias sedimentary series upstream of the basin, and impermeable shale and clay belonging to the Triassic sedimentary series downstream of the basin (Couvreur 1988). The study area is rich in karstic limestone formations that provide scattered springs, playing an important role in watercourse supply, drinking water supply and irrigation. The highly permeable hydrogeological formations in the Tassaout river basin (upstream of the Moulay Youssef dam) are made up of recent quaternary formations in the form of alluvium and carbonate rocks, often characterised by fissures, which can, in most cases, give rise to surface and underground karstic forms. The less permeable formations consist of marl and limestone from the Middle Lias, sandstone from the Upper Lias, conglomerates and marl and sandstone from the Lower Cenomanian (Castany 1982). This class accounts for 56% of the total surface area of the river basin. Impermeable formations are generally associated with outcrops of Lower Triassic doleritic basalt, Cambro-Ordovician schist formations and clay formations (Stéphano Triassic). Lower Triassic doleritic basalt outcrops, Cambro-Ordovician schist formations and clay formations (Stéphano Triassic). This class accounts for 13% of the total surface area of the river basin (Fig. 1).

2.2 *Hydro-Climatic Context*

Average annual rainfall at the Tamsmat station was 514 mm^{-1} over the period 1976–2016. This record was characterized by high variability, with successive wet and dry periods. The average coefficient of variation (CV) was estimated at 0.31, indicating that this basin generally receives moderate rainfall.

There is considerable variability in discharge from one year to the next. The average annual discharge was 8.9 m^3/s^{-1} at the study station over the period 1978–2016. Only 15 years exceed the modulus, but 24 are "water deficit" years.

To study the temporal heterogeneity of the discharge, the *NICHOLSON* rainfall index, and the low-pass filter method of "*HANNING*" were used to highlight hydrological trends and distinguish between periods of deficit and surplus. Changes in hydrological regimes leading to surplus and deficit trends are supported by statistical studies to detect tipping points or break points (Boukrim et al. 2011). These transition

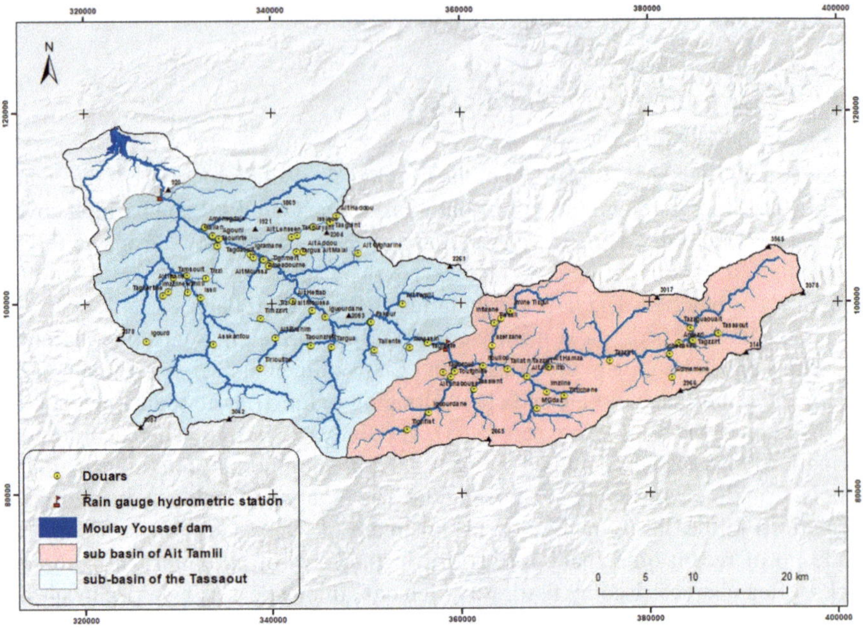

Fig. 1 National and regional localisation of the Tassaout basin (Alia Moulay Youssef dam). *Source* Topographic map of Demnat and Ait Tamlil 1/100000, DEM 2017

tests make it possible to identify the pivotal years of climate change and improve water indicators. This heterogeneity in annual yields is characterised by a continuity of wet and dry periods of varying duration. The *NICHOLSON* rainfall index was produced over four periods. These are two dry periods (1978–1987) and (1999–2006), and two wet periods extending along (1988–1998) and (2007–2016). This indicator shows that there was a significant water deficit over the 1978–1987 period. Maximum values were reached in 1983 and 1984 respectively, at the Tamsemat and Ait Tamlil stations (Fig. 2).

3 Methods

- *Daily discharge volume deficit based on the SPA filter*

Low-flow can be determined and analysed in terms of duration and/or water deficit and defined as events below a threshold Q0 (Yevjevich 1967). The SPA (Sequent Peak Algorithm) was originally developed to size the volume of the reservoir (Tallaksen and Lanen 2004), but it has also been used to calculate a range of daily flow deficits in relation to a given threshold.

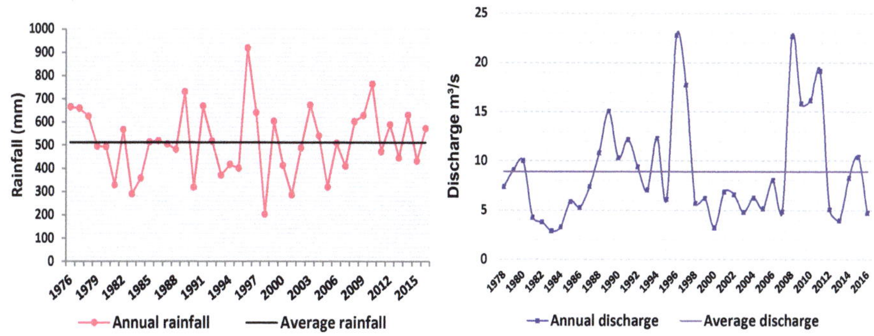

Fig. 2 Annual change in rainfall and discharge at the Tamsmat station (1976–2016)

The information extracted from each event is specified, such as the start (t0) and end (tf) dates of the event, the duration of the event (d = tf − t0) and the volume deficit V for the threshold, (i.e. the sum of the difference between the threshold and the observed flow).

If qt is the daily discharge rate and q0 is the threshold extracted from the flow duration curve, then the following formula is used to determine the deficit volume by applying the SPA algorithm method:

$$S_t = \begin{cases} S_{t-1} + q_0 - q_t, positive \\ 0, negative \end{cases} \quad (1)$$

Low-flow represents an extreme natural event and is generally caused by a prolonged lack of precipitation. Consequently, low-flow levels cannot be considered as absolute, but appear to be linked to "thresholds" that can vary according to the geographical context and the natural characteristics of the basin (Giuntoli and Renard 2010). The notion of severity (or amplitude) describes the minimum discharge threshold reached by the watercourse during low-flow periods (Garcia 2016), but can also refer to aspects linked to the duration of the water deficit, which represents the loss of volume and seasonality at specific periods and determines the time of year when the water shortage occurs (Catalogne 2012) (Fig. 3).

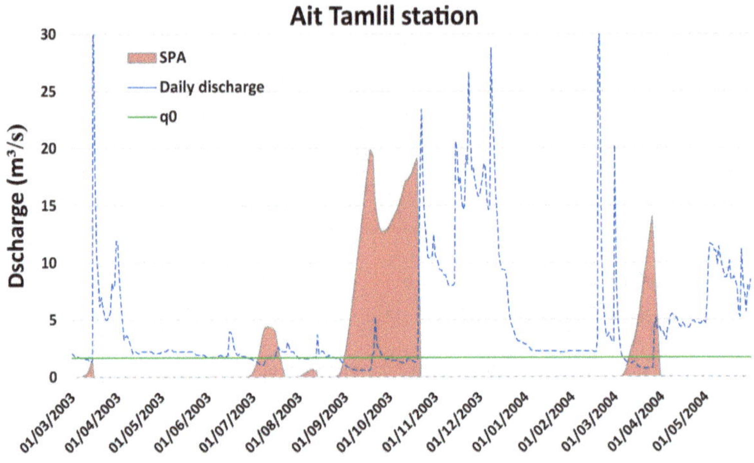

Fig. 3 Example of the SPA algorithm applied to the daily discharge at the Ait Temlil station (2003–2004)

4 Results

4.1 Identification of Deficit Periods Using Low-Flow Indexes Derived from the Flow Duration Curve

To extract the periods of water deficit using the SPA method, we will consider the low-flow indices resulting from the flow duration curve classified as threshold q0. This method gave the results illustrated in Fig. 4.

The null interval in the filter corresponds to a sustained flow. Conversely, low-flow events correspond to positive values in the SPA algorithm. The main advantage of

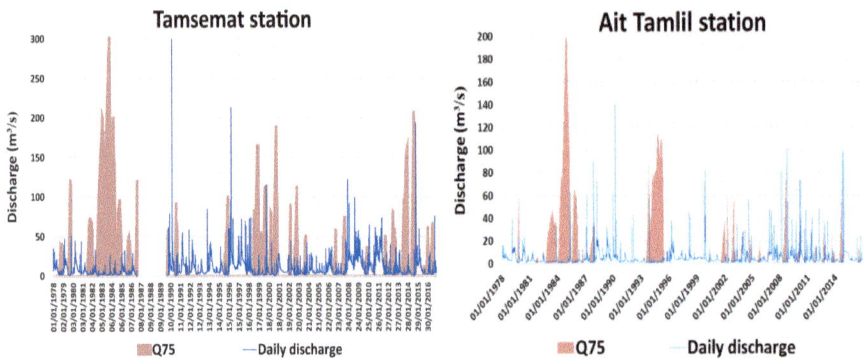

Fig. 4 Cumulative water deficit using the SPA algorithm and the index (Q75) from the flow duration curve

this method is that the histogram reflects the severity of the water deficit, with the highest values corresponding to the most severe low-flow periods.

This choice posed the problem of studying phases that do not reflect low-flow events, so the choice of a higher threshold resulted in several years of deficits attributed to the hydrological drought phenomenon according to the classification proposed by Dracup et al. (1980). On the other hand, very low indices can indicate very severe low-flow conditions, but only for a limited period (Table 1).

For each study chronicle, we extracted the durations, deficit volumes and maximum annual deficits by considering the threshold $Q_0 = Q75$ extracted from the flow duration curve. Severe deficit episodes occurred. These episodes are confined to specific years of severe drought in the two basins studied, and thus reflect a very marked shortage of water. The most significant events occurred in 1983, 1984 and 1985, which were marked by severe climatic and hydrological drought.

The Ait Tamlil station has many episodes (51) with a total of around 4765 days, but the average deficit volume is only 109.2 hm^3 over the record period (1978–2016), while the Tamsemat station only has 48 episodes, with a total of around 4399 days, although the downstream area has suffered a very pronounced deficit volume (247.5 hm^3).

The most severe volume deficit was observed at Tamsemat between 25/6/1982 and 21/05/1984, which is 697 days below the Q75 threshold, with a total volume of 26.2 hm^3. At the Ait Tamlil station, the most severe volume deficit occurred between 4/12/1983 and 30/4/1985, which is. 514 days below the proposed threshold, with a total volume of 17.2 hm^3.

In general, the years with severe low-flow periods correspond to the driest years in the study period.

4.2 Seasonality of Water Deficit Periods

One of the characteristics that distinguishes the low-flow event is its seasonality, as defined by different researchers Catalogne (2012, Chakir 2022, Laaha and Blöschl 2006, Smakhtin 2001).

The results of the seasonality of the water deficit reveal a range of information on the temporal and spatial distribution of low-flow flows. Most water deficit episodes occur during low-water periods, in particular the period from July to September. Water deficit episodes in the Ait Tamlil sub-basin are characterised by prolonged periods. On the other hand, the Tassaout river basin (upstream of the M. Youssef dam) is characterized by shorter periods but a remarkable water deficit. The Q75 index derived from the flow duration curve shows short to moderate periods of water deficit, except in 1983, 1984 and 1994, which were characterized by a severe water deficit (Fig. 5).

Table 1 Cumulative deficit volume using the SPA algorithm and the index (Q75) from the flow duration curve

Tamsemat station					Ait tamlil station				
Period	Start date	End date	Duration	Volume deficit (hm^3)	od	Start date	End date	Duration	Volume deficit (hm^3)
1	14/07/1978	25/12/1978	165	3,67	1	15/07/1979	19/10/1979	97	5,12
2	10/07/1979	19/10/1979	102	10,60	2	23/07/1980	26/08/1980	35	0,34
3	31/10/1980	04/11/1980	5	0,03	3	30/08/1980	03/09/1980	5	0,03
					4	18/07/1981	25/11/1981	131	0,55
4	26/06/1981	17/02/1982	237	6,42	5	29/07/1982	25/08/1982	28	0,09
5	25/06/1982	21/05/1984	697	26,19	6	01/09/1982	29/10/1983	424	4,20
6	01/07/1984	06/01/1985	190	8,43	7	04/12/1983	30/04/1985	514	17,27
7	02/07/1985	16/07/1985	15	0,09	8	12/06/1985	17/03/1986	279	5,77
8	28/07/1985	04/01/1986	161	4,16	9	29/06/1986	01/10/1986	95	1,43
9	14/06/1986	31/08/1986	79	10,53	10	23/05/1987	24/10/1987	155	3,63
10	01/09/1989	26/10/1989	56	4,92	11	22/11/1987	29/11/1987	8	0,11
11	24/06/1990	19/09/1990	88	8,09	12	10/01/1988	15/01/1988	6	0,04
12	27/08/1991	13/09/1991	18	0,53	13	05/08/1990	16/09/1990	43	0,74
13	25/10/1991	05/12/1991	42	1,07	14	31/10/1991	13/11/1991	14	0,01
14	14/09/1992	12/10/1992	29	0,29	15	08/01/1992	17/02/1992	41	0,31
15	20/07/1995	27/12/1995	161	8,78	16	21/02/1992	25/02/1992	5	0,04
16	04/06/1998	15/06/1998	12	0,84	17	09/05/1993	31/05/1995	753	9,90
17	21/06/1998	15/01/1999	209	14,42	18	01/01/1996	11/01/1996	11	0,14

(continued)

Table 1 (continued)

Tamsemat station					Ait tamlil station				
Period	Start date	End date	Duration	Volume deficit (hm^3)	od	Start date	End date	Duration	Volume deficit (hm^3)
18	03/02/1999	15/05/1999	102	4,76	19	17/02/2000	07/04/2000	51	1,28
19	10/06/1999	11/10/1999	124	9,98	20	23/06/2001	01/01/2002	193	3,31
20	01/02/2000	09/04/2000	69	6,96	21	23/01/2002	19/04/2002	87	5,83
21	23/06/2000	01/12/2000	162	16,44	22	14/07/2002	09/02/2003	211	4,79
22	01/02/2002	04/04/2002	63	7,91	23	08/03/2003	16/03/2003	9	0,12
23	07/08/2002	09/01/2003	156	9,86	24	29/06/2003	23/07/2003	25	0,38
					25	02/08/2003	13/08/2003	12	0,06
24	26/06/2003	09/08/2003	45	1,01	26	26/08/2003	22/10/2003	58	1,72
25	16/08/2003	31/10/2003	77	4,38	27	07/03/2004	01/04/2004	26	1,21
26	16/02/2004	19/02/2004	4	0,08	28	17/07/2004	01/08/2004	16	0,47
27	26/08/2005	03/09/2005	9	0,17	29	06/08/2004	03/11/2004	90	2,85
28	13/10/2005	16/10/2005	4	0,12	30	14/07/2005	25/11/2005	135	2,01
29	15/07/2006	10/11/2006	119	5,21	31	04/07/2006	05/09/2006	64	1,05
30	21/11/2006	03/12/2006	13	0,09	32	01/09/2007	22/09/2007	22	1,59
31	22/01/2007	28/01/2007	7	0,14	33	25/03/2008	18/09/2008	178	8,21
32	22/03/2007	29/03/2007	9	0,09	34	30/09/2008	31/10/2008	32	1,57
33	04/04/2007	10/04/2007	7	0,10	35	29/07/2009	28/08/2009	31	0,97
34	13/06/2007	17/10/2007	127	6,56	36	06/10/2009	21/12/2009	77	0,93
35	20/09/2009	24/09/2009	5	0,25	37	27/12/2009	09/01/2010	14	0,52
36	27/10/2009	24/12/2009	59	3,29	38	31/07/2010	15/08/2010	16	1,06

(continued)

Table 1 (continued)

Tamsemat station					Ait tamlil station				
Period	Start date	End date	Duration	Volume deficit (hm³)	od	Start date	End date	Duration	Volume deficit (hm³)
37	19/08/2010	01/09/2010	14	1,49	39	18/08/2010	10/10/2010	54	3
38	01/09/2011	20/11/2011	81	4,51	40	25/02/2011	13/03/2011	17	0,60
39	22/12/2011	26/01/2012	36	1,51	41	21/07/2011	10/08/2011	21	0,30
40	22/02/2012	26/03/2012	34	1,42	42	24/08/2011	27/08/2011	4	0,01
41	10/05/2012	17/05/2012	8	0,03	43	04/09/2011	20/11/2011	78	1,74
					44	14/03/2012	22/03/2012	9	0,02
42	24/05/2012	01/11/2012	162	7,04	45	21/07/2012	28/10/2012	100	2,51
43	10/01/2013	19/01/2013	10	0,28	46	28/06/2013	09/07/2013	12	0,08
44	13/06/2013	05/02/2014	238	14,19	47	17/07/2013	18/01/2014	186	2,64
45	24/05/2014	10/11/2014	171	18,05	48	22/05/2014	20/09/2014	122	5,25
46	04/09/2015	03/10/2015	30	1,14	49	27/09/2014	20/11/2014	55	2,04
47	07/12/2015	04/03/2016	89	5,44	50	03/06/2016	07/09/2016	97	1,37
48	24/05/2016	30/08/2016	99	5,91	51	10/09/2016	28/09/2016	19	0,25
Average			91,6	5,16	Average			93,4	2,14
Max			697,0	26,19	Max			753,0	17,27

5 Discussion

Low-flow events can be determined and analyzed using a discharge threshold, which is used to identify the deficit and its duration. The analysis of low-flow situations based on a discharge threshold has a number of advantages, the most important of which is that it takes into account the duration of hydrological events, unlike other indicators that limit low-flow to a temporal context that can be identical whatever the year (Lang 2007). Several authors, interested in deficit volumes, (Fleig et al. 2006) use thresholds that generally correspond to the indices derived from the flow duration

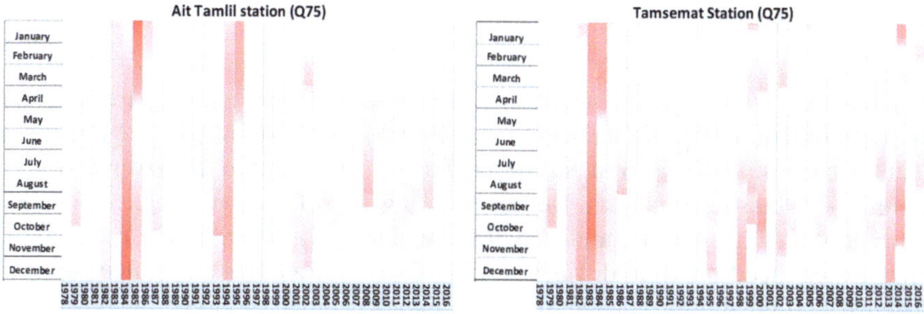

Fig. 5 Seasonality of periods of water deficit according to the Q75 index derived from the flow duration curve

curve classified between 70 and 95%. We might also mention the work of Tallaksen et al. (1997), who analyzed the duration and volume of water deficit according to three different thresholds: 50, 70 and 90%. They also found that the occurrence of droughts over several years rather than in other years was also an important feature in the choice of an appropriate threshold, and that this choice depended on the type of water deficit under consideration rather than a random choice. In the continental context, New Brunswick (Canada) was highlighted in a low-flow study by El-Jabi et al. (2011) by adopting the Q90 index, below which deficit periods were extracted. They stressed that the justification for applying this index was based on hydrological and economic characteristics.

The low-flow study in the Tassaout river basin, based on the Q75 index derived from the flow duration curve (DC), revealed periods of water deficit lasting several months in certain years. The downstream part of the basin (Tamsemat station) had a deficit of 26,2 hm^3 in 1982, 1983 and 1984, while the upstream part (Ait Tamlil station) had a deficit of 17,2 hm^3 in 1983, 1984 and 1985. Based on the extracted threshold, it was found that the total number of days varied from year to year and between the upstream and downstream parts of the basin. The percentage of days below the Q75 threshold represents 34% of the days in the study chronicle (1978–2016) in the Tassaout river basin (upstream of the M. Youssef dam).

The results showed that the Tassaout basin (upstream of the M. Youssef dam) experiences remarkably severe low-flow periods during particularly extreme events. Low-flow occurred frequently over many years (more than 70%), this being due to climatic factors affecting the hydrological regime of the river basins. The occurrence of low-flow on a spatial scale differs between upstream and downstream, mainly due to the physical characteristics of the basin. Low-flow days occur more frequently upstream (Ait Tamlil catchment) than downstream. This is due to frequent low-flow during the winter period, when snow accumulates and holds back large quantities of water. However, the deficit is very pronounced downstream as a result of human activities, in particular the withdrawal of water for irrigated farming (Chakir and El Ghachi 2021, Goumih et al. 2022).

6 Conclusion

Assessing the hydrological conditions in the Tassaout basin in the upstream of the M. Youssef dam involves establishing a reference point for low-flow conditions at Q75. Our investigation into identifying deficit episodes within this river basin is a preliminary statistical analysis. Defining the severity of the hydrological situation relies on selecting appropriate thresholds that signify the extent of the issue. Some years experienced multiple distinct periods. Examination of the low-flow severity unveiled two notably prolonged and voluminous deficient periods: the span from 1982 to 1985, and the interval from 1993 to 1995.

Measuring the volume deficit during low-flow occurrences serves to enhance the management of water resources during periods of scarcity, particularly upstream of dams. It equips water sector authorities with a lucid perspective, enabling them to proactively address crises and strategize effective solutions. This approach involves identifying influencing factors, rationalizing management practices, and cultivating human resilience in the face of climatic adversities and potential disasters. By adopting predictive and preventive management strategies, the aim is to mitigate risks associated with drought and water shortages.

References

Boukrim S, Lahrach, Sadkaoui N, Chaouni A (2011) Impact des changements climatiques sur l'hydrologie du bassin versant de l'Ouergha (Rif-Maroc). Revue scientifique et Technique LJEE 42–55

Castany G (1982) Principes et méthodes de l'hydrogéologie. DUNOD. PARIS

Catalogne C (2012) Amélioration des méthodes de prédétermination des débits de référence d'étiage en sites peu ou pas jaugés. https://theses.hal.science/tel-01547922

Chakir M (2022) Étude statistique des étiages et des tarissements dans le bassin versant de l'oued de Tassaout (Amont du Barrage de Moulay Youssef): Détermination, extraction, modélisation et problématique de la gestion (1978–2020) (Haut Atlas - Maroc)

Chakir M, El Ghachi M (2021) Estimation des prélèvements d'eau pour l'irrigation dans le bassin de la Tassaout (en amont du Barrage de Moulay Youssef) en 2015 : cartographie, suivi et mesure. In: Monitoring I dans le bassin de l'Oum Er-Rbia (Maroc) : retours d'expériences en hydro-climatologie, pp 55–64

Corbonnois J, Humbert J (2000) Ressources et gestion de l'eau dans le Nord-Est de la France. In: L'eau dans les régions françaises, d'un extrême à l'autre. SEDES, pp 219–249

Couvreur G (1988) Essai sur l'évolution morphologique de haut atlas calcaire—Maroc

Dracup JA, Lee KS, Paulson EG Jr (1980) On the definition of droughts. Water Resour Res 16:297–302. https://doi.org/10.1029/WR016i002p00297

El-Jabi N, Ashkar F, Issa M (2011) Application stochastique du phénomène d'étiage. Can J Civ Eng 24:191–200. https://doi.org/10.1139/cjce-24-2-191

Fleig AK, Tallaksen LM, Hisdal H, Demuth S (2006) A global evaluation of streamflow drought characteristics. Hydrol Earth Syst Sci 10:535–552. https://doi.org/10.5194/hess-10-535-2006

Garcia F (2016) Amélioration d'une modélisation hydrologique régionalisée pour estimer les statistiques d'étiage

Giuntoli I, Renard B (2010) Identification des impacts hydrologiques du changement climatique : vers un réseau de référence pour la surveillance des étiages. ONEMA-Irstea

Goumih M, El Ghachi M, Ouakhir H, Ennaji N, Midaoui A (2022) Hydrological response of the Assif Melloul basin in the Central High Atlas (Morocco): monitoring and Creating a rating curve. In: IOP conference series: earth and environmental science, vol 1090(1). https://doi.org/10.1088/1755-1315/1090/1/012003

Laaha G, Blöschl G (2006) A comparison of low flow regionalisation methods -catchment grouping. J Hydrol 323:193–214

Lambert R (1975) Été, étiage et tarissement des cours d'eau dans le bassin garonnais. Revue Géographique De L'est 15:139–151. https://doi.org/10.3406/rgest.1975.1315

Lang C (2007) Etiages et tarissements: vers quelles modélisations? L'approche conceptuelle et l'analyse statistique en réponse à la diversité spatiale des écoulements en étiage des cours d'eau de l'Est français

Larue J-P, Giret A (2004) L'assèchement de cours d'eau dans le bassin de la Maine entre 1989 et 1992. Norois. Environnement, aménagement, société, pp 117–133. https://doi.org/10.4000/norois.944

Raml QA, Arzek AS (2018) Evaluating the qualitative characteristics of groundwater and its suitability for human use in the Kirkuk district. J Educ Sci Stud 2

Roche MF (1986) Dictionnaire français d'hydrologie de surface avec équivalents en anglais, espagnol, allemand. Dunod, Paris, New York

Smakhtin VU (2001) Low flow hydrology: a review. J Hydrol 240:147–186. https://doi.org/10.1016/S0022-1694(00)00340-1

Tallaksen L, van Lanen HAJ (2004) Hydrological drought. Elsevier, Processes and estimation methods for streamflow and groundwater

Tallaksen LM, Madsen H, Clausen B (1997) On the definition and modelling of streamflow drought duration and deficit volume. Hydrol Sci J 42:15–33

Yevjevich V (1969) An objective approach to definitions and investigations of continental hydrologic droughts. VUJICA YEVJEVICH: fort Collins, Colorado State University, 1967, 19 p. (Hydrology paper no. 23). J Hydrol 7:353–353. https://doi.org/10.1016/0022-1694(69)90110-3

Measurement and Analysis of NO_2 Concentration by Differential Optical Absorption Spectroscopy: Towards Enhanced Air Quality Monitoring

Lotanna Ucheagu, Mbadiwe S. Benyeogor, Kolawole I. Agbaogun, Daodu E. Tunde, Olusegun I. Lawal, Chibuike Orazulike, Andrew O. Benyeogor, Amos Odju, Tobore L. Igbigbi, and Augustina C. Amaechi

Abstract The effective monitoring of nitrogen dioxide (NO_2) assumes pivotal significance in fostering sustainable environments, primarily due to its substantial impact on air quality. This study constitutes a significant stride in this direction by

L. Ucheagu
Institute of Environmental Physics (IUP) University of Bremen, Bremen, Germany
e-mail: ucheagu@uni-bremen.de

M. S. Benyeogor (✉) · K. I. Agbaogun
Institute of Physics, University of Münster, Münster, Germany
e-mail: mbenyeog@uni-muenster.de

K. I. Agbaogun
e-mail: kiagbaogun@uni-muenster.de

D. E. Tunde · O. I. Lawal
Advanced Aerospace Engines Laboratory, National Space Research and Development Agency (NASRDA), Oka Akoko, Ondo State, Nigeria

C. Orazulike
Department of Hydro Sciences, Technische Universität Dresden, Dresden, Germany
e-mail: cor001@un-ihe.org

UNESCO-IHE Delft Institute for Water Education, Delft, Netherlands

A. O. Benyeogor · A. Odju
Department of Business Information Systems, Central Michigan University, Mount Pleasant, MI 8859, USA
e-mail: Benye1a@cmich.edu

A. Odju
e-mail: odju1a@cmich.edu

T. L. Igbigbi
City of Cleveland Department of Public Utilities, Cleveland, Ohio State, USA
e-mail: Tobore_igbigbi@ClevelandWater.com

A. C. Amaechi
International Institute of Tropical Agriculture (IITA), Ibadan, Nigeria
e-mail: a.amaechi@cgiar.org

M. Azrour et al. (eds.), *Sustainable and Green Technologies for Water and Environmental Management*, World Sustainability Series,
https://doi.org/10.1007/978-3-031-52419-6_12

harnessing the capabilities of Differential Optical Absorption Spectroscopy (DOAS) to measure and comprehend NO_2 levels, thereby shedding comprehensive light on its vertical distribution within the troposphere. The DOAS method also offered a unique advantage in terms of identifying and quantifying trace gases with unknown vertical columns in the designated region. These intricate tasks are accomplished by leveraging UV and visible wavelength range spectroscopy, which makes use of scattered sunlight as the fundamental light source. Our work delves into the technical intricacies underpinning DOAS while emphasizing the broader context of its utility. These include the use of pixel-based gas profile analysis, statistical methods, and Python scripts to analyze the measured gas concentration data. By capitalizing on these advanced techniques, a comprehensive understanding of trace gas distribution and behavior is achieved, thereby making a substantial contribution to the enhancement of environmental monitoring strategies. In essence, our work offers both technical and analytical insights that support sustainable practices for monitoring and managing air quality.

Keywords Air quality · Nitrogen dioxide · DOAS · Trace gas · Zenith-sky

1 Introduction

Air pollution is a pressing environmental concern, and nitrogen dioxide (NO_2) plays a significant role in contributing to this issue (Kellogg and Schneider 1978). It has adverse effects on the environment and human health, warranting urgent attention and mitigation measures (Fuentes and Rowe 1998). Effective monitoring of NO_2 is crucial (Vrekoussis et al. 2022), necessitating advanced techniques like Differential Optical Absorption Spectroscopy (DOAS) (Platt and Stutz 2008), known for its precision in measuring atmospheric NO_2 concentrations using UV and visible wavelength range spectroscopy with scattered sunlight. Despite the importance of monitoring NO_2, there is a need for comprehensive data on its concentration and vertical distribution. This study addresses this gap by analytically applying DOAS to assess NO_2 levels and vertical distribution in a tropospheric area, while also identifying unknown trace gases and estimating their vertical columns. Hence, valuable insights into air quality management are provided, contributing to informed environmental protection policies.

2 Literature Review

The NO_2 is a critical trace gas in the nitrogen cycle (Vrekoussis et al. 2022). It plays a significant role in various processes, including nitrification and denitrification, influencing the transformation of nitrogen compounds in the atmosphere and ecosystem. However, monitoring NO_2 levels is essential for understanding its impact on air quality. The DOAS technique has emerged as a precise method to measure the concentration of atmospheric NO_2, necessitating effective data analysis. This measures the specific narrowband absorption spectra of trace gases in the UV and visible spectral regions. The technique is based on the principle of absorption spectroscopy, which involves the interaction of light with matter. Here, light of a specific wavelength is absorbed by a sample of a gas, and the amount of absorption is proportional to the concentration of the gas in the sample. DOAS applies this principle by measuring the differential absorption of light at different wavelengths, providing a sensitive and accurate method for determining the concentration of trace gases in the atmosphere (Lancaster et al. 1998).

The Beer-Lambert-Law is used to describe the relationship between the attenuation of incoming radiation and the properties of the medium. This law provides the theoretical basis for absorption spectroscopy and DOAS. By measuring the attenuation of radiation at different wavelengths, DOAS can determine the concentration and total amount of trace gases in the atmosphere. Since DOAS measures the sky brightness as a function of wavelength at different solar zenith angles (SZA), accounting for extinction by molecular absorption of J species, as well as Mie and Rayleigh scattering (Lancaster et al. 1998), the Lambert-Beer law can be written as

$$
\begin{aligned}
I(\lambda, \Theta) = \alpha(\lambda, \Theta) I_0(\lambda) \exp\Big[-\int \sum \sigma_{j}(\lambda)\rho_j(s) \\
+ \sigma_{Mie}(\lambda)\rho_{Mie}(s) + \sigma_{Ray}(s)ds \Big]
\end{aligned}
\quad (1)
$$

where σ_j is the absorption cross-section of species j, ρ_x is the density of absorber or scatterer x, I_0 is the radiance outside the atmosphere, and the integral is taken along the light path s. The factor α determines the intensity of the scattered light and depends on the density of scatterers and the phase function and is a smooth function of wavelength.

Defining the slant column as an integral of the absorber column along the light path ($SC_j = \int \rho_j(s)ds$) and assuming that the absorption cross-sections do not vary significantly along the light path leading to:

$$
\begin{aligned}
I(\lambda, \Theta) = \alpha(\lambda, \Theta) I_0(\lambda) \exp\Big[-\sum_{j=1}^{J} \sigma_j(\lambda) SC_j \\
+ \sigma_{Mie}(\lambda) SC_{Mie} + \sigma_{Ray}(\lambda) SC_{Ray} \Big]
\end{aligned}
\quad (2)
$$

In the DOAS method, the terms varying smoothly with wavelength are approximated by a low-order polynomial including extinction by scattering (Richter and Burrows 2002; Eisinger et al. 1997; Richter et al. 1999). Therefore, Eq. (2) can be rewritten as:

$$
\ln\left(\frac{I(\lambda, \Theta)}{I_0(\lambda)} \right) = -\sum_{j=1}^{J} \sigma'_j(\lambda) SC_j + \sum_{p} b_p^* \lambda^p \quad (3)
$$

where $\sigma_j'(\lambda)$ represents the portion of the absorption cross-section that varies rapidly with wavelength. In the final analysis, Eq. 3 is solved for many wavelengths in a linear least squares analysis with the slant columns SC_j and the polynomial coefficients b_p^* as parameters. The vertical column of the gas j can be calculated from the slant column SC_j knowing the airmass factor AMF of the gas. Airmass factors are defined as the ratio of the slant column and the vertical column of absorber j:

$$\text{AMF}, j(\lambda, \Theta) = \frac{SC_j(\lambda, \Theta)}{VC_j(\lambda, \Theta)} \tag{4}$$

The DOAS method's limitation in measuring extraterrestrial spectra with ground-based instruments necessitates the use of the measured intensity at Θ_o (high sun) as I_0. This choice is made because the light path through the atmosphere is shorter at that solar zenith angle. Therefore, only the difference in absorption between the two measurements, given in Eq. 5 as

$$\Delta SC_j = SC_j(\Theta) - SC_j(\Theta_0) \tag{5}$$

can be determined (Richter and Burrows 2002). Besides, with the assumptions that the vertical column in the background measurement is the same as in the actual measurement, and the differences are only due to changes in the light path, the vertical column can be calculated with Eq. 6, as

$$\begin{aligned} \Delta SC_j = SC_j(\Theta) - SC_j(\Theta_0) = VC_j(\lambda) \cdot \text{AMF}_j(\lambda, \Theta) \\ -VC_j(\lambda) \cdot \text{AMF}_j(\lambda, \Theta_0). \end{aligned} \tag{6}$$

Therefore,

$$VC_j(\lambda) = \frac{\Delta SC_j}{\text{AMF}_j(\lambda, \Theta) - \text{AMF}_j(\lambda, \Theta_0)} \tag{7}$$

In the present study, the measurement of the NO_2 concentration in the cell includes a background measurement conducted without the cell, and the light path within the cell corresponds to the length of the cell. When the cell is exposed to sunlight, the NO_2 photolytic cycle is initiated as follows:

$$NO_2 + hv \rightarrow O^* + NO \tag{8}$$

$$O^* + O_2 + M \rightarrow O_3 + M \tag{9}$$

$$O_3 + NO \rightarrow NO_2 + O_2 \tag{10}$$

The concentration of NO_2 inside the cell decreases continuously until an equilibrium is reached between NO_2 and O_3.

DOAS is effective in measuring various trace gases, e.g., NO_2, SO_2, O_3, and *HCHO*, with impacts on health and the environment. Studies by Richter and Eisinger

(Eisinger et al. 1997; Richter et al. 1999) explored BrO variations in Bremen, revealing bromine conversion to sulfates and aerosols. Richter and Burrows (Richter and Burrows 2002) analyzed GOME data, uncovering NO_2 in Africa's free troposphere and outflow regions due to biomass burning and lightning. Chai et al. (Chai et al. 2009) studied NO_2 emissions using 4D-Var, reducing nitrogen oxide emissions. Expounding on this cited literature, notably the contributions by Richter and Eisinger, The present paper emphasizes the analytical aspect of DOAS measurement and data processing, aiming to provide valuable insights for environmental management. This includes understanding the behavior of trace gases (with a focus on NO_2), the sources and impacts of pollutants, and the dynamics of atmospheric composition changes over time.

3 Methodology

In the present study, experimental DOAS-based measurements of NO_2 concentration are conducted in both indoor and outdoor settings. Inside the laboratory, sky brightness measurements are taken using the DOAS instrument positioned on the laboratory building's roof between 10 a.m. and 12:30 p.m., with stable but cloudy sky conditions. Outside the laboratory, complete data on measured intensities with and without the cell, along with measurements without the cell at different horizon angles, are collected. Essential data on trace gases are also gathered to calculate the required quantities.

The adopted experimental setup for DOAS consists of several key components, as shown in Fig. 1. Here, a telescope with two ports allows for zenith-sky observations and off-axis observations with an elevation angle range of $-5°$ to $30°$. The incoming light is directed through a mirror into a quartz fiber bundle that carries it to the grating spectrometer. Inside the spectrometer, the different wavelengths of light are separated. The setup includes a cooled Charge-Coupled Device (CCD) detector that produces an electric signal upon interaction with photons. This detector stores and displays data for an image, where each pixel's intensity corresponds to a specific color in the color spectrum. A shutter controls the exposure time of the CCD detector, while a lens focuses the light on the CCD. To calibrate the spectrometer, a calibration lamp is utilized. Data acquisition is performed using the AMAX OMA program running on a computer. As part of the DOAS system, these components enhanced accurate measurements of trace gases.

To conduct measurements, the experiment followed a series of steps. Initially, 212 measurements were taken at the zenith (90°) for 90 min, consisting of 106 measurements with the NO_2 cell and 106 without the cell. These two measurements were rapidly performed in series, with an exposure time of 0.025s, automatically selected by the software. A Vernier with a resolution of $\sigma = 0.05$ cm was used to measure the cell length as well. In the second phase, measurements were taken without the cell at various elevation angles, including one at 90°, and 15 measurements between 1°

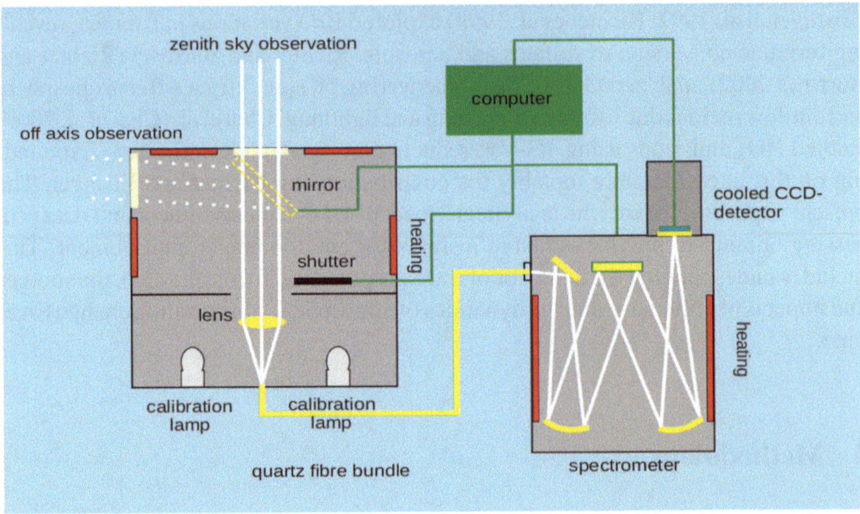

Fig. 1 Sketch of DOAS setup (*Source* A. Richter, Introduction to Measurement Techniques in Environmental Physics, 2009)

and 15° towards the horizon. The exposure time for these measurements remained 0.025 s. Finally, a dark measurement was performed with an exposure time of 0.025 s while the zenith entrance of the telescope was covered with a black cloth and a box.

4 Data Analysis and Results

Here, the NO_2 concentration for each measurement time-stamp is calculated using the first set of measurements taken at an elevation angle of 90°. The measurements without the cell are denoted as I_0, and those with the cell are labeled as I. The adopted methods of data analysis are discussed as follows.

Since the data is stored by the detector as pixels, the number of pixels must be converted to wavelengths by applying the wavelength calibration formula:

$$\lambda(i) = \sum_{j=0}^{2} a_j \left(\frac{i-1}{N-1}\right) \tag{11}$$

where N is the pixel number and a_j are calibration coefficients typical for NO_2 ($a_0 = 429.494$, $a_1 = 93.112$, $a_2 = -6.050$). With the calibration formula, the corresponding wavelengths are calculated.

A Python script was implemented with a defined loop to perform the following steps for each measurement:

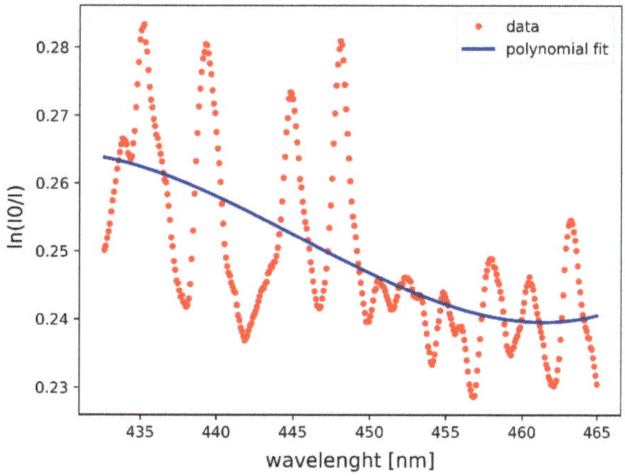

Fig. 2 Polynomial fit for the first measurement of NO_2 content in the cell in the selected fitting window. The y-axis shows the logarithm of the ratio between I_0 (no cell) and I (with cell), and the x-axis represents the wavelength. The orange line is the 3rd-order polynomial fit

- The dark current measurement was subtracted from both I and I_0.
- The data were cut and restricted to the fitting window [432.5 − 465] nm.
- The optical density (OD) of the measurement was calculated by taking $\ln(I_0/I)$.
- A polynomial of order 3 was fitted to the optical density (OD) (second term on the right side of Eq. 3), as shown in Fig. 2.
- The polynomial fit was subtracted from the OD to obtain the differential optical depth (DOD):

$$\text{DOD} = \ln\left(\frac{I_0(\lambda)}{I(\lambda, \theta)}\right) - \sum_p b_p \lambda^p. \tag{12}$$

- As expressed in Eq. 13 and plotted in Fig. 3, the slant column (SC) was determined by fitting the given differential cross-section to the measured DOD as they are related linearly.

$$\text{DOD} = \sigma_{\text{NO}_2} \cdot \text{SC}_{\text{NO}_2} + \epsilon \tag{13}$$

- The NO_2 concentration inside the cell was derived as:

$$C = \frac{\text{SC}}{l} \tag{14}$$

where l is the length of the cell.

The result was plotted as shown in Fig. 4, and the time dependence of the NO_2 concentration inside the cell was obtained. To estimate the atmospheric NO_2 change during the measurement time, we replicated the analysis steps using the measurement without the cell only, taking the first measurement without the cell as I_0. The results are presented in Fig. 5.

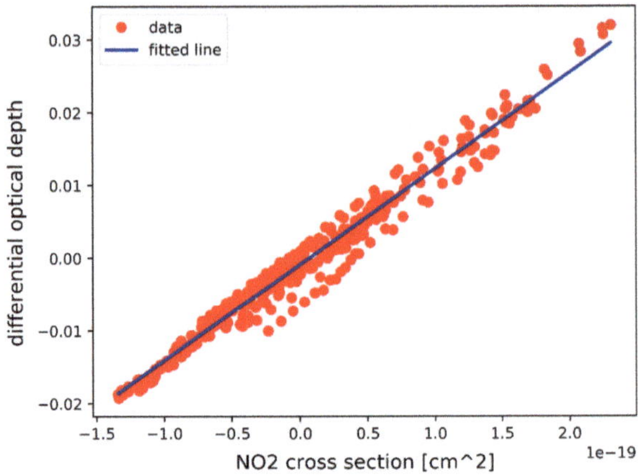

Fig. 3 Linear fit between cross section and the differential term for the first measurement. The slope of the fit gives the NO_2 slant column

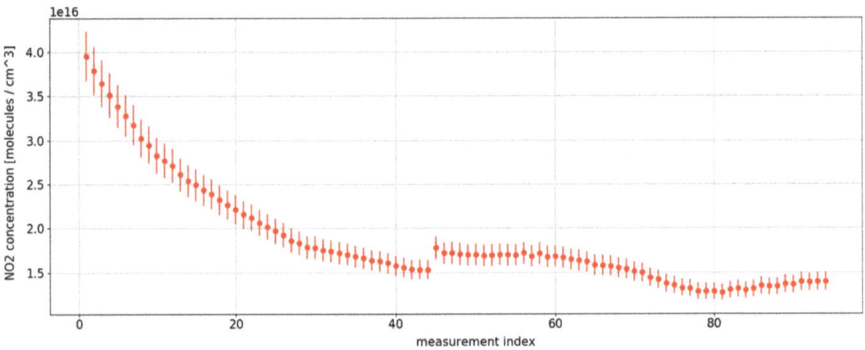

Fig. 4 With the Cell: NO_2 concentration [molecules/cm^3] during the measurement

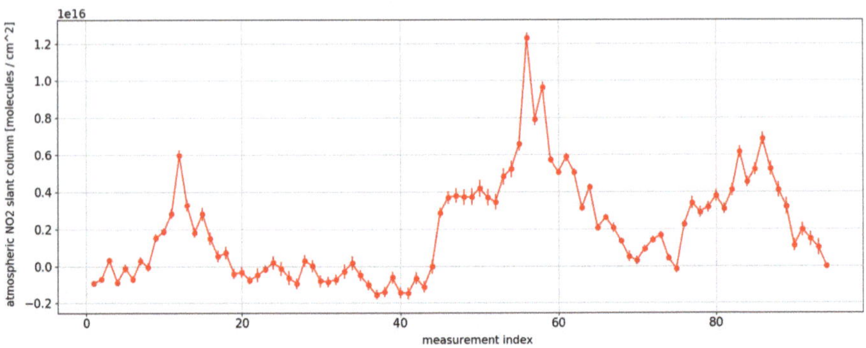

Fig. 5 Without the Cell: NO_2 concentration [molecules/cm^3] during the measurement

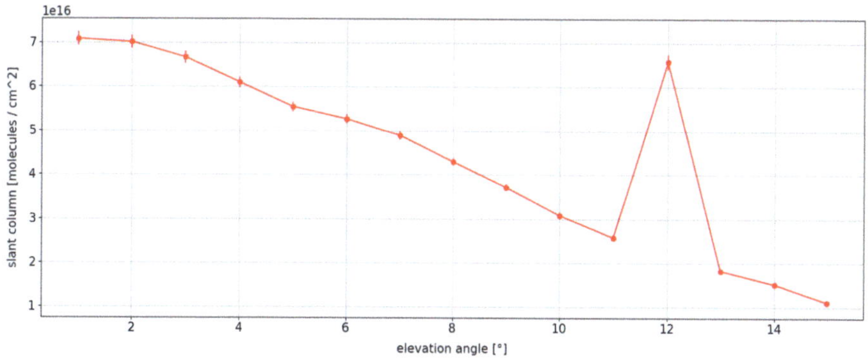

Fig. 6 Atmospheric NO_2 slant column at different elevation angle (signal for angle 12° can be ignored)

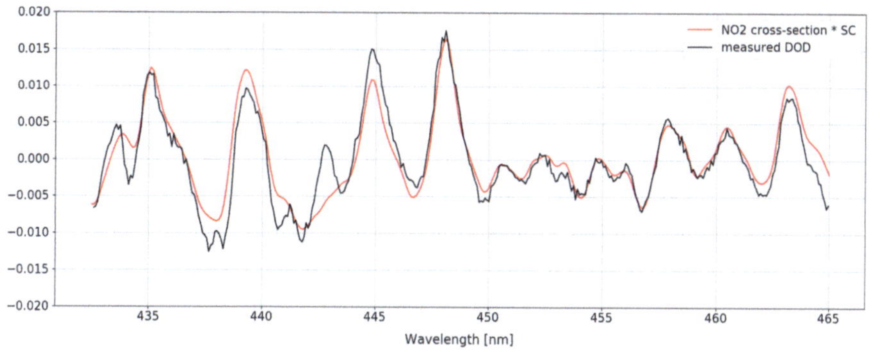

Fig. 7 Comparison between σ_{NO_2} and the measured DOD for measurement at 1°

4.1 *Off-Axis Measurements*

In order to determine the slant column of NO_2, the same analysis has been performed over the measurements at different elevation angles using the atmospheric measurements at 1° to 15° elevation as I and the zenith-sky measurement as I_0. The results of these are shown in Figs. 6, 7 and 8, for the respective angles of elevation.

4.2 *Simulated Trace Column over the Test Location*

In the new fitting window [336–357] nm, the same analysis was carried out using two simulated atmospheric measurements taken at different solar zenith angles (20° and 80°). Measurement at the zenith angle 20° was considered as I_0 and at 80° as I. To identify the trace gas in the simulated atmospheric spectra, the model spectrum

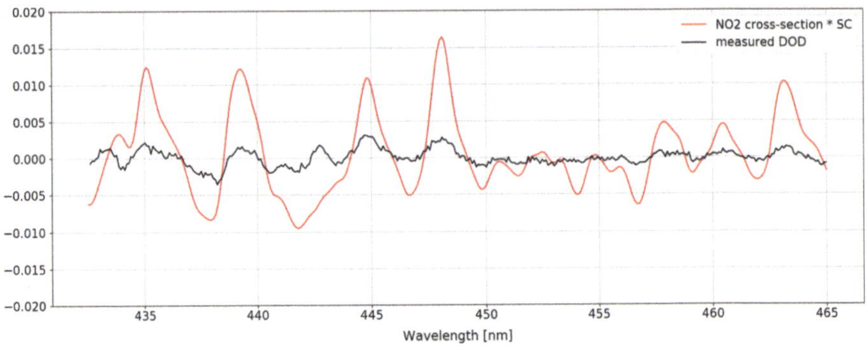

Fig. 8 Comparison between σ_{NO_2} and the measured DOD for measurement at 15°

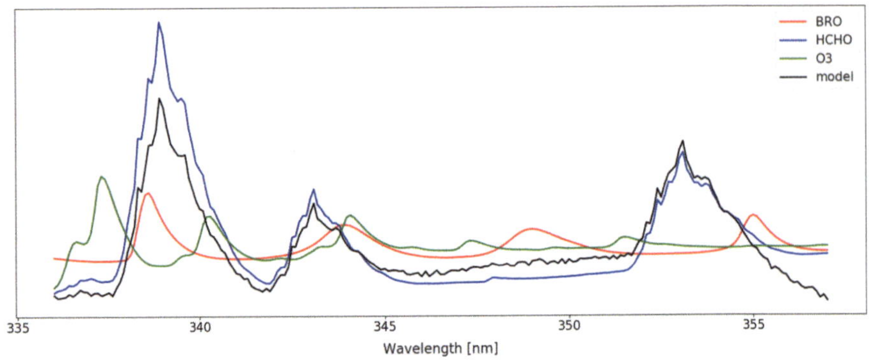

Fig. 9 Model DOD versus scaled BrO, HCHO, and O_3 cross sections (Y-axis is unitless due to scaling for comparison)

was compared with the cross sections of three gases: O_3 (ozone), BrO (bromine monoxide), and HCHO (formaldehyde). Since the slant column SC is wavelength independent, according to Eq. 13, DOD and cross sections must show a spectrum with the same shape (SC acts like a scaling factor). Therefore, by plotting the model spectrum and three given cross-sections, the trace gas was identified. The difference between slant columns at 80° and 20° ΔSC was calculated using Eq. 13. According to Eq. 7, the vertical column (VC) was obtained from the ΔSC, knowing the airmass factor of the identified trace gas at 80° and 20°. According to Fig. 9, the only gas that exists in the model is the *HCHO*.

Therefore,

$$VC = \frac{SC(80°) - SC(20°)}{AMF(80°) - AMF(20°)}. \quad (15)$$

Knowing that for HCHO AMF(80°) = 2.0984 and AMF(20°) = 1.7039, the vertical column is then calculated as:

$VC_{HCHO} = (9.2 \pm 0.25) \times 10^{15}$molecules/cm^2

4.3 Error Analysis

The slant column's error in each step is equal to the standard deviation of the slope in the linear fit in Eq. 13, which is calculated by the covariance function ("cov" command) for a linear fit in Python. The error of the NO_2 concentration in the cell is obtained by the formula below:

$$\sigma_c = \sqrt{\left(\frac{\sigma_{SC}}{l}\right)^2 + \left(\frac{SC}{l} \cdot \sigma_l\right)^2} \tag{16}$$

where the error σ_l is calculated taking into account that l is measured as the subtraction of two lengths, the total cell height $l_1 = (12.17 \pm 0.05)$ cm and the internal edge $l_2 = (2.14 \pm 0.05)$ cm, measured with a Vernier ($\sigma_V = 0.05$ cm):

$$l = l_1 - l_2 \text{ therefore, } \sigma_l = \sigma_V \sqrt{2}. \tag{17}$$

Also, the error of the vertical column is calculated by:

$$\sigma_{vc} = \left| \frac{1}{\text{AMF}(80) - \text{AMF}(20)} \right| \cdot \sigma_{SC}. \tag{18}$$

4.4 Discussion of Result

Figure 4 shows that the concentration of NO_2 in the cell reduced from an initial value of $(3.94 \pm 0.28) \times 10^{16}$ molecules/cm^3 to $(1.4 \pm 0.09) \times 10^{16}$ molecules/cm^3 during the measurement time. Due to a technical problem, an interruption occurred in the middle of the measurement, which is observed as a jump in Fig. 4. According to Fig. 5, atmospheric NO_2 fluctuated during the measurement between around 1.2×10^{16} and -0.2×10^{16} molecules/cm^2. To make a comparison between atmospheric NO_2 and the concentration of NO_2 in the cell, we must multiply the cell concentration by a factor of 10 given by the cell length to account for the conversion from molecules/cm^3 to molecules/cm^2. Thus, during the measurement, the fluctuation of atmospheric NO_2 is always less than the cell signal by one order of magnitude. Therefore, the measurements were not considerably affected by atmospheric NO_2 changes. The atmospheric NO_2 slant column at different elevation angles is portrayed in Fig. 6. It is observed that the slant column decreased with the increase in the elevation angle, suggesting a higher concentration of NO_2 at low altitudes, and a longer light path at low elevation angles will result in a greater slant column.

The plots in Figs. 7 and 8 reveal a decrease in the correlation between the measured DOD and cross-section as the elevation angle increases. The weaker agreement between the measured DOD and σ_{NO_2} at 15° suggests the presence of other gases besides NO_2 significantly contributing to radiation absorption at that altitude. On the contrary, the strong agreement at 1° indicates that NO_2 is the primary gas that absorbs at that altitude in the fitting window. Simulated atmospheric measurements identified HCHO as the only trace gas present. The vertical column of HCHO over Bremen was determined to be $(9.2 \pm 0.25) \times 10^{15}$ molecules/cm^2, consistent with the values reported by SCHIAMACHY(Wittrock 2006) and GOME2 (Vrekoussis 2010; Liu 2019).

5 Conclusion

This study underscores the importance of DOAS for the analysis of trace gases, providing altitude-specific measurements and non-intrusive long-term monitoring. It reveals that NO_2 dominates radiation absorption at lower altitudes, while other gases may contribute at higher altitudes. Moreover, atmospheric NO_2 fluctuations have minimal impact on NO_2 measurements in the cell. The correlation between the DOD and cross-section decreases with elevation angle, suggesting other gases' influence. In addition, simulated measurements identifying the presence of HCHO gas are crucial to characterizing the atmosphere's chemical profile and assessing its toxicity level. The DOAS method, complemented by the demonstrated analytics, can facilitate air quality monitoring, serving as a valuable support platform for environmental protection strategies.

Acknowledgements Experiments and data collection were conducted at IUP, University of Bremen, Bremen, Germany.

References

Chai T, Carmichael GR et al (2009) Regional NOx emission inversion through a four-dimensional variational approach using SCIAMACHY tropospheric NO_2 column observations. Atmos Environ 43(32):5046–5055

Eisinger M, Richter A, Ladstätter-Weißenmayer A, Burrows JP (1997) DOAS zenith sky observations: 1. BrO measurements over Bremen (53°n) 1993–1994. J Atmos Chem 26(1):93–108

Fuentes JMC, Rowe JG (1998) The effect of air pollution from nitrogen dioxide (NO_2) on epiphytic lichens in Seville, Spain. Aerobiologia 14:241–247

Kellogg WW, Schneider SH (1978) Global air pollution and climate change. IEEE Trans Geosci Electron 16(1):44–50

Lancaster DG, Richter D, Curl RF, Tittel FK (1998) Real-time measurements of trace gases using a compact difference-frequency-based sensor operating at 3.5 μm. Appl Phys B 67:339–345

Liu S, Valks P, Pinardi G, De Smedt I, Yu H, Beirle S, Richter A (2019) An improved total and tropospheric NO_2 column retrieval for GOME-2. Atmos Meas Tech 12:1029–1057

Platt U, Stutz J (2008) Differential optical absorption spectroscopy: principles and applications. Physics of earth and space environments. Springer, Berlin

Richter A, Burrows J (2002) Tropospheric NO_2 from GOME measurements. Adv Space Res 29(11):1673–1683

Richter A, Eisinger M, Ladstätter-Weißenmayer A, Burrows JP (1999) DOAS zenith sky observations. 2. seasonal variation of BrO over Bremen (53°n) 1994–1995. J Atmos Chem 32:83–99
Vrekoussis M, Pikridas M et al (2022) Local and regional air pollution characteristics in Cyprus: a long-term trace gases observations analysis. Sci Total Environ 845:157315
Vrekoussis M, Wittrock F, Richter A, Burrows JP (2010) GOME-2 observations of oxygenated VOCs: what can we learn from the ratio glyoxal to formaldehyde on a global scale? Atmos Chem Phys 10:10145–10160
Wittrock F (2006) The retrieval of oxygenated volatile organic compounds by remote sensing techniques. In: HCHO from SCIAMACHY measurements. https://www.iup.uni-bremen.de/doas/hcho_from_scia.htm

Tomato Growth Promotion by Trichoderma Asperellum Laboratory-Made Bioproduct

Hanane E. L. Kaissoumi, Fadoua Berbera, Najoua Mouden, Abdelatif OuazzaniChahdi, Amina Ouazzani Touhami, Karima Selmaoui, Rachid Benkirane, and Allal Douira

Abstract The endophytic fungus Trichoderma asperellum known as biological control agent has also capacity to stimulate plant growth. Tomato plants were treated with the moroccan T. asperellum laboratory-made biofungicidal and biostimulant product slurry at a concentration of 107 conidia.mL^{-1} by dipping before planting and by fertigation with different volumes, 5 L, 10 L, 15 L and 20 L, three times every 20 days during cultivation. The tomato plants grew well compared to the control plants. The aerial part length, numbers of leaves, flowers and fruits respectively increased with time varying from 48.29 to 55.64/34.28 cm; 10.36 to 12.56/8.6; 6.86 to 9.72/5.2; 0 to 0.26/0 after the first application, from 79. 06 to 91.09/ 51.83 cm; 27.6 to 32.43/22.53; 23.9 to 42.4/9.8; 3.84–5.88/2.74 after the second application and from 95.55 to 112.12/70.53 cm; 39.92 to 45.62/22.48; 17.08 to 58.99/7.94; 14.32 to 23.2/12.14 after the third application. By the end of the trial, the root part length and the fresh weights of the aerial and root parts attained 53/48 cm; 555.96/238.3 g and 31.76/20.06 g respectively. T. asperellum was able to colonize the roots, stems and leaves of the tomato plants, with significant re-isolation percentages reaching 90%, 85% and 66.66%. The T. asperellum based bioproduct has shown its ability to promote the growth of tomato plants, the gain percentages of the different agronomic parameters are very important which can be multiplied by 2, 3 or even 5, as in some cases.

H. E. L. Kaissoumi (✉) · A. OuazzaniChahdi · A. O. Touhami · K. Selmaoui · R. Benkirane · A. Douira
Laboratoire des Productions Végétales, Animales et Agro-Industrie, Département de Biologie, Faculté des Sciences, Université Ibn Tofail, Kenitra, Morocco
e-mail: elkaissoumihanane1216@gmail.com

F. Berbera
Laboratoire d'Analyse Médicale, CHP Moulay Abdellah Mohammedia, Ministère de la Santé, Mohammedia, Morocco

N. Mouden
Laboratoire de Chimie Moléculaire et Molécules de l'Environnement, Faculté Pluridisciplinaire de Nador, Université Mohammed 1er Oujd, Oujda, Morocco

M. Azrour et al. (eds.), *Sustainable and Green Technologies for Water and Environmental Management*, World Sustainability Series,
https://doi.org/10.1007/978-3-031-52419-6_13

Keywords Trichoderma Asperellum · Promotion · Growth · Tomato · Agriculture · Farming

1 Introduction

Tomatoes are one of the most important early crops. In 2006, this sector produced 1,398,831 tonnes on an area of 14,781 hectares (FAOSTAT 2023). Tomatoes are cultivated throughout the world, and are a valuable source of potassium, antioxidants, ascorbic acid, vitamin A, lycopene and tocopherols needed to maintain good human health (Capobianco-Uriarte et al. 2021). In fact, tomatoes constitute the most popular vegetable consumed throughout the African continent (Rwomushana et al. 2019). In 2019, over 21 million tonnes of tomatoes have been produced across Africa, with an increase in production of more than 60% in the last 20 years (FAOSTAT 2021). Growing tomatoes is also key to socio-economic progress, as it represents a significant source of income to rural farmers in several African developing countries (Arah 2015; Mulugeta et al. 2020). It also offers employment to local women, who represent over 60% of the workforce in this crop (Rwomushana et al. 2019). However, despite the competitiveness of tomato, it faces biotic constraints such as the emergence of new pests and diseases involving the excessive use of chemicals causing the development of resistance and rendering some phytosanitary treatment products ineffective (Elame et al. 2019). In addition, pathogenic fungi are viewed as the greatest menace to the species of plant hosts, constituting the primary factor in the reduction of plant development (about 65%) (Zhang and Ma 2017; Almeida et al. 2019; Hewedy et al. 2020a, b). In addition, the continual exposition under stress and to climate change around the world has made plants more vulnerable to disease (Attia et al. 2022). Today, tomato crops are subject to many infections due to a range of causes, such as fungal bacteria, viruses and disorders of physiology, resulting in symptom disorders involving fruit blotching, fruit rots, leaf wilting and leaf burns (Jones et al. 2014; Elshafie et al. 2017).

In the context of reducing environmentally unfriendly agricultural inputs, new alternatives are being developed. Among these, the use of beneficial microorganisms seems to be a promising strategy.

Biological biocontrol agents are increasingly crucial and, in many instances, can complement or even provide a substitute for licensed chemicals when fungal antagonisms are significant (Whipps et al. 2001). Numerous biopesticides and biofertilizers have become commercially established, most of them founded on beneficiogenic genus Trichoderma symbionts (Woo et al. 2006). Hence, Trichoderma strains are key anti-pathogenic biocontrol factors (Pal and Gardener 2006; Contreras-Cornejo et al. 2016; Zhang et al. 2022). Their use as biofungicides against pathogens such as Botrytis cinerea, Fusarium spp, Pythium spp, Rhizoctonia solani and Sclerotium rolfsii on economically important crops is frequent (Mohiddin et al. 2010; Olowe et al. 2022; Sharma et al. 2022).

The Trichoderma spp. has been the subject of much research that has contributed to the study of biological control using fungi (Ortiz and Orduz 2001; Heraux et al. 2005). It includes important species for antibiotic and enzyme production (Howell 2003; Viesturs et al. 1996) and biological control activity against fungi and nematodes (Brunner et al. 2005; Sahebani and Hadavi 2008; Paulitz and Belanger 2001; Darles 2013; Szabó et al. 2012; Poveda et al. 2020). It also contributes to the induction of systemic acquired resistance in plants through endophytism. Trichoderma species can also promote plant growth and development (Chang et al. 1986; Souza et al. 2008; Gravel et al. 2007; Shanmugaiah et al. 2009; Singh et al. 2016).

Trichoderma strains have been extensively described as promoters of crop development. This characteristic is isolate-specific as opposed to species-specific, as different isolates display differing degrees of vegetative distinctiveness. The most common indication of plant growth stimulation is the increase in root and/or shoot biomass, however alterations in plant development and morphology have also been noted. The extent to which growth is enhanced may vary widely due to a combination of various determining parameters, such as the type of crop, the conditions under which it is grown, the level of inoculum and the nature of the treatment formulation. A large number of Trichoderma-based bioinoculants are now available commercially, with blends of different strains becoming more common due to the improved coherence of their efficiencies (Stewart and Hill 2014).

Several works have shown that Trichoderma fungal species are the most promising microorganisms for improving the growth of tomato plants by enhancing nutrient uptake and preventing them against biotic and abiotic environmental stresses (Tucci et al. 2011; Uddin et al. 2018; Ruiz-Cisneros et al. 2018; Shivam et al. 2019; Khoshmanzar et al. 2020; Sani et al. 2020; Morán-Diez et al. 2020; Wang et al. 2021).

Thus, the aim is to optimize the growth of tomato plants by introducing a biological product based on Trichoderma asperellum in a greenhouse tomato crop.

2 Materials and Methods

2.1 Introduction Biofungicidal and Biostimulant Product

Laboratory-made liquid formulation product by adding conidia of Trichoderma asperellum KU987252 (Ouazzani et al. 2019).

Preparation of the seedlings. Seeds of tomato variety "Mouna" disinfected with Sodium hypochlorite (5%) and washed several times at the tap water were germinated in plastic pots, filled with a wet peat and incubated in a greenhouse for four weeks.

Green-house trials. The sandy soil of Maamora (Chliyeh et al. 2014) never cultivated was brought from the grounds and distributed in the greenhouse in five ridges with black plastic mulch with a drip irrigation system T.TAPE.

Treatment of seedlings before planting. The roots of the seedlings with their peat were soaked in a basin containing 2 L of a slurry product based on Trichoderma asperellum conidia adjusted to a concentration of 107 conidia.mL^{-1} for 45 min. Control seedlings were soaked in distilled water. The seedlings were planted in the greenhouse (25 seedlings/ridge) and irrigated every 3 days with well water.

Treatment of plants after planting. The product slurries at a concentration of 107 conidia.mL^{-1} of T. asperellum at different volumes 5 L, 10 L, 15 L and 20 L were applied by injection into the drip line of the four ridges. The control ridge was treated with well-water. The treatments were applied three times every 20 days during the cultivation.

Evaluation of agronomic parameters of treated plants. Fifteen days after the different treatments, the length of aerial parts was measured with a steel tape measure and the numbers of leaves, flowers and fruits were counted in 5 randomly sampled tomato plants.

Four months after plantation, the plants were dug up, the roots were washed under a stream of tap water and drained on absorbent paper overnight under ambient laboratory conditions. Plants were cut at the collar, the height of the aerial part and the length of the roots were measured with a steel tape measure. The fresh weights of the aerial and root parts of the plants were weighed by a precision balance and the dry mass was also weighed after drying at 70 °C for 48 h. The gain percentage for each of the parameters is calculated in relation to the control treated with water.

Re-isolation from tomato plants. Roots, stems, and leaves of tomato plants from different treatments were cut into fragments of 1–2 mm. The fragments were disinfected with 95° alcohol for five minutes and aseptically washed with sterile distilled water, then dried between two sheets of sterile filter paper.

The fragments of the different sections were sterilely deposited in Petri dishes containing PSA medium (Potato Sucrose Agar: 200 g potato, sucrose: 20 g, Agar–agar: 15 g, distilled water: 1000 mL) with three fragments per dish. The plates were incubated in the dark at 25 °C for 4 days.

The reisolation percentage of Trichoderma asperellum, from the the different sections (PRf %) of the tomato plants was calculated according to the following formula:

$$PRf(\%) = \frac{Nf}{TNF} \times 100$$

Nf: Number of fragments showing a colony of T. asperellum.

TNf: Total number of fragments.

Statistical analysis. Data processing consisted of ANOVA order I analysis of variance and comparison of means using the LSD tests at the 5% level.

3 Results and Discussion

The biostimulant effect of the Laboratory-made liquid formulation biofungicide biostimulant product based on Trichoderma asperellum was evaluated in tomato plants grown under greenhouse. The bioproduct induced a better stimulation response of the agronomic growth parameters significantly higher than that of the untreated controls (Fig. 1).

The three treatments with the different volumes of the bioproduct slurry stimulated the development of tomato plants. The aerial part length, the average numbers of leaves, flowers and fruits of the treated tomato plants are very significant compared to the control. They varied respectively between 48.29–55.64 cm/34.28 cm, 79.06–91.09 cm/51.83 cm and 95.55–112.12 cm/70.53 cm with the gain percentages from 140.86 to 162.04%,156.12 to 187.26%, 153.47–158.96%; 10.36–12.56/8.6, 27.6–32.43/22.53, 39,92–45.62/22,48 with the gain percentages from 120.04 to 146.04%, 122.5 to 143,94%, 177.58–203.02%; 6.86–9.27/5.2, 23.9–42.4/9.8, 17.08–58.99/7.94 with the gain percentages from 131 to 186.92%, 243,38 to 431.77%, 215.11 to 742.94%; 0–0.26/0, 3.84–5.88/2.74, 14.32–23.2/12.14 with the gain percentages from 140 to 214.59%—117.95–191.10% (Table 1).

At the end of the trials, root length showed a slight increase for the 10 L and 20 L slurry volume treatments over the control, 52.33 and 53 cm/48 cm with percentage gains of 109.02 and 110.41% while the other treatments were not different from the control. On the other hand, the fresh weights of the root and aerial parts showed a

Fig. 1 Effect of different volumes of *Trichoderma asperellum* product slurry on the development of tomato plants

Table 1 Time evolution of agronomic growth parameters of tomato plants after treatments with different volumes of Trichoderma asperellum product slurry

Treatments	Volumes (L)	APL (cm)	GP_{APL} (%)	LAN	GP_{LAN} (%)	FAN	GP_{FAN} (%)	FrAN	GP_{FrAN} (%)
1st treatment	0	34.28^{e}	–	8.6^{c}	–	5.2^{c}	–	0	–
	5	55.64^{a}	162.04^{a}	12.56^{a}	146.04^{a}	9.28^{a}	178.46^{a}	0.16	–
	10	48.29^{d}	140.86^{d}	10.56^{b}	122.79^{b}	6.86^{b}	131.92^{b}	0.26	–
	15	50.54^{c}	147.43^{c}	10.36^{b}	120.04^{b}	7.04^{b}	135.38^{b}	0.16	–
	20	52.72^{b}	153.79^{b}	11.16^{b}	129.76^{b}	9.72^{a}	186.92^{a}	0	–
2nd treatment	0	51.83^{c}	–	22.53^{d}	–	9.82^{e}	–	2.74^{c}	–
	5	89.24^{a}	173.18^{a}	30.48^{b}	135.2^{b}	42.4^{a}	431.77^{a}	5.88^{a}	214.59^{a}
	10	80.92^{b}	156.12^{b}	27.6^{c}	122.5^{c}	29.4^{c}	299.38^{c}	3.84^{b}	140.14^{b}
	15	79.06^{b}	187.26^{b}	29.14^{b}	129.33^{b}	23.9^{d}	243.38^{d}	3.97^{b}	144.89^{b}
	20	91.09^{a}	175.74^{a}	32.43^{a}	143.94^{a}	38.47^{b}	391.75^{b}	5.39^{a}	196.7^{a}
3rd treatment	0	70.53^{d}	–	22.48^{c}	–	7.94^{d}	–	12.14^{c}	
	5	95.55^{c}	135.47^{c}	45.64^{a}	203.02^{a}	40.48^{b}	509.82^{b}	23.2^{a}	191.10^{a}
	10	100.8^{b}	142.91^{b}	39.92^{b}	177.58^{b}	41.60^{b}	523.92^{b}	14.32^{bc}	117.95^{b}
	15	112.12^{a}	158.96^{a}	44.8^{a}	199.28^{a}	17.08^{c}	215.11^{c}	17.28^{b}	142.33^{b}
	20	102.73^{b}	145.65^{b}	44.61^{a}	198.44^{a}	58.99^{a}	742.94^{a}	21.81^{a}	179.65^{a}

Values of each treatment within a column having the same letter do not differ significantly ($P < 0.05$) according to LSD test

APL: Aerial part length; GP_{APL}: APL Gain percentage; LAN: Leave average number per plant; GP_{LAN}: LAN percentages of gains; FAN: Flower average number per plant; GP_{FAN}: FAN percentages of gains; FrAN: Fruit average number per plant; GP_{FrAN}: FrAN percentages of gains.

significant increase compared to the control, varying respectively between 314.46–555.96 g/238.3 g and 22.63–31.76 g with percentage gains of 131.95–233.30% and 112.81–158.32% (Table 2).

The re-isolation of T. asperllum was positive from the organs of tomato plants treated with different volumes of the bioproduct slurry while it was negative for the control. Roots showed the highest re-isolation percentages reaching 90%, followed by stems 85% and leaves 66.66% (Table 3).

These results are in agreement with the work of Windham et al. (1986) who demonstrated that the addition of Trichoderma asperullum to a previously autoclaved soil increased the weight of the roots as well as the aerial part of tomato and tobacco by 213–291% compared to the untreated control. Similarly, the work of Baker et al. (1984) showed that radish plant weights increased by over 274% after treatment with Trichoderma asperellum strain T-95.

Caron et al. (2020) reported an improvement in height and fresh weight of the aerial part of tomato plants in the presence of Trichoderma spp. Similarly, Kleifeld and Chet (1992) showed that Trichoderma asperellum inoculated into sandy soil increased dry biomass of cucumber plants by 43% compared to the control. On the other hand, corn seeds treated with Trichoderma asperellum showed a significant

Table 2 Effect of Trichoderma asperellum product treatments on tomato plant biomass at the end of the trial

Volumes	Growth agronomic parameters					
	RPL (cm)	PG_{RPL} (%)	APW (g)	PG_{APW} (%)	RPW (g)	PG_{RPW} (%)
0 L	48^{b}	–	238.3^{e}	–	20.06^{d}	–
5 L	47.33^{b}	98.60^{c}	460.1^{b}	193.07^{b}	31.5^{a}	157.02^{a}
10 L	52.33^{a}	109.02^{a}	314.46^{d}	131.95^{d}	22.63^{c}	112.81^{c}
15 L	48.33^{b}	100.68^{b}	431.56^{c}	181.09^{c}	25^{b}	124.62^{b}
20 L	53^{a}	110.41^{a}	555.96^{a}	233.30^{a}	31.76^{a}	158.32^{a}

Values within a column affected by the same letter do not differ significantly ($P < 0.05$) according to the LSD test

RPL: Root Part Length; PG_{RPL}: RPL percentages of gains; APW: Aerial Part Weight; PG_{APW}: APW percentages of gains; RPW: Root Part Weight; PG_{RPW}: RPW percentages of gains

Table 3 Re-isolation of *Trichoderma asperellum* from different organs of tomato plants treated with the bioproduct at the end of the trial (%)

Organs	Volumes				
	0 L	5 L	10 L	15 L	20 L
Leaf	0^{c}	49.99^{b}	66.66^{a}	55.55^{ab}	61.11^{a}
Stem	0^{c}	85^{a}	75^{a}	66.66^{b}	73.33^{ab}
Root	0^{c}	78.88^{a}	66.66^{b}	90^{a}	88.88^{a}

Values within a line affected by the same letter do not differ significantly ($P < 0.05$) according to the LSD test

increase in stem and root fresh weight even in the presence of Phytium ultimum in the soil (Harman et al. 2004).

Trichoderma asperellum improved tomato seed growth and effectively countered wilt damage due to the phytopathogen F. oxysporum through increasing leaf number, shoot and root length (cm) and fresh and dry weights (g); in addition, Trichoderma protects tomato berries against F. oxysporum postharvest infection (Sehim et al. 2023).

Trichoderma species commonly provide the growth and yield of plants, as well as root penetration (Harman et al. 2004; Mouria et al. 2007).

Trichoderma spp. increase plant growth as a biofertilizer and promote plant defense functions (Harman et al. 2004; Shoresh et al. 2010; Harman 2011). Using Trichoderma has been shown to significantly improve the growth of salad lettuce, bell pepper and tomatoes (Vinale et al. 2006; Tančić-Živanov et al. 2020). Treating corn crops using Trichoderma harzianum strain T-22 demonstrated two rows of increased root length compared with control plants (Harman et al. 2004).

The studies conducted by Srivastava (2004) and Mouria et al. (2007) have revealed that the colonisation of plants' roots with Trichoderma strains and their competition in the rhizosphere soil are both responsible for improving plant roots' growth and development.

A number of factors have been suggested as possible explanations for enhanced growth, including the control of secondary pathogens, improvements in nutrient absorption, enhanced carbohydrate metabolism and photosynthesis, as well as the synthesis of phytohormones. There is good evidence that microbially-produced indole acetic acid (IAA) may play a role in this regard, although it is more likely that Trichoderma stimulates growth by affecting the imbalance of hormones such as IAA, gibberellic acid and ethylene (Stewart and Hill 2014).

4 Conclusion

The application of the biofungicide product based on Trichoderma asperullum is beneficial for the growth and production of treated tomato plants whatever the volume of the slurry brought to the ground level compared to the control. Thus, this product can be applied alone as a biostimulant during the tomato biological crop.

References

Almeida F, Rodrigues ML, Coelho C (2019) The still underestimated problem of fungal diseases worldwide. Front Microbiol 10:214. https://doi.org/10.3389/fmicb.2019.00214

Arah IK (2015) An overview of post-harvest challenges facing tomato production in Africa. In: African Studies Association of Australasia and the Pacific (AFSAAP) 37th annual conference Africa: diversity and development. Dunedin, New Zealand, pp 1–21

Attia MS, Abdelaziz AM, Al-Askar AA, Arishi AA, Abdelhakim AM, Hashem AH (2022) Plant growth-promoting fungi as biocontrol tool against *Fusarium* wilt disease of tomato plant. J Fungi 8:775. https://doi.org/10.3390/jof8080775

Baker R, Elad Y, Chet I (1984) The controlled experiment in the scientific method with special emphasis on biological control. Phytopathology 74:1019–1021

Brunner K, Zeilinger S, Ciliento R, Woo SL, Lorito M, Kubicek CP, Mach RL (2005) Improvement of the fungal biocontrol agent *Trichoderma atroviride* to enhance both antagonism and induction of plant systemic disease resistance. Appl Environ Microbiol 71:3959–3965

Capobianco-Uriarte MDIM, Aparicio J, De Pablo-Valenciano J, Casado-Belmonte MDP (2021) The European tomato market. An approach by export competitiveness maps. PLoS ONE 16(5):e0250867. https://doi.org/10.1371/journal.Pone.0250867

Caron J, Laverdière L, Thibodeau PO, Bélanger RR (2020) Utilisation d'une souche indigène de *Trichoderma harzianum* contre cinq agents pathogènes chez le concombre et la tomate de serre au Québec. Phytoprotection 83:73–87

Chang YC, Chang YC, Baker R, Kleifeld O, Chet I (1986) Increased growth of plants in the presence of the biological control agent *Trichoderma harzianum*. Plant Dis 70:145–148

Chliyeh M, Ouazzani Chahdi A, Selmaoui K, Ouazzani Touhami A, Filali Maltouf A, El Modafar C, Moukhli A, Oukabli A, Benkirane R, Douira A (2014) Effect of *Trichoderma* and arbuscular mycorrhizal fungi againt *verticillium* wilt of tomato. Int J Recent Sci Res 5(2):449–459

Contreras-Cornejo HA, Macias-Rodriguez L, Del-Val E, Larsen J (2016) Ecological functions of *Trichoderma* spp. and their secondary metabolites in the rhizosphere: interaction with plant. FEMS Microbiol Ecol 92:fiw036. https://doi.org/10.1093/femsec/fiw036
Darles B (2013) Evaluation de l'efficacité de diffèrent produits de biocontrôle, pour lutter contre la fusariose du Dipladenia. Rapport de stage, (Bioingénierie) option Biotechnologie végétales, Université Paul Sabatier, Toulouse, pp 35
De Souza JT, Bailey BA, Pomella AWV, Erbe EF, Murphy CA, Bae H, Hebbar PK (2008) Colonization of cacao seedlings by *Trichoderma stromaticum*, a mycoparasite of the witches' broom pathogen, and its influence on plant growth and resistance. Biol Contr 46:36–45
Elame F, Lionboui H, Wifaya A, Mokrini F, Mimouni A, Azim K (2019) Revue Marocaine des Sciences Agronomiques et Vétérinaires 7(4):595–599
Elshafie HS, Sakr S, Bufo SA, Camele I (2017) An attempt of biocontrol the tomato-wilt disease caused by *Verticillium dahlia* using *Burkholderia gladioli* pv. *agaricicola* and its bioactive secondary metabolites. Int J Plant Biol 8:7263. https://doi.org/10.4081/pb.2017.7263
FAOSTAT (2021, November 3). https://www.fao.org/faostat/en/#data/QCL
FAOSTAT (2023) Crops and livestock products, Tomatoes (Morocco). https://www.fao.org/faostat/en/#data/QCL. Consulté le 20/04/2023
Gravel V, Antoun H, Tweddell RJ (2007) Growth stimulation and fruit yield improvement of greenhouse tomato plants by inoculation with *Pseudomonas putida* or *Trichoderma atroviride*: possible role of indole acetic acid (IAA). Soil Biol Biochem 39:1968–1977
Harman GE (2011) *Trichoderma*—not just for biocontrol anymore. Phytoparasitica 39:103–108
Harman GE, Petzoldt R, Comis A, Chen J (2004) Interactions between *Trichoderma harzianum* strains T22 and maize inbred line mo 17 and effects of these interactions on diseases caused by *Pythium ultimum* and *Colletotrichum graminicola*. Phytopathology 94(2):147–153
Heraux FMG, Hallett SG, Ragothama KG, Weller SC (2005) Composted chicken manure as a medium for the production and delivery of *Trichoderma* virens for weed control. Hort Sci 40:1394–1397
Hewedy OA, Abdel-lateif KS, Bakr RA (2020a) Genetic diversity and biocontrol efficacy and indigenous *Trichoderma* isolates against *Fusarium* wilt of pepper. J Basic Microbiol 60:126–135. https://doi.org/10.1002/jobm.201900493
Hewedy OA, Abdel KS, lateif MF, Seleiman AS, Albarakaty FM, El-Meihy RM (2020b) Phylogenetic diversity of *Trichoderma* strains and their antagonistic potential against soil-borne pathogens under stress conditions. Biology 9:189. https://doi.org/10.3390/biology9080189
Howell CR (2003) Mechanisms employed by *Trichoderma* species in the biological control of plant diseases: the history and evolution of current concepts. Plant Dis 87:4–10
Jones JB, Zitter TA, Momol TM, Miller SA (2014) Compendium of tomato diseases and pests. APS Press, United States
Khoshmanzar E, Aliasgharzad N, Neyshabouri MR et al (2020) Effects of *Trichoderma* isolates on tomato growth and inducing its tolerance to water-deficit stress. Int J Environ Sci Technol 17:869–878. https://doi.org/10.1007/s13762-019-02405-4
Kleifeld O, Chet I (1992) *Trichoderma harzianum*—Interaction with plants and effect on growth response. Plant Soil 144:267–272
Mohiddin FA, Khan MR, Khan SM, Bhat BH (2010) Why *Trichoderma* is considered super hero (super fungus) against the evil parasites? Plant Pathol J 9:92–102. https://doi.org/10.3923/ppj.2010.92.102
Morán-Diez ME, Tranque E, Bettiol W, Monte E, Hermosa R (2020) Differential response of tomato plants to the application of three *Trichoderma* species when evaluating the control of pseudomonas syringae populations. Plants 9:626. https://doi.org/10.3390/plants9050626
Mouria B, Ouazzani-Touhami A, Douira A (2007) Effet de diverses souches du *Trichoderma* sur la croissance d'une culture de tomate en serre et leur aptitude à coloniser les racines et le substrat Phytoprotection 88(3):103–110

Mulugeta T, Muhinyuza JB, Gous-Meyer R, Matsaunyane L, Andreasson E, Alexandersson E (2020) Botanicals and plant strengtheners for potato and tomato cultivation in Africa. J Integr Agric 19(2):406–427

Olowe OM, Nicola L, Asemoloye MD, Akanmu AO, Babalola OO (2022) *Trichoderma* : potential bio-resource for the management of tomato root rot diseases in Africa. Microbiol Res 257:126978. https://doi.org/10.1016/j.micres.2022.126978

Ortiz A, Orduz S (2001) *In vitro* evaluation of *Trichoderma* and *Gliocladium* antagonism against the symbiotic fungus of the leaf-cutting and Attacephalots. Mycopathologia 150:53–60

Ouazzani Chahdi A, Ouazzani Touhami A, Khirallah W, Benkirane R, Douira A (2019) Production, formulation et recyclage d'un produit biofongicide et biostimulant à base de *Trichoderma asperellum.* MA 41534 Date de publication, June 31. https://patentregister.ompic.ma/SearchPatent/searchByDepot?typeNum=AP&numDepot=41534&count=0&lang=FR

Pal KK, Gardener BM (2006) Biological control of plant pathogens. American Psychopathological Society, St. Paul, MN

Paulitz TC, Belanger RR (2001) Biological control in greenhouse systems. Annual Rev Phytopathol 39:103–133

Poveda J, Abril-Urias P, Escobar C (2020) Biological control of plant-parasitic Nematodes by filamentous fungi inducers of resistance: *Trichoderma*, Mycorrhizal and Endophytic Fungi. Front Microbiol 11:992. https://doi.org/10.3389/fmicb.2020.00992

Ruiz-Cisneros MF, Ornelas-Paz JJ, Olivas-Orozco GI, Acosta-Muñiz CH, Sepúlveda-Ahumada DR, Pérez-Corral DA, Rios-Velasco C, Salas-Marina MA, Fernández-Pavía SP (2018) Effect of *Trichoderma* spp. and phytopathogenic fungi on plant growth and tomato fruit quality. Revista Mexicana de Fitopatologia 36(3):444–456

Rwomushana I, Beale T, Chipabika G, Day R, Gonzalez-Moreno P, Lamontagne Godwin J, Makale F, Pratt C, Tambo J (2019) Evidence note: tomato leafminet (Tuta absoluta): impacts and coping strategies for Africa. CABI Work. Paper 12:1–56

Sahebani N, Hadavi N (2008) Biological control of the root-knot nematode *Meloidogyne javanica* by *Trichoderma harzianum.* Soil Biol Biochem 40:2016–2020

Sani Md NH, Hasan M, Uddain J, Subramaniam S (2020) Impact of application of *Trichoderma* and biochar on growth, productivity and nutritional quality of tomato under reduced N-P-K fertilization. Ann Agric Sci 65(1):107–115

Sehim AE, Hewedy OA, Altammar KH, Alhumaidi MS, Abd-Elghaffar RY (2023) *Trichoderma asperellum* empowers tomato plants and suppresses *Fusarium oxysporum* through priming responses. Front Microbiol 14:1140378

Shanmugaiah V, Balasubramanian N, Gomathinayagam S, Manoharan PT, Rajendran A (2009) Effect of single application of Trichoderma viride and Pseudomonas fluorescens on growth promotion in cotton plants. Afr J Agric Res 4(11):1220–1225

Sharma A, Salwan R, Kaur R, Sharma R, Sharma V (2022) Characterization and evaluation of bioformulation from antagonistic and flower inducing *Trichoderma asperellum* isolate UCRD5. Biocatal Agric Biotechnol 43:102437. https://doi.org/10.1016/j.bcab.2022.102437

Shivam M, Rai D, Rai B, Dubey S (2019) Growth promotion effect of *Trichoderma* Isolates on tomato seedlings. J Plant Dis Sci 14(2):115–118

Shoresh M, Harman GE, Mastouri F (2010) Induced systemic resistance and plant responses to fungal biocontrol agents. Ann Rev Phytopath 48:21–43

Singh V, Upadhyay RS, Sarma BK, Singh HB (2016) Seed bio-priming with *Trichoderma asperellum* effectively modulate plant growth promotion in pea. Int J Agric, Environ Biotechnol 9(3):361–365

Srivastava VK (2004) *Trichoderma* spp- a boon for better crop health. Pest 28(8):40–45

Stewart A, Hill R (2014) Chapter 31—Applications of *Trichoderma* in plant growth promotion. In: Gupta VK, Schmoll M, Herrera-Estrella A, Upadhyay RS, Druzhinina I, Tuohy MG (2014) Biotechnology and Biology of Trichoderma. Elsevier, pp 415–428. ISBN 9780444595768. https://doi.org/10.1016/B978-0-444-59576-8.00031-X

Szabó M, Csepregi K, Gálber M, Virányi F, Fekete C (2012) Control plant-parasitic nematodes with *Trichoderma* species and nematode-trapping fungi: the role of chi18-5 and chi18-12 genes in nematode egg-parasitism. Biol Contr 63(2):121–128

Tančić-Živanov S, Medić-Pap, S, Danojević D, Prvulović D (2020) Effect of *Trichoderma* spp. on growth promotion and antioxidative activity of pepper seedlings. Braz Arch Biol Technol 63. https://doi.org/10.1590/1678-4324-2020180659

Tucci M, Ruocco M, De Masi L, De Palma M, Lorito M (2011) The beneficial effect of *Trichoderma* spp. on tomato is modulated by the plant genotype. Mol Plant Pathol 12(4):341–354. https://doi.org/10.1111/j.1364-3703.2010.00674.x

Uddin MN, Rahman UU, Khan W et al (2018) Effect of *Trichoderma harzianum* on tomato plant growth and its antagonistic activity against *Phythium ultimum* and *Phytopthora capsici*. Egypt J Biol Pest Contr 28:32. https://doi.org/10.1186/s41938-018-0032-5

Viesturs U, Leite M, Treimanis A, Eremeeva T, Apsite A, Eisimonte M, Jansons P (1996) Production of cellulases and xylanases by *Trichoderma viride* and biological processing of lignocellulose and recycled paper fibers. Appl Biochem Biotechnol 57:349–360

Vinale F, Marra R, Scala F, Ghisalberti EL, Lorito M, Sivasithamparam K (2006) Major secondary metabolites produced by two commercial *Trichoderma* strains active against different phytopathogens. Lett Appl Microbiol 43:143–148

Wang R, Chen D, Khan RAA, Cui J, Hou J, Liu T (2021) A novel *Trichoderma asperellum* strain DQ-1 promotes tomato growth and induces resistance to gray mold caused by *Botrytis cinerea*. FEMS Microbiol Lett 368(20):fnab140

Whipps JM, Lumsden RD (2001) Commercial use of fungi as plant disease biological control agent: status and prospects. In: Butt T, Jackson C, Magan N (eds) Fungal biocontrol agents progress, problems and potential. CAB Publishing, Wallingford, pp 9–22

Windham MTY, Baker R (1986) A mechanism for increased plant growth induced by *Trichoderma spp*. Phytopathology 76:518–521

Woo SL, Scala F, Ruocco M, Lorito M (2006) The molecular biology of the interactions between *Trichoderma* spp., phytopathogenic fungi, and plants. Phytopathology 96:181–185

Zhang Y, Ma LJ (2017) Deciphering pathogenicity of Fusarium oxysporum from a phylogenomics perspective. Adv Genet 100:179–209. https://doi.org/10.1016/bs.adgen.2017.09.010

Zhang Y, Xiao J, Yang K, Wang Y, Tian Y, Liang Z (2022) Transcriptomic and metabonomic insight into the biocontrol mechanism of *Trichoderma asperellum* M45a against watermelon Fusarium wilt. PLoS ONE 17:e0272702. https://doi.org/10.1371/journal.pone.0272702

Integrated Control of Strawberry Anthracnose by Trichoderma Asperellum–Pyraclostrobin/Boscalid Combination

Hanane El Kaissoumi, Fadoua Berbera, Najoua Mouden, Amina Ouazzani Touhami, Karima Selmaoui, Rachid Benkirane, and Allal Douira

Abstract The different Trichoderma asperellum—Pyraclostrobin + Boscalid preventive treatment programs were applied before inoculation with Colletotrichum gloeosporioides, to the aerial parts of Camarosa strawberry plants. They reduced anthracnose symptoms for 80 days, The disease severity reduction percentages on leaves were ranged from 66.66 to 99.1%. The numbers of symptomatic flowers, green and red strawberries are zero to low compared to the inoculated control, the reduction percentages varied between 94.1–100%, 92.25–100% and 81.35–100%. The C. gloeosporioides inoculum on leaves and petioles of treated plants was low to nil compared to plants inoculated with the pathogen, the re-isolation percentages fluctuated from 0 to 25%/100% and 0 to 33%/100%. The combined treatments provided better development of the aerial and root parts than the inoculated plants. The aerial perpendicular diameter growth, root length development and their fresh weights reached respectively 16.66/7.66 cm, 35.33/14 cm, 8.6/3.3 g and 21.83/9.16 g. The Trichoderma asperellum—Pyraclostrobin + Boscalid combination provided integrated protection of strawberry plants against anthracnose and compatibility between biological control agent and fungicide.

Keywords Strawberry · Colletotrichum Gleoesporioides · Anthracnose · Trichoderma Asperellum · Agriculture · Farming

H. E. Kaissoumi (✉) · F. Berbera · A. O. Touhami · K. Selmaoui · R. Benkirane · A. Douira
Laboratoire des Productions Végétales, Animales et Agro-Industrie, Département de Biologie, Faculté des Sciences, Université Ibn Tofail, Kenitra, Morocco
e-mail: elkaissoumihanane1216@gmail.com

F. Berbera
Laboratoire d'Analyse Médicale, CHP Moulay Abdellah Mohammedia, Ministère de la Santé, Mohammedia, Morocco

N. Mouden
Laboratoire de Chimie Moléculaire et Molécules de l'Environnement, Faculté Pluridisciplinaire de Nador, Université Mohammed 1er Oujd, Oujda, Morocco

M. Azrour et al. (eds.), *Sustainable and Green Technologies for Water and Environmental Management*, World Sustainability Series,
https://doi.org/10.1007/978-3-031-52419-6_14

1 Introduction

Strawberry cultivation is one of the most widespread horticultural crops in the world. In Morocco, this culture plays a very important socio-economic role in the perimeter of Gharb-Loukkos. The cultivated area and production practices have changed significantly, from 750 hectare with 31.000 tons harvested in 1995–3400 hectares during the 2020–2021 campaign, with an average production of 102.000 tons (Anonymous 2023).

Although strawberry fields have had a good yield for several years in the Gharb-Loukkos region, strawberry producers face many challenges and the strawberry plants were subject to several phytosanitary constraints always cause considerable losses, such as climatic conditions and insufficient storage techniques. It can be attacked by several fungal diseases, including gray mold (Feliziani and Romanazzi 2016; Dowling et al. 2017), powdery mildew (Carisse et al. 2013), angular leaf spot (Roach et al. 2016) and anthracnose (Forcelini et al. 2018; El Kaissoumi et al. 2018).

Colletotrichum spp, the agents responsible for strawberry anthracnose, represent a highly aggressive pathogen which can affect all aspects of the strawberry crop, from leaves and stems to petioles, vines, roots and crowns, especially in hot, humid climates (Buddie et al. 1999). Furthermore, over a period of only seven days, the infection of plant crowns results in more frequent seedling/plant withering and death due to insufficient water, causing substantial losses (Cai et al. 2011). Strawberry anthracnose is an important limiting condition for strawberry production around the world (Hou et al. 2022).

Colletotrichum gloeosporioides is among the most aggressive species of this genus and causes economically significant losses on the strawberry plant. The disease can be observed on runners, leaves (Kaissoumi et al. 2018; Tellier et al. 2021), flowers, ripening fruits (Carisse et al. 2013; Kaissoumi et al. 2018; Food and Rural Affairs (OMAFRA) 2009; Paulus 1990; Freeman 2008; Kim et al. 2016), green fruits (Maas 1998), roots and crown (Tellier et al. 2021; Freeman 2008; Peres et al. 2005).

Thus, to deal with these diseases, it is important to choose methods that optimize costs and minimize adverse effects (Ministry of Agriculture, Fisheries and Food of Quebec (MAPAQ) 2019). The most important control strategy for strawberry anthracnose is the use of agrochemicals (Li et al. 2021). A number of different fungicides, such as benzimid-azoles, strobilurins and imazalil, are frequently prescribed for the control of strawberry anthracnose (Li et al. 2021; Xu et al. 2014).

The most economical and effective programs for the control of anthracnose disease in strawberries combine cultural, biological, chemical, genetic and physical different preventive methods. Indeed, the integrated pest management, allows to obtain a sustainable production system that takes into account economic, health and environmental concerns (Vincent and Panneton 2001; Carvalho 2006; Nicolopoulou-Stamati et al. 2016).

The objective of this work is to evaluate different strategies of integrated protection against strawberry anthracnose by alternating biological control by Trichoderma asperellum and chemical control by a fungicide combining Strobulirins and carboxamides chemical families.

2 Material and Methods

2.1 *Plant Material*

Healthy looking Camarosa variety bare root strawberry plants are planted in pots containing Maamora soil characterized by a pH = 7.53, 0.7% of organic matter, 0.05% of Nitrogen, 0.239% of P2O5, 0.15 meq.100^{-1} g of K2O, 0.20 meq.100^{-1} g of Mg, 7351.5 mg.Kg^{-1} of Ca (Chliyeh et al. 2014) and placed in a greenhouse and watered daily until the 4–5 leaf stage required for inoculation.

Pathogen

An isolate of Colletotrichum gloeosporioides (Coll3) isolated from the collar of Camarosa variety strawberry plant in 2014 and knowns for its high pathogenicity (Kaissoumi et al. 2018) was cultured on PSA medium (Potato Sucrose Agar: Potato = 200 g, sucrose = 15 g, Agar–Agar = 20 g, 1000 ml distilled water). The incubation is carried out at 24 °C in the dark for 7 days.

Antagonist Agent

Trichoderma asperellum strain KU987252 was isolated from compost. Its antagonistic and biofungicidal power against various fungal pathogens and biostimulant effect on tomato and strawberry are known (Abdellatif et al. 2019). It was cultured on PSA medium at 28 °C in the dark for 7 days.

Fungicide Tested

The specialty SIGNUM WG (Boscalid 26.7% and Pyraclostrobin 6.7%) is approved in Morocco against gray mold in strawberries at a registered dose of 150 g/hl (ONSSA (National Food Safety Office) 2022).

Preparation of the Inoculum

The surfaces of the cultures of the strain of T. asperellum and C. gloeosporioides seven days old and loaded with conidia, are immersed in sterile distilled water and then scraped with a metal spatula. The resulting spore suspension is filtered through muslin cloth. The concentration was determined using a Malassez plate. The conidial suspension is adjusted to the concentration of 107 conidia mL^{-1} for the biocontrol agent and 105 conidia mL^{-1} for the pathogen, with distilled water to which 0.05% Tween 20 has been added and 0.05% gelatin.

2.2 Inoculation and Treatments

The aerial parts of strawberry plants were inoculated and/or treated at the 4–5 leaf stage. The inoculation was made by spraying the vegetative part with 30 mL of the conidial suspension of Colletotrichum gloeosporioides. Treatment by specialty SIGNUM WG, was carried out by spraying 30 mL of the product mixture at the approved dose of 501 ppm. The biological treatment was carried out by spraying 30 mL of the suspension of Trichoderma asperellum. The healthy control was treated by spraying 30 mL of distilled water to which 0.05% Tween 20 and 0.05% gelatin were added.

The chemical and biological treatments were carried out according to time every 20 days preventively 48 h before inoculation by Colletotrichum gloeosporioides. Three repetitions are carried out for each treatment.

Several application trials were carried out as follows in Table 1. The plants inoculated are placed in special black plastic sacks humidified with diluted distilled water during the first three days to ensure an atmosphere with saturated humidity. The containers are then stored in a greenhouse from March 13 to May 13, 2017.

Table 1 The different control and treatment programs applied over time on strawberry plants

Treatments	Time (Days)			
	0	20	40	60
Control 1	Water	Water	Water	Water
Control 2	*C. gleoesporioides*	*C. gleoesporioides*	*C. gleoesporioides*	*C. gleoesporioides*
Control 3	*T. asperellum*	*T. asperellum*	*T. asperellum*	*T. asperellum*
Control 4	SIGNUM WG	SIGNUM WG	SIGNUM WG	SIGNUM WG
Trial 1	SIGNUM WG/ Coll3	SIGNUM WG/ Coll3	SIGNUM WG/ Coll3	SIGNUM WG/ Coll3
Trial 2	*T. asperellum*/ Coll3	*T. asperellum*/ Coll3	*T. asperellum*/ Coll3	*T. asperellum*/ Coll3
Trial 3	SIGNUM WG/ Coll3	*T. asperellum*/ Coll3	SIGNUM WG/ Coll3	*T. asperellum*/ Coll3
Trial 4	*T. asperellum*/ Coll3	SIGNUM WG/ Coll3	*T. asperellum*/ Coll3	SIGNUM WG/ Coll3
Trial 5	*T. asperellum*/ Coll3	SIGNUM WG/ Coll3	SIGNUM WG/ Coll3	*T. asperellum*/ Coll3
Trial 6	*T. asperellum*/ Coll3	*T. asperellum*/ Coll3	SIGNUM WG/ Coll3	*T. asperellum/ Coll3*
Trial 7	*T. asperellum*/ Coll3	SIGNUM WG/ Coll3	*T. asperellum*/ Coll3	*T. asperellum*/ Coll3

Table 2 Combination efficiency Trichoderma asperellum-Pyraclostrobin/ Boscalid on the reduction of the severity of anthracnose on the leaves of strawberry plants (expressed in %).

0	No symptoms
1	Less than 0.5% of the limbus showing symptoms
2	0.6–5% of limbus with symptoms
3	6–15% of limbus with symptoms
4	16–30% of limbus with symptoms
5	31–50% of limbus with symptoms
6	51–80% of limbus with symptoms
7	81–100% of limbus with symptoms

Scoring of Results

The development of leaf symptoms was followed as a function of time every 20 days for two months, the diseased leaf area was estimated according to the Stover scale modified by Gauhl et al. (1995) (Table 2).

The disease severity index (SI) was calculated according to the Eq. (1).

$$IS(\%) = \frac{\sum nb}{(N-1)\times T} \times 100 \quad (1)$$

n: Number of sheets for each degree of the scale.

b: Degree of arbitrary scale representing severity.

N: Number of degrees used in the scale.

T: Total number of sheets evaluated.

The percentage reduction in disease severity (RM%) was calculated according to the Eq. (2).

$$RM(\%) = \frac{ISti - ISTr}{ISti} \times 100 \quad (2)$$

ISti: Disease severity index of plants inoculated with C. gleoesporioides.

ISTr: Disease severity index of plants treated with T. asperellum and/or SIGNUM WG then inoculated with C. gleoesporioides.

Every 20 days and before the next treatment, the percentage reductions in the numbers of flowers, green strawberries and diseased red strawberries are estimated in the strawberry plants treated with T. asperellum and/or SIGNUM WG then inoculated with C. gleoesporioides compared to the number of flowers, green strawberries and red strawberries estimated in plants inoculated with C. gleoesporioides.

At the end of the tests, the plants were separated from the soil and washed under a stream of water in order to remove the adherent soil particles. The aerial part of each plant was cut at the level of the collar, the lengths of the aerial and root parts are measured using a double decimeter. The fresh weights of both parts were measured using a precision balance. The dry weights were also weighed using a precision balance, after drying in an oven at 70 °C for 24 h.

The growth rates of each of the parameters, the fresh and dry lengths and weights of the aerial and root parts, are calculated in relation to those of the healthy control.

Reisolation

After two months of treatment, re-isolation was carried out on the leaves and petioles of the strawberry plants of the various trials. They were cut into 2 cm segments and disinfected separately with 95° alcohol for 2 min. The fragments were then rinsed several times with sterile distilled water, dried on sterile filter paper, subcultured on PSA medium and incubated in the dark at 22 °C. After seven days of incubation, the observation of the colonies developed around the fragments of leaves and petioles was made under an optical microscope in order to identify the fungal species isolated and the percentage of re-isolation (Pr%) was calculated according to the Eq. (3).

$$Pr(\%) = \frac{NsX}{NT} \times 100 \tag{3}$$

Ns X: Number of segments containing fungal species X.

NT: Total number of segments.

Statistical Analysis

The statistical processing of the data concerned the analysis of variance and the LSD test at the 5% threshold. The percentages are transformed into Arcsin $\sqrt{P}$.

3 Results and Discussion

The effectiveness of the various integrated pest management strategies by alternating treatments with Trichoderma asperellum and the Pyraclostrobin/Boscalid association was evaluated on the reduction of anthracnose symptoms caused by Colletotrichum gleoesporioides on strawberries.

The protection levels of the Trichoderma asperellum strain and the chemical preventive treatment on strawberry plants in the presence of the pathogen were significantly significant and reduced anthracnose disease for 80 days. The reduction percentage in disease severity on the leaves of strawberry plants inoculated with C. gleoesporioides and preventively treated with the alternation of the Trichoderma asperellum-Pyraclostrobin/Boscalid combination compared to those treated with either T. asperellum alone or SIGNUM WG alone varied respectively between 95.70 and 66.66%/96.24%–88.97%; 89.5 and 75.00%/68.39%–78.91%; 99.1 and 73.45%/95.81%–77.82%; 92.81 and 78.43%/72.72%–79.78% after 20, 40, 60 and 80 days of treatment (Table 3).

Table 3 Combination efficiency Trichoderma asperellum-Pyraclostrobin/Boscalid on the reduction of the severity of anthracnose on the leaves of strawberry plants (expressed in %)

Treatments	Time (days)			
	20	40	60	80
Trial 1	88.97[b]	78.91[b]	77.82[b]	79.78[c]
Trial 2	96.24[a]	68.39[b]	95.81[b]	72.72[d]
Trial 3	66.60[b]	75.0[c]	73.95[b]	78.43[c]
Trial 4	83.47[c]	82.4[b]	86.35[c]	84.78[b]
Trial 5	95.70[a]	89.5[a]	99.1[a]	92.81[a]
Trial 6	82.64[c]	70.2[c]	87.46[a]	80.31[b]
Trial 7	88.56[b]	88.85[a]	94.40[b]	91.24[a]

Values assigned by the same letter do not differ significantly

Similarly, this combination of biological control agent and phytosanitary product reduced the number of flowers, green strawberries and red strawberries showing symptoms of anthracnose similar to treatments with T. asperellum alone or SIGNUM WG alone. The disease reduction percentages varied from 88.5 to 100% in the flowers, from 93.5 to 100% in the green strawberries and from 78.47 to 100% in the red strawberries during the entire cultivation period (Table 4).

The re-isolation carried out at the end of the tests from the leaves and petioles of the strawberry plants which had undergone treatment with the combination. T asperellum—Pyraclostrobin/Boscalid before inoculation with C. gleoesporioides made it possible to show that the inoculum of Trichoderma asperellum was able to persist on the aerial parts and reduce the development of C. gleoesporioides in the presence of the fungicide. The re-isolation percentages of T. asperellum varied in trials from 4 to 7 between 75 and 100% at the level of the leaves and between 33.33 and 66.66% at the level of the petioles. Whereas in test 3, they are 33.33% at the level of the leaves and 100% at the level of the petioles. Maintaining T. asperellum in the aerial parts of strawberry plants was important, the re-isolation percentages were 100% and 66.66% respectively in the absence and presence of C. gleoesporioides (Table 5).

The inoculum of C. gleoesporioides was greatly reduced on the aerial parts of strawberry plants treated in the various trials with the combination T. asperellum—Pyraclostrobin/Boscalid. The re-isolation percentages varied from 0 to 25% on the leaves and from 0 to 33.33% on the petioles, whereas they are 100% in the plants inoculated with C. gleoesporioides. The treatment with T. asperellum alone was able to maintain the inoculum of C. gleoesporioides at a low level, the re–isolation percentage is 33.33% in comparison with the fungicide alone for which it is 66.66% (Table 5).

Treatment with Trichoderma asperellum first, followed by the phytosanitary product in trials 4, 5, 6 and 7, made it possible to protect the aerial and root parts against Colletotrichum gleoesporioides and to allow them good development, the agronomic parameters measured varied respectively between 11.33 and 12.66 cm,

Table 4 Combination efficiency Trichoderma asperellum- Pyraclostrobin/Boscalid on reduction in numbers of diseased flowers, green strawberries and red strawberries in treated strawberry plants compared to those inoculated with Colletotrichum gleoesporioides (expressed in %)

Treatments	Time (days)											
	20			40			60			80		
	FL	GS	RS	FL	GS	RS	FL	GS	RS	FL	GS	RS
Trial 1	100[a]	100[a]	77.14[d]	100[a]	88.31[b]	79.81[d]	100[a]	100[a]	100[a]	100[a]	95.35[b]	100[a]
Trial 2	100[a]	100[a]	92.86[b]	100[a]	90.26[c]	91.28[b]	100[a]	100[a]	100[a]	100[a]	92.25[c]	84.46[d]
Trial 3	88.35[c]	97.24[b]	84.14[c]	100[a]	100[a]	100[a]	92.72[b]	100[a]	93.41[c]	100[a]	100[a]	100[a]
Trial 4	88.35[c]	100[a]	87.31[c]	100[a]	100[a]	92.53[b]	100[a]	100[a]	100[a]	100[a]	100[a]	93.78[b]
Trial 5	100[a]	100[a]	96.82[a]	100[a]	100[a]	85.06[c]	100[a]	100[a]	93.41[c]	94.01[b]	100[a]	87.57[c]
Trial 6	94.1[b]	100[a]	100[a]	88.72[b]	93.5[b]	78.47[d]	100[a]	100[a]	86.29[d]	100[a]	94.83[b]	81.35[e]
Trial 7	100[a]	94.46[b]	90.48[b]	100[a]	100[a]	92.53[b]	100[a]	100[a]	96.58[b]	100[a]	100[a]	93.78[b]

FL: Flowers, GS: Green strawberries, RS: Red Strawberries
Values assigned by the same letter do not differ significantly

Table 5 Reisolation of fungal species from the aerial part of strawberry plants at the end of the tests (expressed in %)

Treatments	Leaves			Petioles		
	Colletotrichum gloeosporioides	*Trichoderma asperellum*	*Various*	*Colletotrichum gloeosporioides*	*Trichoderma asperellum*	*Various*
Control 2	100[a]	–	–	100[a]	–	–
Control 3	–	100[a]	–	-	100[a]	–
Trail 1	66.66[b]	–	33.33 Rzp	33.33[b]	–	66.66 Asp
Trail 2	33.33[c]	66.66[c]	–	33.33[b]	66.66[b]	–
Trail 3	25[d]	33.33[d]	41.76 Pen	0[d]	100[a]	–
Trial 4	0[e]	100[a]	–	33.33[b]	66.66[b]	–
Trial 5	25[d]	75[b]	–	33.33[b]	66.66[b]	–
Trial 6	0[e]	100[a]	–	0[d]	33.33[c]	66.66 Rzp
Trial 7	0[e]	100[a]	–	25[c]	33.33[c]	41.76 Alt

Values within a column affected by the same letter do not differ significantly ($P < 0.05$) according to LSD test
Rzp: *Rhizopus* sp., Asp: *Aspergillus* sp., Pen: *Penicillium* sp., Alt: Alternaria sp.

28.33 and 35.33 cm, 5.63 and 8.6 g, 11.33 and 21.83 g. In contrast, in trial 3, treatment with the fungicide followed by the biological control agent had little effect on plant development in the presence of C. gleoesporioides. Indeed, the inoculation of the aerial parts of the strawberry plants by Colletotrichum gleoesporioides caused a significant reduction in the various agronomic parameters compared to control 1 (water control). The perpendicular length of the aerial part, the length of the root part; their fresh weight are respectively 7.66 cm/15 cm, 14 cm/33 cm, 3.3 g/12.7 cm and 9.16 g/17.9 cm. However, at the plants treated with the strain of Trichoderma asperellum, a significant stimulation of the parameters 17.16 cm, 37.33 cm, 6.83 g and 22.1 g (Table 6).

Anthracnose is an important disease diagnosed in strawberry growing regions around the world. Yield losses are reported following the development of this disease, fruits as well as foliage, petioles, stolons, flowers, collars and roots can be infected by this disease. Generally, the damage observed is caused by Colletotrichum, when the environmental conditions of the strawberry fields are favorable, i.e. humid and hot weather (between 20 and 25 °C) (Anonymous 2018).

In Morocco, farmers mainly use chemical control with different active ingredients (Fenhexamid, Boscalid + Pyraclostrobin, Thiophanate-methyl, Cyprodinil + Fludioxonil, Fluopyram + Trifloxystrobin, Difenoconazole, Myclobutanil, Mancozeb, Penconazole, Triadimenol). The most widely used are Myclobutanil and Triadimenol against Oïdium and Fenhexamid, the Boscalide + Pyraclostrobin combination and Mancozeb against gray rot. The combination Fluopyram + Trifloxystrobin

Table 6 Effect of treatments combining Trichoderma asperellum and the Pyraclostrobin/Boscalid association against anthracnose on the development of strawberry

Treatments	Growth agronomic parameters			
	PLA (cm)	RPL (cm)	APW (g)	RPW (g)
Control 1	15^{ab}	33^{b}	12.7^{a}	17.9^{a}
Control 2	7.66^{c}	14^{d}	3.3^{c}	9.16^{b}
Control 3	17.16^{a}	37.33^{a}	6.83^{bc}	22.1^{a}
Control 4	12.66^{ab}	32.33^{b}	5.83^{bc}	11.5^{b}
Trial 1	13.66^{ab}	23.36^{c}	12.2^{a}	14.6^{ab}
Trial 2	16.66^{a}	28.33^{b}	3.56^{c}	13.1^{b}
Trial 3	12.66^{ab}	20.33^{cd}	3.16^{c}	8.93^{b}
Trial 4	12.66^{ab}	31.66^{b}	8.13^{b}	21.83^{a}
Trial 5	11.33^{b}	32^{b}	6^{c}	12.8^{b}
Trial 6	12^{ab}	28.33^{bc}	5.63^{bc}	11.33^{b}
Trial 7	12^{ab}	35.33^{ab}	8.6^{b}	19.76^{a}

Values within a column affected by the same letter do not differ significantly ($P < 0.05$) according to LSD test.
PLA: Aerial part length; RPL: Root Part Length; APW Aerial Part Weight; RPW: Root Part Weight

is used against both pathogens. The plots receive between 7 and 15 treatments per cycle (Tanji et al. 2014).

However, the application of chemical control suffers from several deviations. Indeed, like any selection pressure, it causes the emergence of resistant genotypes (Kranthi et al. 2001; Pray et al. 2002; Shelton et al. 2002), pushing farmers to overdose, to increase the frequency of use sometimes without respecting the conditions of application (stages of the pest, preventive or curative treatment, time before harvest, etc.) and to use very toxic products thus aggravating the situation. This situation is harmful both for public health and the environment and for the budget of the farmer, in addition to the rejection of fruits exported abroad following non-compliance with the doses of pesticides authorized by the importing countries (Sellami et al. 2015).

The massive use of chemical phytosanitary products has mainly led to the appearance of resistant bio-aggressors. The wrong answer to combat these is to increase the quantity and frequency of application of the phytosanitary product used. Moreover, biopesticides represent one of the most promising alternatives to this dependence since they present numerous ecological advantages. Indeed, used in an integrated control strategy, alone or in combination with chemical pesticides, they make it possible to conquer harmful organisms and limit the appearance of cases of resistance (Sellami et al. 2015; Khalequzzaman et al. 2016).

So, it has become necessary to establish new strategies aimed at reducing the impact of phytosanitary products on public health and the environment, mainly through the development of integrated pest management and the use of biological pesticides.

The most effective methods of controlling anthracnose disease in strawberries must combine different methods that achieve a sustainable production system that takes into account economic, health and environmental concerns. Indeed, biological control through the use of micro-organisms provides effective phytosanitary protection given the natural ubiquity of microbiological agents in ecosystems, their great variety, their easy dissemination, their specificity of action and at the same time presents a low risk to human health and the environment (Cavero et al. 2015).

Indeed, integrated pest management can always be defined as a pest and disease control strategy. It is based on the implementation of all methods and techniques that are economically, ecologically and toxicologically acceptable with the aim of reducing the importance of harmful organisms to a level below the threshold of economic harm, giving priority natural limiting factors (Mrabet et al. 2022).

The microorganisms used in microbiological control belong to several taxa, namely viruses, bacteria, microfungi, nematodes and protozoa (Noel et al. 2018).

Trichoderma was originally exploited as a biological control agent against soil diseases and damping off (Adandonon et al. 2006; Sawant 2014). Subsequently, the efficacy of Trichoderma was demonstrated for the management of plant leaf diseases (Abdel-kader and El-tayeb 2012). In addition, Trichoderma harzianum can stimulate the defense mechanisms of treated plants and thus increase their protection (Hibar et al. 2005; Meyer et al. 1998; Windham and Baker 1986; Yedidia et al. 1999; Fenibo et al. 2020; Sood et al. 2020; Daami-Remadi 2001; Howell 2003). Several mechanisms of action involved in the biocontrol of pathogens have been identified in Trichoderma (Sood et al. 2020), namely mycoparasitism, secretion of volatile and diffusible substances and competition for space and nutrients (Sood et al. 2020; Daami-Remadi 2001; Howell 2003; Wang et al. 2015). In addition, Trichoderma harzianum can stimulate the defense mechanisms of treated plants and thus increase their protection (Hibar et al. 2005; Meyer et al. 1998; Windham and Baker 1986; Yedidia et al. 1999).

This mechanism is generally possible thanks to the various organic and inorganic molecules secreted by Trichoderma, including specific proteins regulating different physiological processes (Alfiky and Weisskopf 2021). Under stress conditions, Trichoderma enhances plant uptake and assimilation of essential soil macronutrients such as nitrogen (Fiorentino et al. 2018), phosphorus (Meng et al. 2019) and trace micro-nutrients, such as iron (Chen et al. 2019). This role of nutrient regulator provided by Trichoderma is essential for the defense of plants against bioaggressors (Aesan et al. 2020).

Indeed, the use of biological pesticides or biopesticides has several advantages such as the reduction of the risks of development of resistance, the high specificity towards target organisms, the most effective action even at very low quantities, as well as the reduction of pollution since they are generally biodegradable (Sellami et al. 2015). Several reasons justify the choice of these biological pesticides or biopesticides. Indeed, they restrict or eliminate the use of chemical pesticides, are less toxic than chemical pesticides, reduce the risks of developing resistance, have a greater specificity of action and are effective at very low quantities (Xu et al. 2011). In addition, they improve the quality of life of agricultural workers, do not require a required delay before harvesting, offer consumers healthy products which have better press with consumers. They degrade rapidly, thus reducing the risk of pollution (Yovo 2007). The products considered as biopesticides by European and global regulatory agencies are of various origins. They can be classified into three main categories; depending on their nature: microbial biopesticides; plant biopesticides and animal biopesticides (Chandler et al. 2011; Leng et al. 2011).

4 Conclusion

The results obtained during this study showed that the treatment by the combination of the biological fight and the chemical fight on the strawberry plants under greenhouse from the flowering to the harvest made it possible to reduce the incidence of the disease of the anthracnose and stimulate growth. Thus, the use of biological control as an alternative method to chemical control by beneficial microorganisms can be of major scientific and agronomic interest by reducing the number of treatments with phytosanitary products while giving a similar result of protection against diseases in most of their beneficial influence on the development of the plant.

References

Abdel-kader MH, El-tayeb TA (2012) Field implementation using chlorophyll derivatives with sunlight for malaria, filaria and dengue fever vectors control in infested African swamps. Malar J 11(1):42

Abdellatif OC, Amina OT, Wafae K, Rachid B, Allal D (2019) Production, formulation and recycling of a biofungicide and biostimulant product based on Trichoderma asperellum. MA 41534. Publication date 31 May 2019. https://patentregister.ompic.ma/SearchPatentsearchByDepot?typeNum=AP&numDepot=41534&count=0&lang=FR

Adandonon A, Aveling TAS, Labuschagne N, Tamo M (2006) Biocontrol agents in combination with Moringaoleifera extract for integrated control of Sclerotium-causedcowpeadamping-off and stem rot. Eur J Plant Pathol 115(4):409–418

Aesan TE, Oancea AO, atefan LM, Mãnoiu VS, Ghiurea M, Răut I, ConstantinescuAruxandei D, Toma A, Bira AF, Pomohaci CM, Oancea F (2020) Effects of foliartreatment with a Trichoderma plant biostimulant consortium on Passiflora caerulea yield and quality. Microorganisms 8(13):2–27. https://doi.org/10.3390/microorganisms8010123

Alfiky A, Weisskopf L (2021) DecipheringTrichoderma–plant–pathogen interactions for better-development of biocontrol applications. J Fungi 7(1):1–18. https://doi.org/10.3390/jof7010061

Anonymous (2018) RAP Leader in integrated pest management. Strawberry technical sheet. Anthracnose in strawberries. (Phytosanitary Warning Network). https://fraisesetframboisesduquebec.com/wp-content/uploads/2015/02/Survol-des-pratiques-et-des-recherches-sur-la-fraise-biologique-ici-et-ailleurs_2018.pdf

Anonymous (2023) Interproberries Morocco: Moroccan interprofessional federation of fruit berries. Moroccan berries sector. In: 4th international strawberry congress, Antwerp, Belgium, 21–22 September 2022. https://www.iscbelgium.com. Accessed 10 Feb 2023

Buddie AG, Martínez-Culebras P, Bridge PD, García MD, Querol A, Cannon PF, Monte E (1999) Molecular characterization of Colletotrichum isolates derived from strawberry. Mycol Res 103:385–394

Cai L, Giraud T, Zhang N, Begerow D, Cai G, Shivas RG (2011) The evolution of species concepts and species recognition criteria in plant pathogenic fungi. Fungal Divers 50:121–133

Carisse O, Lefebvre A, Van der Heyden H, Roberge L, Brodeur L (2013) Analysis of incidence-severity relationships for strawberry powdery mildew as influenced by cultivar, cultivar type, and production systems. Plant Dis 97(3):354–362

Carvalho FP (2006) Agriculture, pesticides, food security and food safety. Environ Sci Policy 9(7):685–692

Cavero PAS, Hanada RE, Gasparotto L, Neto RAC, de Souza JT (2015) Biological control of banana black Sigatokadisease with Trichoderma. Crop Protect 45(6):951–957. Trichoderma. https://doi.org/10.1590/0103-8478cr20140436951

Chandler D, Bailey AS, Tatchell GM, Davidson G, Greaves J, Grant WP (2011) The development, regulation and use of biopesticides for integrated pest management. Philos Trans Royal Soc B: Biol Sci 366(1573):1987–1998

Chen M, Liu Q, Gao SS, Young AE, Jacobsen SE, Tang Y (2019) Genomemining and biosynthesis of a polyketide from a biofertilizer fungus that can facilitate reductive iron assimilation in plant. Proc National Acad Sci USA 116(12):5499–5504. https://doi.org/10.1073/pnas.1819998116

Chliyeh M, Ouazzani Chahdi A, Selmaoui K, Ouazzani Touhami A, Filali Maltouf A, El Modafar C, Moukhli A, Oukabli A, Benkirane R, Douira A (2014) Effect of Trichoderma harzianum and Arbuscular mycorrhizal fungi against Verticillium wilt of tomato. Int J Recent Sci Res 5(2):449–459

Daami-Remadi M (2001) Biological control of Fusarium spp. pathogens responsible for dry rot of potato tubers. Dissertation for Advanced Studies in Plant Protection and Environment. Tunisia: Higher School of Horticulture and Livestock of Chott Mariem, 72 p

Dowling ME, Hu MJ, Schnabel G (2017) Identification and characterization of Botrytis fragariae isolates on strawberry in the United States. Plant Dis 101(10):1769–1773

El Kaissoumi H, Mouden N, Chliyeh M, Benkirane R, OuazzaniTouhami A, Douira A (2018) Comparative pathogenicity of Colletotrichumspp. against different varieties of strawberry plants (Fragariaananassa) widely grown in Morocco. Acta Phytopathologica and EntomologicaHungarica 53(2):1–20

Feliziani and Romanazzi (2016) Postharvest decay of strawberry fruit: etiology, epidemiology, and disease management. J Berry Res 6:47–63. https://doi.org/10.3233/JBR-150113,IOSPress

Fenibo E, Ijoma G, Matambo T (2020) Biopesticides in sustainable agriculture: current status and prospects. Preprints, pp 1–43

Fiorentino N, Ventorino V, Woo SL, Pepe O, De Rosa A, Gioia L, Romano I, Lombardi N, Napolitano M, Colla G, Rouphael Y (2018) Trichoderma-based biostimulants modulaterhizospheremicrobial populations and improve N uptakeefficiency, yield, and nutritionalquality of leafyvegetables. Front Plant Sci 9(June):1–15. https://doi.org/10.3389/fpls.2018.00743

Forcelini BB, Rebello CS, Wang NY, Peres NA (2018) Fitness, competitive ability, and mutation stability of isolates of Colletotrichum acutatum from strawberry resistant to QoI fungicides. Phytopathology 108(4):462–468

Freeman S (2008) Management, survival strategies, and host range of Colletotrichum acutatum on strawberry. HortScience 43(1):66–68

Gauhl F, Pasberg-Gauhl C, Vuylsteke D, Ortiz R (1995) Multilocational evaluation of black Sigatoka resistance in banana and plantain. IITA Research Guide 47. 2nd edn. Training Program, International Institute of Tropical Agriculture (IITA), Ibadan, Nigeria, 59 p

Hibar K, Daami-Remadi M, Khiareddine H, El Mahjoub M (2005) In vitro and in vivo inhibitory effect of Trichoderma harzianum on Fusarium oxysporum. sp. radicis-lycopersici. Biotechnol, Agron, Soc Environ 9(3):163–171

Hou SF, Liu JJ, Xu TF, Li XF, Li S, Wang HQ (2022) Simultaneous detection of three crown rot pathogens in field-grown strawberry plants using a multiplex PCR assay. Crop Prot 156:105957

Howell CR (2003) Mechanisms employed by Trichoderma species in the biological control of plant diseases: the history and evolution of current concepts. Plant Dis 87(1):4–10
Khalequzzaman KM, Uddin MK, Hossain MM, Hasan MK (2016) Effect of fungicides in controlling wilt disease of cumin. Malays J Med Biol Res 3:69–74
Kim MS, Jin JS, Kwak YS, Hwang GS (2016) Metabolic response of strawberry (Fragaria × ananassa) leaves exposed to the angular leaf spot bacterium (Xanthomonas fragariae). J Agric Food Chem 64(9):1889–1898
Kranthi KR, Jadhav D, Wanjari R, Kranthi S, Russell D (2001) Pyrethroid resistance and mechanisms of resistance in field strains of Helicoverpaarmigera (Lepidoptera: Noctuidae). J Econ Entomol 94(1):253–263
Leng J, Lan X, Liu Y, Du S (2011) Shape-memory polymers and their composites: stimulus methods and applications. Prog Mater Sci 56(7):1077–1135
Li X, Jing T, Zhou D, Zhang M, Qi D, Zang X, Zhao Y, Li K, Tang W, Chen Y, Qi C, Wang W, Xie J (2021) Biocontrol efficacy and possible mechanism of Streptomyces sp. H4 against postharvest anthracnose caused by Colletotrichum fragariae on strawberry fruit. Postharvest Biol Technol 175:111401
Maas JL (1998) Compendium of strawberry diseases. Second edition. The American Phytopathological Society Press, St Paul, Minnesota, USA, 98 pp
Meng X, Miao Y, Ma Q, Liu L, Guo K, Wei D, Liu R, Qirong S (2019) TgSWO from Trichodermaguizhouense NJAU4742 promotes growth in cucumber plants by modifying the rootmorphology and the cellwall architecture. MicrobialCellFactories 18(1):1–15. https://doi.org/10.1186/s12934-019-1196-8
Meyer G De, Bigirimana J, Elad Y, Höfte M (1998) Induced systemic resistance in Trichodermaharzianum T39 biocontrol of Botrytis cinerea. Eur J Plant Pathol 104(3):279–286p
Ministry of Agriculture, Fisheries and Food of Quebec (MAPAQ) (2019) Diagnostic portrait of the strawberry and raspberry industry in Quebec. MAPAQ, Quebec, QC, Canada. https://www.mapaq.gouv.qc.ca/fr/Publications/Portraitdiagnosectoralfraises_framboises.pdf
Mrabet R, Bahri H, Zaghouane O, Chiekh M'hamed H, El-Areed SRM, Abou El-Enin MM (2002) Chapter 6: Adoption and spread of conservation agriculture in North Africa. In: Advances in Conservation Agriculture. Volume 3: Adoption and Spread, Kassam A (ed). Burleigh Dodds, Cambridge, UK. ISBN-13: 9781786764751
Nicolopoulou-Stamati P, Maipas S, Kotampasi C, Stamatis P, Hens L (2016) Chemical pesticides and human health: The urgent need for a new concept in agriculture. Front Public Health 4(148)
Noel J-P, Lytle M, Cascio C, Wallace MT (2018) Disrupted integration of exteroceptive and interoceptive signaling in autism spectrum disorder. Autism Res 11(1):194–205
ONSSA (National Food Safety Office) (2022) Approval of chemical inputs—ONSSA, online service. Trade name: SIGNUM WG. http://eservice.onssa.gov.ma/IndPesticide.aspx. Last updated 16 Nov 2022
Ontario Ministry of Agriculture, Food and Rural Affairs (OMAFRA) (2009) OMAFRA, Guelph, ON, Canada. Integrated crop management. http://www.omafra.gov.on.ca/IPM/french/strawberries/diseases-anddisorders/index.html
Paulus AO (1990) Fungal diseases of strawberry. J HortScience 25(8):885–889p
Peres NA, Timmer LW, Adaskaveg JE, Correll JC (2005) Lifestyles of Colletotrichum Acutatum. Plant Dis 89(8):784–796
Pray CE, Huang J, Hu R, Rozelle S (2002) Five years of Bt cotton in China—the benefits continue. Plant J 31(4):423–430
Roach JA, Verma S, Peres NA, Jamieson AR, van de Weg WE, Bink MCAM, Bassil NV, Lee S, Whitaker VM (2016) FaRXf1: a locus conferring resistance to angular leaf spot caused by Xanthomonas fragariae in octoploid strawberry. Theor Appl Genet 129(6):1191–1201
Sawant IS (2014) Trichoderma-foliarpathogen interactions. Open Mycol J 8(1):58–70. https://doi.org/10.2174/1874437001408010058
Sellami S, Tounsi S, Jamoussi K (2015) Biological control, an alternative to chemical phytosanitary products. Sfax Biotechnology Center, Biopesticides Laboratory, Tunisia 19(5):2286–5314p

Shelton AM, Zhao JZ, Roush RT (2002) Economic, ecological, food safety, and social consequences of the deployment of Bt transgenic plants. Ann Rev Entomol 47(1):845–881

Sood M, Kapoor D, Kumar V, Sheteiwy MS, Ramakrishnan M, Landi M, Fabrizio A, Anket S (2020) Trichoderma: the "Secrets" of a multi-talented. Seedlings 9(762). https://doi.org/10.3390/plants9060762

Tanji A, Benicha M, Mamdouh M (2014) Strawberry production technique, Results of surveys at Loukkos. Transfer of Technology in Agriculture MADER/DERD_PNTTA (Ed.), Ministry of Agriculture and Rural Development, Monthly Information and Liaison Bulletin of the PNTTA. No. 201: 9 pp

Tellier S, Breton A, Van der Heyden H (2021) Technical data sheet—Strawberry, Anthracnose in Strawberry. Phytosanitary Warning Network (RAP), MAPAQ, Quebec, QC, Canada. https://www.agrireseau.net/documents/Document_97707.pdf

Vincent C, Panneton B (2001) Physical control methods as alternatives to pesticides. Vertigo—Electron J Environ Sci 2(2)

Wang M, Ma J, Fan L, Yu K, Fu C, Gao J, Chen Y, Li J (2015) Biological control of southern corn leafblight by Trichodermaatroviride. Biocontr Sci Tech 25(10):1133–1146. https://doi.org/10.1080/09583157.2015.1036005

Windham MTY, Baker RA (1986) mechanism for increased plant growth induced by Trichoderma spp. Phytopathology76:518–521

Xu J, Cui W, Cheng JJ, Stomp AM (2011) Production of high-starch duckweed and its conversion to bioethanol. Biosyst Eng 110(2):67–72

Xu XF, Lin T, Yuan SK, Dai DJ, Shi HJ, Zhang CQ, Wang HD (2014) Characterization of baseline sensitivity and resistance risk of Colletotrichum gloeosporioides complex isolates from strawberry and grape to two demethylation-inhibitor fungicides, prochloraz and tebuconazole. Australas Plant Pathol 43:605–613

Yedidia I, Benhamou N, Chet I (1999) Induction of defense responses in cucumber plants (CucumissativusL.) by the biocontrol agent Trichoderma harzianum. Appl Environ Microbiol 65(3):1061–1070

Yovo OPD (2007) Search for genes modifying the clinical expression of sickle cell disease. Doctoral dissertation

A Systematic Study on Implementation of Smart Devices for Sustainable Environment

Bhushan Nirmal, Manan Shah, Mourade Azrour, Jamal Mabrouki, and Azidine Guezzaz

Abstract Smart devices are interconnected devices that collect and exchange data wirelessly through sensors, routers, etc. Industries use IoT to optimize their functioning to reduce environmental impacts and make a more sustainable approach. Some of the objectives would be: increasing resource efficiency by monitoring patterns and optimizing the usage, monitoring the Environment is beneficial for the air, water, and soil quality, and decisions can be taken based on the real-time data that the smart objects generate. In this study topic, we will discuss how IoT technology is going to help the environment through smart objects (i.e. sensors, routers) in maintaining the resources that are not renewable, water resource management, Green transportation like electric buses, intelligent traffic management, and many more topic's through which IoT technology can reduce the dependency of the resources in real-time. The overall output of the IoT technology with the use of smart objects/devices concerning the area of application would be enhancing resource efficiency, lower gas emissions, enhancing waste management such as recycling, cost savings, real-time data present while decision-making, and Sustainable infrastructure such as smart building. Some of the real-time examples are Alexa, Connected cars, and Fitness watches.

B. Nirmal
Department of Computer Engineering, LJ Institute of Technology, Ahmedabad, Gujarat, India

M. Shah (✉)
Department of Chemical Engineering, School of Energy Technology, Pandit Deendayal Energy University, Gandhinagar, Gujarat 382426, India
e-mail: manan.shah@spt.pdpu.ac.in

M. Azrour
STI Laboratory, Faculty of Sciences and Techniques, IDMS Team, Moulay Ismail University of Meknes, Errachidia, Morocco

J. Mabrouki
Faculty of Sciences, Mohammed V University in Rabat, Rabat, Morocco

A. Guezzaz
Higher School of Technology, Cadi Ayyad University, Marrakesh, Morocco

M. Azrour et al. (eds.), *Sustainable and Green Technologies for Water and Environmental Management*, World Sustainability Series,
https://doi.org/10.1007/978-3-031-52419-6_15

Keywords IoT · Smart devices · Environment

1 Introduction

IoT and Smart devices are once that are playing and will play a very big and important role in a Sustainable Environment. IoT and Smart devices refer to the network of interconnected devices that are physical in form, embedded with sensors, software, and other technologies, which enable them to collect data and exchange data over the internet. When IoT is combined with Smart devices and advanced analytics, it can revolutionize how we manage and conserve natural resources, reduce waste, and minimize environmental effects. Smart Devices play a very vital role in predicting and analyzing the air quality, water quality, temperature, humidity, and soil condition, considering this in the case of natural resources that need to be conserved and protect ecosystems. The importance of IoT and Smart devices for a sustainable environment cannot be elaborated in words. These technologies play a vital role in addressing the environmental challenges we face today and in building a more sustainable future. Some of the key reasons would be Data-Driven decision-making, Resource efficiency, Environmental monitoring and Conservation, Sustainable Agriculture, and many more. Therefore IoT and Smart devices help the environment through advanced technologies by various devices for their respective purpose for which the devices are designed.

In IoT and Smart devices there have been numerous advanced technologies that help make a sustainable environment Many people have recently adopted the trend of "living green," which includes using green IT tools and services online, particularly the G-IoT, also known as the Green IoT, expanding the use of environmentally friendly building materials and roofing materials, utilizing renewable energy sources, raising awareness of the value of recycling, etc. (Memic et al. 2022). Our homes are an additional setting in which this technology is used. Around 75% of the primary resources that are accessible on a global scale—energy, fossil fuels, raw materials, food, and water—are consumed by them. According to estimates, the pace of consumption will rise from 40 billion tons in 2010 to 90 billion tons by 2050 (Hajer et al. 2018). Urban cities are the major contributors to global Greenhouse Gases (GHG) (Bibri et al. 2023). Additionally, artificial intelligence is a technological advancement that is opening up new possibilities for these applications of smart cities. In addition to the Internet of Things and artificial intelligence, blockchain technology has the potential to expand the use of smart cities by generating new designs and other innovations (Ahmed et al. 2022). Cities are viewed as highly complex systems with a vast number of interconnected populations, transportation, communication, services, and business infrastructure, as well as improvements to urban residents' quality of life (Alahi et al. 2023). IoT makes it possible for gadgets to interact and serve smart cities. IoT concepts are used in information and communication technologies (ICT) solutions by numerous national governments and business companies to manage the concept of smart cities (Albino et al. 2021, Alahi et al. 2023).

Talking about Smart devices in terms of IoT technology, are interconnected devices or objects that have sensors, software, and network connectivity through which they operate and collect and exchange data. The importance of Smart devices for a sustainable environment is immense. Considering an example of smart devices in providing education during the pandemic times. IoT and AI can be used to improve the educational experience of students. For instance, AI-powered chatbots can be utilized to respond to student inquiries and provide individualized learning experiences (Kamruzzaman et al. 2023; Shah et al. 2021). Our home is a significant additional place where smart gadgets can be used. The linking of numerous household systems, appliances, and devices through devices and sensors is what smart homes do (Ponte et al. 2018). Smart devices make routine or the process much easier and more efficient in terms of using resources and how the process is performed. Many smart devices are used in our day-to-day life such as a smart thermostat, a communicating device that is pre-installed in almost all our electronic devices such as speakers, mobile phones, and smartwatches. Examples would be Amazon's Alexa, Apple's Siri, Microsoft's Cortona, etc.

Various smart devices are used to create a sustainable environment. One of the areas where smart devices with the use of IoT technology are used is Smart cities. Urban regions that use technology to boost the effectiveness and sustainability of their infrastructure are known as smart cities. Information and communication technologies (ICT) when combined (Tang et al. 2016; Ali and Khan 2023) and the Internet of Things (Awan et al. 2022; Ali and Khan 2023), Smart cities maximize resource use, reduce waste and pollution, and improve resident quality of life. Environmental monitoring systems continuously test the quality of air, water, and soil to identify the pollution, and human activities impact and support conservation efforts. Due to increasing urbanization, people purchase vehicles more than their usage, thus creating air pollution, traffic problems, parking congestion, etc. To promote the development of the economy, and sustainable environment and ensure the management of smart mobility (Hamamurad et al. 2022; Mavlutova et al. 2023). Thus variety of devices are helpful in making the environment sustainable.

In this study topic, we will discuss mainly the devices that are used to make the environment sustainable. We would mainly focus on some of the devices such as smart thermostats, smart waste management, smart fire alarms, smart agriculture devices, and smart waste management. Consider these devices because these are the devices that make our day-to-day life much easier and make it more efficient. Thus considering these activities, we shall consider these devices for further study.

2 Smart Devices

In this section, we will about the devices which are involved in making the environment sustainable. Some of the devices would be Energy Management Devices which are used to optimize our homes. Renewable Energy Devices such as solar panels, solar inverters, and wind turbines. Smart Lightning such as Smart LED bulbs and motion sensing lights are used in our homes or anywhere where the going and coming of humans is frequent. Water Management Devices are Smart irrigation systems that manage water levels. Home automation systems such as Smart home hubs (e.g., Samsung Smart Things) and Voice Assistants (e.g., Amazon Echo, Google Home). Transportation and Mobility devices which include using EV vehicles in public transport. Air Quality Monitoring Devices which are the smart purifiers and measure the quality of the air. Remote work and Communication Devices such as video conferencing system and Smart office setup that optimize lighting, and climate control. Sustainable Appliances are smart refrigerators, washing machines, and various other devices. Smart Agricultural devices are used to monitor the water levels for the proper yield of crops. Smart Fire Alarm for detecting if there is any smoke detected. Smart Thermostats are used in homes and industrial areas where electronic devices are connected to the smart thermostat so that the devices do not consume power when it is not needed. Thus the above-listed devices have contributed to creating an environment sustainable. Next, we will see a Flow Chart for IoT and Smart devices for sustainable development (Fig. 1).

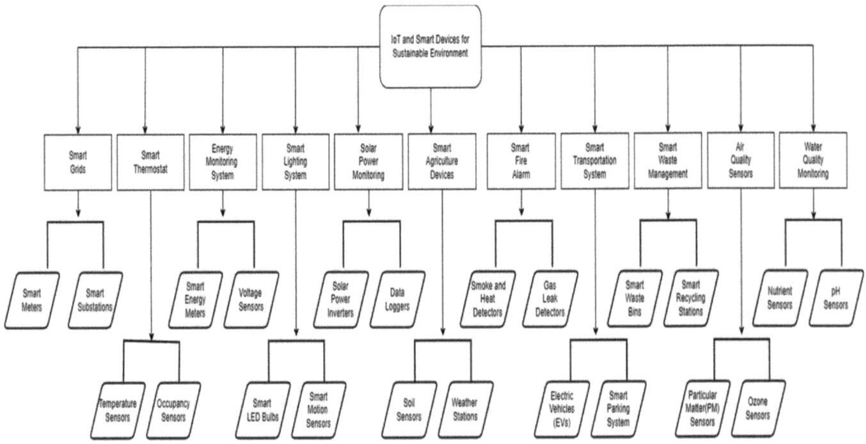

Fig. 1 IoT smart devices for sustainable development

2.1 Smart Agriculture Devices

In this part, we will talk about Smart Agriculture Devices. IoT technology has significantly improved farm management in the realm of agriculture. This technology enables all agricultural machinery and tools to be connected to make informed decisions about irrigation and fertilizer supply (Kumar and Periasamy 2021; Said Mohamed et al. 2021). Monitoring environmental elements is not the only effective way to boost crop output (Suma et al. 2017). Numerous other factors also contribute to a greater degree of productivity loss. In the field, various sensors like temperature, moisture, and PIR sensors are used. These sensors' data collection is connected to the microcontroller through RS232. Thus various factors are involved that collect data and monitor it. Technology is widely employed today and is typically used to reduce or optimize human effort. Technology is advancing in agriculture in an effort to increase crop quality and yield. The term "smart agriculture" is still new to most farmers (Sushanth and Sujatha 2018; Simo et al. 2022). Smart agriculture refers to the use of technologies such as the Internet of Things (IoT), wireless sensor networks, positioning systems, and Artificial Intelligence (AI) on farms (Naresh and Munaswamy, 2019; Simo et al. 2022).

Considering Smart Agriculture there are various fields wherein we use the technology which makes the productivity more than the traditional methods of farming. Communication technology is also to be involved for the sensors and other devices to stay connected and work properly. According to a study on communication technologies for IoT (Alam et al. 2018; Chettri and Bera 2020; Quy et al. 2022), communication technologies must gradually advance IoT device evolution if IoT is to be integrated into the smart agriculture sector. They are crucial to the advancement of IoT systems. The current communication methods can be divided into three categories: protocol, spectrum, and topology.

Numerous wireless communication protocols have been put out for the field of smart agriculture. By interacting, exchanging data, and making decisions based on these protocols, gadgets in a smart agricultural system may monitor and control farming conditions, increase yields, and boost production efficiency. Based on the communication range, the conventional, low-power communication protocol numbers used in smart agriculture may be split into short-range and long-range categories (Quy et al. 2022). Devices which are included in this communication range are:

- Short-range: NFMI (near-field magnetic induction) (Pal and Kant 2019; Quy et al. 2022), Bluetooth (Collotta et al. 2018; Quy et al. 2022), ZigBee (Bacco et al. 2018; Quy et al. 2022), terahertz (Z-Wave) (Gente et al. 2016; Afsharinejad et al. 2017, Quy et al. 2022), and RFID (Wang et al. 2019; Quy et al. 2022).
- Long-range: LoRa (Ali et al. 2021; Quy et al. 2022), Sigfox (Joris et al. 2019; Quy et al. 2022), and NB-IoT (Narrowband IoT) (Popli et al. 2019; Quy et al. 2022).

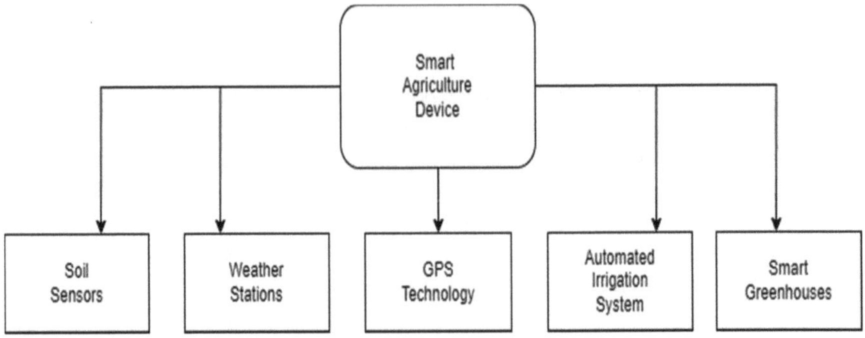

Fig. 2 Classification of smart agriculture devices

An overview of which devices are involved in smart agriculture devices for sustainable development (Fig. 2).

These are some of the various devices which are involved in smart agricultural devices. Starting with soil sensors, these sensors continuously monitor the soil. They collect the data and then forward it where the data is stored, After that real-time decisions are taken upon what measures are to be taken for the betterment of the soil quality. Due to this, there is low use of pesticides, and thus quality is maintained. After that comes the GPS technology which again has various uses in farming. For Precision farming GPS technology would enable the farmers to precisely know the location of the crops, what are the soil types, and other features on their field. Another system that uses the technology is the Auto-steering system which helps farmers to guide farm machinery and where human involvement is minimal, thus increasing accuracy leading to efficient farming and harvesting. Next, we would talk about weather stations where the sensors are located at different locations. These sensors collect weather data which include some of the key meteorological parameters such as temperature, humidity, precipitation, wind speed and direction, solar radiation, and pressure.

The next device is an automated irrigation system that performs action based on data that is collected from various sensors, actuators, and controllers. Using an automated irrigation system you can perform precision watering i.e. it can deliver the right amount of water directly to the root zone thus resulting in reduced usage of water compared to the traditional methods where water is given manually. Through this device, real-time monitoring can be done through the soil moisture sensor which continuously measures the moisture content of the soil. Another benefit of using this system is saving labor costs, thus reducing the amount of time in other activities and focusing mainly on farming.

Smart greenhouses are the ones in which a particular environment is created for farming. The roof and walls of a greenhouse are often constructed from transparent materials like glass or plastic. Plants are cultivated in a controlled environment in greenhouses, where moisture levels, soil nutrients, light levels, temperature, etc. are all regulated. Because of this, greenhouse technology allows anyone to grow any plant, at any time, by creating the right environmental conditions (Tripathy et al. 2021; Quy et al. 2022).

Soil Sensors: a pH sensor that creates a map of where lime should be applied spatially. The significance of space–time modeling for PA is developed since crop development is not only spatially variable but also highly dynamic in time. In the case study, soil moisture is modeled spatially and temporally using space–time geostatistics and a combination of geographical and temporal variables (Oliver et al. 2013). Thus soil sensors must be used to maintain the quality of the soil and create better and organic farming.

Weather Stations: Prior to mainstreaming climate adaptation into agricultural development strategies and plans, it is important to understand the level of knowledge that local farmers have about the climate and climate risk. The local knowledge and experiences of farmers in Burkina Faso, Chad, and Niger are examined in this research in relation to how they are coping with climate change and variability between 1960 and 2010. Elements used include rainfall, temperature, and climate extreme occurrences. If used on a larger scale, this data will highlight "best management practices" and provide lessons learned for increased adoption and accomplishments that can assist in speeding up adaptation to climate change.

In my opinion, Smart agriculture devices play a vital for farmers in terms of irrigation or cost saving. There is no doubt that setting up this technology is going to cost them but the benefits which the technology provides are immense. Their labor cost is decreased, the crop that can cause harm due to the weather conditions is decreased to a much greater extent, and the soil sensors predict what type of crop should be in a particular area of the farm. All in all Smart Agriculture devices are the paths through which a farmer achieves better farming.

3 Smart Fire Alarm

In this part of the study, we will study in detail how smart fire alarms are useful when used in any particular area. Intelligent fire detection systems, often known as smart fire alarms, are cutting-edge safety solutions created to detect and respond to fires more precisely, rapidly, and effectively than conventional fire alarm systems. These systems go beyond the simple function of sounding an alarm when smoke or fire is detected since they include a variety of advanced sensors, data processing tools, and communication possibilities. Early fire detection and warning systems are separated into smoke detectors and infrared detectors based on the various operating methods (Xu et al. 2019; Li et al. 2022). Generally speaking, smoke alarms can only be activated following the continual combustion of flammable materials (Fu et al. 2019; Li et al. 2022), illustrating the necessity for time for smoke to accumulate and the consequently lengthy response time, often exceeding 100 s (Zhang et al. 2020; Evans and Stroup 1986; Qualey, 2000; Li et al. 2022). Fire displays a variety of physical characteristics when it burns. To address these many qualities, scientists and engineers have created vision-based fire detectors (VFDs), as well as fire detectors that are sound-sensitive, flame-sensitive, temperature-sensitive, gas-sensitive, or solid-sensitive (Hong et al. 2019; Avazov et al. 2022). Sensors pick up on the chemical characteristics of smoke, setting off an alarm. This strategy, nevertheless, might cause erroneous warnings. Until the smoke is in close enough proximity to the sensors to trigger them, the alarm won't sound. The flame's smoke and temperature were noticed by these detection systems, which were developed as parts of conventional alarm systems (Avazov et al. 2022).

Thus for the detection of false warnings and monitoring in real time, VFDs are used. A distinctive non-contact fire detector is the VFD. Based on the flame data or shape of the fire's smoke, the VFD can detect a fire from a distance. With the development of computer vision and machine learning, VFDs have become increasingly popular and have been in use for a long time. Furthermore, compared to conventional fire detection technology, VFDs are more dependable and well-made. VFDs have thereby greatly enhanced the detection of fires in huge regions, including forests, warehouses, and shopping centers. The prevention of emergencies requires the early detection of fires. Achieving a high detection rate with a low false alarm rate is the main goal of VFDs. VFDs can also be used for both daytime and nocturnal surveillance (Avazov et al. 2022). The proponents put together the required sensors, electrical parts, microcontrollers, and other electrical modules to develop a functioning smart fire alarm system that could detect smoke, detect temperature spikes, detect the presence of fire, broadcast a GPS coordinate to the fire department, transmit and receive data through RF signals using long-range radio frequency (RF) module and SMS technology, and be functional (Sansolis et al. 2021).

The newly created SFMSS used radiofrequency technology to automatically monitor the Smart Fire Alarm systems installed in particular urban establishments. It processed SMS alerts from a smart fire alarm device that broadcast GPS locations in response to an alarm. To locate the site of the fire occurrence, the system software automatically transformed the GPS coordinate into a legible address location and plotted it on a city map (Sansolis et al. 2021).

A general overview of the benefits of smart fire alarms in a sustainable environment:

- Advanced algorithms on smart fire alarms can discriminate between real fire occurrences and false alarms brought on by things like dust or cooking smoke. By reducing false alarms, they save resources and cut down on carbon emissions from responding cars while also reducing the unnecessary deployment of emergency services.
- Faster fire detection allows for the containment and extinction of fires before they spread, minimizing property damage and the release of dangerous chemicals into the environment.
- Building automation systems and some intelligent fire alarm systems are interconnected. These systems can turn down unneeded HVAC and electrical systems in the case of a fire, helping to save energy and lessen the emergency's impact on the environment.

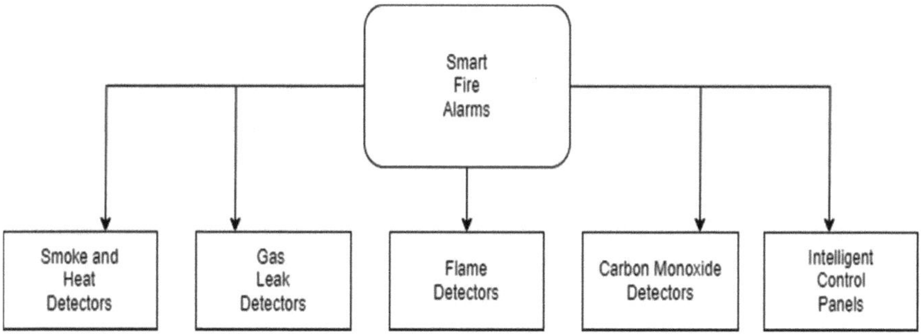

Fig. 3 Applications of smart fire alarms

- Better sustainability planning can be done using the data produced by smart fire alarms. Building owners and managers can spot patterns in fire events, enhance evacuation protocols, and decide on fire safety and sustainability solutions after doing their research.

A Flowchart of which devices that come under Smart Fire Alarm (Fig. 3):

The design of smoke and heat detectors makes it possible for them to identify fire's early warning indicators. Early fire detection helps lessen the effects of larger fires, including substantial damage and environmental effects. As there is early detection, the manual method of putting out a fire requires the use of water, but by employing these detectors, we may use less water and, more importantly, spare the lives of people or other living things.

Once more, gas leak detectors are utilized to identify gas leaks at an early stage. It can be avoided, which can lower the likelihood of mishaps, fires, explosions, etc. Air pollution is avoided because the risk of injury is reduced. The reduction of greenhouse emissions is the key consideration while employing gas leak detectors. Air pollution is caused by greenhouse gases, thus utilizing detectors to catch it early can help maintain a sustainable ecosystem.

A fire that breaks out in a residential or industrial area could seriously harm both the environment and the local populace. Flame detectors are employed to catch it early. The flame detectors, which are also used in forests, are highly useful because they alert us to any signs of fire right away and allow us to stop a significant amount of carbon dioxide and other pollutants from entering the environment. If any flame is seen, the appliances can be turned off to prevent further damage. An ongoing load of electric appliances also has the potential of short-circuiting and catching fire.

The detection of carbon monoxide, a potentially dangerous gas, is another type of sensor covered by smart fire alarms. An odorless, colorless gas called carbon monoxide is created when combustion is not complete. The region is evacuated to protect people from potentially dangerous causes when it reaches a specific level. Installing the detectors is required in industrial and residential areas, which is part of maintaining a sustainable environment.

In essence, intelligent control panels are a management system that acts as a hub for the management and control of multiple systems. The HVAC (Heating, Ventilation, and Air Conditioning) systems, lighting, and other energy-consuming devices can all be monitored by these panels. Better sustainability planning can be done using the data produced by smart fire alarms. Building owners and managers can spot patterns in fire events, enhance evacuation protocols, and decide on fire safety and sustainability solutions after doing their research. Since many intelligent control panels are remotely accessible, security staff or building operators may keep an eye on the system's status and react to alarms from a single place. Particularly in bigger or multi-site facilities, this remote capacity can result in quicker reaction times and more effective resource management.

Case Study:

(1) For any early hazard detection, such as fire or escaping carbon monoxide, smoke detectors and carbon monoxide detectors are employed. In this study, issues with the Integrated Qualifications System (ZSK), a recently developed market credential for technicians who install and repair smoke and carbon monoxide detectors, will be discussed. In many nations, such as the EU, it is required to install smoke and carbon monoxide detectors in houses; however, at the time, the only legal need in Poland is that the maker of these goods acquire a certificate. Theoretical research methods were employed throughout the research process, including synthesis, generalization, inference, comparison, and analogy as well as investigation of literature and legal documents. Smoke detectors produce a sound when they detect smoke in order to retain the essential alarm functionality. The current review of the specifications for installing these detectors demonstrates how various the approaches are to using this kind of equipment. However, in some nations, including the EU, installing these devices is required (Garlińska et al. 2021).
(2) With the rapid development of China's economy, the number and size of large space buildings which include stadiums, large shopping malls, etc. have increased. As the buildings are large in area, with complex structures, property and equipment, and greater fire loads, the primary focus is on fire prevention and safety. One of the fire detection devices used in fire safety design for large spaces is the spot type flame detector (IR flame detector and UV/IR flame detector), which, like video image type flame detectors, has a longer detection distance and a wider protecting range. It is used to detect naked fire sources in high and large architectural spaces. With this adoption of high and large spaces in architecture and buildings, structure and their functionalities require prevention against fire and should be focused on preventive measures, early detection, and timely control (Xu and Zhang 2013).

All in all, in my opinion, rules and regulations should be defined in every residential, commercial or industrial area where there are people in greater numbers. Firstly Smart fire alarms prevent false alarms which reduces the use of resources.

Secondly, by improving the accuracy of the Smart device, effective use of the emergency services can be done. Lastly, smart fire alarms play a vital role in increasing safety while promoting sustainability, thus making this an important device for a safer world.

4 Smart Thermostat

A smart Thermostat or Thermometer is a device that also contributes to the sustainable environment. The most recent report from the International Panel on Climate Change (IPCC) calls for global carbon neutrality by 2050 in order to keep the rise in global temperature to less than 1.5 °C beyond preindustrial levels. A rise in the global temperature above this point is thought to have disastrous consequences (Masson-Delmotte et al. 2018; Alhamayani et al. 2021). Given the history of the European building stock, the majority of residences have modified their room layout and partition design over time. Additionally, the original building materials utilized have deteriorated, resulting in lower thermal characteristics. Additionally, the outdated heating apparatus causes discomfort and uses more energy than is necessary. The only way to address the issue of uncomfortable existing buildings that either negatively impact inhabitants' health and well-being or motivate them to take actions that could jeopardize the building's energy efficiency is to improve building performance (Nicol 2007, Seri et al. 2021). An advantage of the smart thermostat is monetary and energy savings. Traditional thermostats allow you to set the temperature to a single setting that must be altered manually, which is not an efficient use of energy. We'll call this intellectual development at zero. Traditional programmable versions enable planning of the ideal temperature for a room at specific times of the day, making it feasible to lower it while no one is there, resulting in significant power savings (first level of intellectualization). For instance, the Siemens REA23 thermostat (Siemens REA23 manual 2020) operates in a self-learning mode that automatically adjusts the PID controller's adaptive proportional-integral parameters. But rather than being determined automatically based on an analysis of the user's activity, the temperature regulation is carried out according to a timetable specified by the user. Smart thermostats build and modify schedules automatically and dynamically based on user activity and presence while taking into account user-defined priorities, allowing for the highest level of cost-effectiveness (second level of intellectualization). Some thermostat models research people's preferences over time in order to create their own schedules and be flexible. Other approaches dynamically react to the user's actions rather than predicting the timetable based on past activity. Models with automatic scheduling take into account a user's habits, responses to a series of straightforward questions, and the current weather (Nazeer et al. 2019; Jamil et al. 2019; Harikrishnan and Sivagami 2017; Kadir et al. 2011; Korolev et al. 2020).

The main purposes of smart thermostats are to increase energy efficiency and support a sustainable environment. Remote control, learning algorithms, energy consumption tracking, geofencing, interoperability with smart homes, voice control, and many more technologies can help achieve this. Thermostats with remote control capabilities let users manage their heating and cooling systems from their mobile devices or websites. Algorithms for learning where machine learning is applied, and through it, the system adopts users' preferences and everyday activities. It also enables energy tracking, whereby the user receives reports of usage so they can understand the current information and energy utilization. When a resident is not at home and the system is aware of their arrival or departure time, this is known as geofencing. The system can then automatically alter the temperature to provide comfort while preserving energy. This gadget is easily connected to other smart home components including lighting, window coverings, and security systems, enabling energy-saving activities. The user can change the temperature by speaking commands when voice assistants like Amazon Alexa or Google Assistant are integrated.

An overview of which devices are included in Smart thermostats (Fig. 4).

Smart Thermostat Control Unit: This device contains features including a remote control that you may use to change the temperature in your home and operate from your mobile device or the company's website. You may adjust heating and cooling schedules and conserve energy in accordance with your preferences thanks to the control device. The devices' excellent HVAC system compatibility ensures broad use in a variety of homes and environments. It can also work with other smart home appliances like sensors, locks, and smart lights to create a smart home ecosystem.

Temperature Sensor: The primary component of the smart thermostat would be temperature sensors. For instance, if it is placed in a room, it helps to keep the

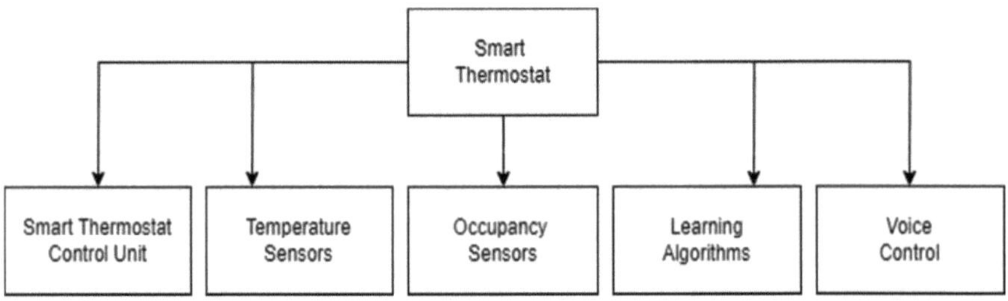

Fig. 4 Devices included in smart thermostats

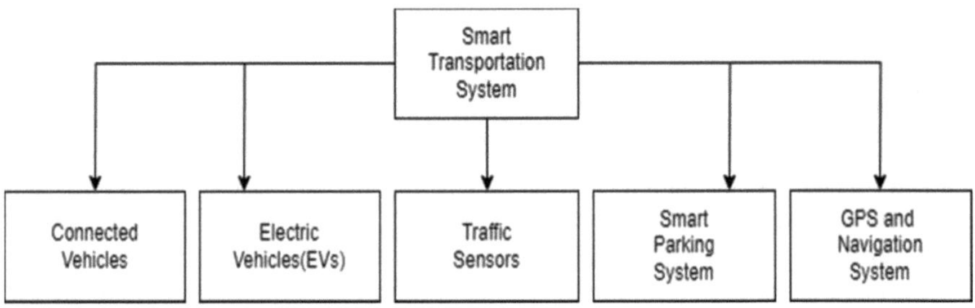

Fig. 5 Various devices in smart transportation systems

temperature comfortable by maintaining the room's temperature. We can divide it into zones and attach a temperature sensor to a smart thermostat to prevent overcooling or overheating. Additionally, energy savings are achieved by cutting back on overall energy use and raising the effectiveness of the heating and cooling system. This sensor allows for customization, enabling the user to set temperature in various locations to their preferred level of comfort.

Occupancy Sensor: A clever approach to improving sustainability is to include occupancy sensors in smart thermostats. By altering temperature settings according to whether a room is inhabited or vacant, occupancy sensors can improve the effectiveness of heating and cooling systems. Automating temperature adjustments based on occupancy, provides user convenience. The sensors pick up on the pattern over time and become better at determining whether a space is occupied or not.

Learning Algorithms: Strong tools for boosting sustainability in residential and commercial settings are learning smart thermostats. Thermostats can use these learning algorithms, sometimes known as "adaptive learning" or "smart learning," to optimize heating and cooling systems for comfort while maximizing energy economy. As a result, this is a smart thermostat upgrade that includes learning capabilities. The feedback and recommendation feature on these devices allows users to receive feedback on their energy usage and suggestions for setting improvements.

Voice Control: By making it simpler for consumers to operate and optimize their heating and cooling systems, voice control devices can improve the usability and convenience of smart thermostats. This may provide a hands-free control that restricts physical interaction with the thermostat. Access to information such as temperature, humidity, etc. may be quicker because we are able to provide voice requests. These devices give users feedback on energy usage and cost savings because they are powered by learning algorithms.

Case Study:

(1) Remote patient monitoring has been made possible by current technological breakthroughs. The goal of this case study is to demonstrate how distant motion sensors may be used to track and measure people as they move around the house. An 8-participant research was conducted to validate the use of remote motion sensors to assess movement in the home. A big database containing information from smart home thermostats was investigated to compare the signs of sleep, physical activity, and sedentary behavior. According to the results, there was a substantial positive linear association between the total number of sensors triggered and the number of steps study participants took inside, with a Spearman rank correlation coefficient of 0.8 (P 0.001) being significant. The findings demonstrate that remote motion sensor data from a smart thermostat solution are a workable alternative to traditional survey data collection methods (Sahu et al. 2020).
(2) The study makes use of a smart thermostat to assess how weekdays and seasons affect sleeping habits and indoor time. It serves as an example of how data from smart home thermostats may be used to assess household sleeping habits and time spent at home. Data from the 2018 Ecobee is used. Donate your Data Set,

> In 481 North American houses with a minimum of 300 days of data, at least six sensors and a maximum occupant count of four were chosen. The findings demonstrate that weekdays have a substantial impact on sleep characteristics. Additionally, results for waking hours and weekends were also displayed. The findings show that Internet of Things data can be utilized to help public health officials comprehend fluctuations in sleep indicators brought on by widespread occurrences like pandemics and climate change. (Jalali et al. 2022).

Overall, I think a smart thermostat is a terrific addition to the ecology of smart homes. It is advantageous due to its many features that allow us to reduce energy use and, consequently, the cost of energy consumption. It can also be embedded with various gadgets that improve the functionality of a smart thermostat, such as a learning algorithm. It makes a wonderful smart home device since it allows people to access their comfort in line with their preferences without having to move physically.

5 Smart Transportation System

We shall talk about smart transportation systems in this topic. With more people migrating into cities than ever before, the modern world is seeing unprecedented urbanization. This population transition has increased air pollution, traffic congestion, and the demand on the transportation system. Big data from mobile Internet and Internet of Things (IoT) technologies is the key source of information for smart transportation, and it can answer four common questions: who, where, when, and what they did (Ganin et al. 2019; Huang et al. 2022a, b). These technologies can thus precisely track the unique traits and behavior patterns of travelers. Smart transportation can mine and analyze big data on transportation, depict a depiction of individual trips, realize comprehensive data integration in the transportation field, and support traffic planning and management using machine learning and other artificial intelligence (AI) technologies (Huang et al. 2022a, b; Cheng et al. 2022; Chung 2021; Ahmed et al. 2022).

Modern society has come to accept ideas like Industry 4.0 and the Internet of Things (IoT) (Liyakat, 2022). The initial objective of Industry 4.0, with an emphasis on the manufacturing sector, was the integration of physical production systems with the digital economy (Guevara and Cheein 2020). A much wider range of topics are now covered by Industry 4.0, such as smart transportation management, smart cities, and smart healthcare. It is crucial to emphasize that innovation, the adoption of new business models, and the creation of new revenue sources are some of Industry 4.0's primary goals. To create new business models that provide new services, IoT must be used (Vaijnath et al. 2022; Liyakat 2022; Banoth et al. 2022; Liyakat 2022).

In order to take use of the potential offered by urbanization, it is necessary to develop cities and human communities that are environmentally benign and commercially efficient (Anthopoulos 2017; Pertheban and Oqbah 2022). Barns (2018) suggested that a smart city be one with programs that have been proven to work and an active presence in at least five of the criteria listed below. Smart citizens, smart infrastructure, smart mobility, and smart governance are a few examples. Additionally, transportation network companies (TNCs) have created information system platforms to support the growth of smart transportation, such as the urban brain system suggested by Alibaba, Tencent, and Huawei (Zhang et al. 2017; Huang et al. 2022a, b). The Apollo initiative, which was started by Baidu and offers solutions for intelligent transportation, autonomous driving, and autonomous cars, is a good example (Zhang et al. 2018; Huang et al. 2022a, b). The development of transportation systems can benefit greatly from smart transportation products and system features. These new technologies have several advantages, including more thorough applications, larger monitoring regions, precise oversight, quicker building, and more efficient management. The top-level design still has a number of problems, including typical requirements and system designs (Wang et al. 2015; Huang et al. 2022a, b).

Traditional vehicular methods are adequate for the amount of traffic that is observed annually. Intelligent Transportation System (ITS) is one of the better alternatives put out. The potential for vehicular networks is greater, and the ITS can be created and implemented more quickly. These networks are currently dealing with a serious issue as a result of the rise in demand for transportation-related services, which must be safe, more efficient, less expensive, and more sustainable. Automakers anticipate releasing a variety of car types with vehicle-to-vehicle (V2V) connections as a result of new technologies. Autonomous vehicles, sometimes referred to as connected vehicles or connected automobiles, are predicted to be on the road within a decade (Hassan et al. 2023).

Devices that are included in smart transportation systems.

Connected Vehicles: Traditional vehicular methods are adequate for the amount of traffic that is observed annually. Intelligent Transportation System (ITS) is one of the better alternatives put out. By increasing transportation efficiency, lowering emissions, and promoting safety, connected vehicles are essential in fostering a more sustainable world. With the help of communication between the infrastructure and the vehicles, real-time data on traffic conditions is made available. Additionally, they offer drivers feedback on their driving style, including acceleration, braking, and speed. The user can escape congestion because the vehicles have access to real-time traffic data. It offers connectivity for electric vehicle battery monitoring and charging optimization.

Electric Vehicles (EVs): Electric vehicles (EVs) are a key component of the environmental sustainability movement. They have a number of important advantages that help lessen the impact on the environment and address problems with conventional internal combustion engine (ICE) vehicles. Since EVs are powered by batteries and have no emissions, they help to reduce air pollution, especially in cities where bad air quality is a significant problem. The running costs of EVs are lower than those of cars with internal combustion engines, which is another benefit. Resulting in a variety of advantages.

Traffic Sensors: Since they allow for continuous monitoring of the movement of vehicles on roads and highways utilizing sensors like sensors, lidar, and inductive loop detectors, these sensors are very beneficial. The sensors can also detect when there has been an incident that has backed up traffic, such as an accident. As a result of fewer traffic delays, fuel consumption and emissions may be reduced. These sensors increase the dependability and efficiency of public transportation systems, such as buses and trains, which reduces the demand for private vehicles on the road.

Smart Parking System: A sustainable urban environment must include a smart parking system. By increasing parking operations' efficiency, lessening traffic congestion, lowering vehicle emissions, and boosting urban living in general, it seeks to address a number of sustainability-related issues. This system improves space usage by using infrared or ultrasonic sensors to track the occupancy of parking spaces in real-time. For urban planning, the system also offers data. City planners can use this data to get important insights that will guide their decisions concerning traffic control, infrastructure for transit, and urban growth.

GPS and Navigation System: The technology offers route optimization, which computes the quickest and most fuel-efficient routes based on real-time traffic data analysis. If we as individuals could view the current traffic conditions in any GPS program, we would choose public transportation over driving our own car. The real-time data allows us to get knowledge about the locations of fueling stations and charging stations, thereby increasing the use of electric vehicles. It is helpful for commercial vehicles as well because it offers real-time traffic data and cuts down on fuel and travel time.

Case Study:

(1) The primary goal of a smart city, as an ITS (intelligent transportation system) driven by data, is to gather and combine various data inputs. The idea behind the smart city bus is to gather and manage public transportation in order to enhance the quality of service offered to passengers and the efficiency of resource allocation to the routes. The platform comprises an information system run by bus operators that can track real-time vehicle movements along with precise measurements from IoT devices put at buses and stations and comments from passengers. In this experiment, the number of people passing through the bus's doors or exits was counted using Time-Of-Flight sensors (Bouloukakis et al. 2023).

(2) This study illustrated the digitalization of city infrastructure, particularly the transportation system. Big data is gathered to track trends and the direction of development as well as to improve the production and service sales processes. In the study scenario, an algorithm is described that uses big data to make strategic judgments about the transportation system. For instance, when it comes to passenger transportation, a business will combine a variety of data to forecast the most popular routes, including GPS and weather information, employee productivity indicators, passenger place of residence/work, additional movements, and transportation expenses. Therefore, by employing the given algorithm, smart cities need smart transportation services (Bulatova 2023).

6 Smart Waste Management

A major problem brought on by the ever-growing population is the rise in trash production. Food waste, material waste, waste from people, waste from worthless items, waste from industries, etc. are all examples of this type of trash. If not handled properly, such trash is dangerous for the environment as a whole. Therefore, waste management is a crucial issue that demands careful consideration and prompt solutions (Gade and Aithal 2021). Human activities account for the majority of the trash produced (Miezah et al. 2015; Gade and Aithal 2021). It should be mentioned that waste management aims to control, collect, process, and recycle garbage in addition to disposing of it properly. The term "waste management" refers to a broad range of activities, including tracking the garbage that is produced in a city from the point at which individuals produce it through waste collection, waste transportation, and depositing it at a waste depository where it may be disposed of in a landfill, burned, or repurposed (Pardini et al. 2020; Gade and Aithal 2021). In waste management operations preventing waste is regarded as the preferred course of action, followed by material reuse, recycling, the creation of energy from waste, and waste disposal, which is regarded as the least preferable course of action (Brancoli et al. 2020; Gade and Aithal 2021). As we have observed repeatedly when dustbins overflow, the concerned parties are unable to receive the information promptly, and as a result, a filthy condition develops around the neighborhood. At the same time, bad odors spread due to waste and the city's unsightly appearance, which encourages air pollution and some dangerous diseases that are easily transmissible. It makes the environment unattractive and the people's living conditions unclean. This results in the spread of several fatal diseases and illnesses in humans. To prevent this, we are preparing to create solid waste management using smart bins. People will benefit from a higher level of living as a result of this (Rathod et al. 2019).

It is important to note that the waste issue we see in industrial operations is mostly a result of technological IoT improvements (Sabir et al. 2020; Aloui et al. 2023). Fortunately, technological developments also offer creative answers that can assist us in overcoming such obstacles when they appear. Best practices for industrial waste management are radically altering as a result of the Internet of Things. First off, we can greatly increase the effectiveness of waste collection and recycling by utilizing IoT services and solutions. The entire waste management process may be automated, enhanced, and interpreted in novel ways while being more affordable. As operators are aware of the number of containers that must be collected each day and the number of trucks that must travel to collect containers, smart containers save time and effort (Sosunova and Porras 2022; Aloui et al. 2023). They also save time when classifying waste by type, making it simpler for staff to send waste to the appropriate authorities for recycling. The ability to gather enormous volumes of data and update them in real time using inexpensive sensors is the primary advantage of IoT. All of the data generated by these low-cost sensors, recording devices, and monitoring tools is sent to the cloud for storage and analysis. AI algorithms can then examine the data and make useful inferences. We'll examine the implementation of

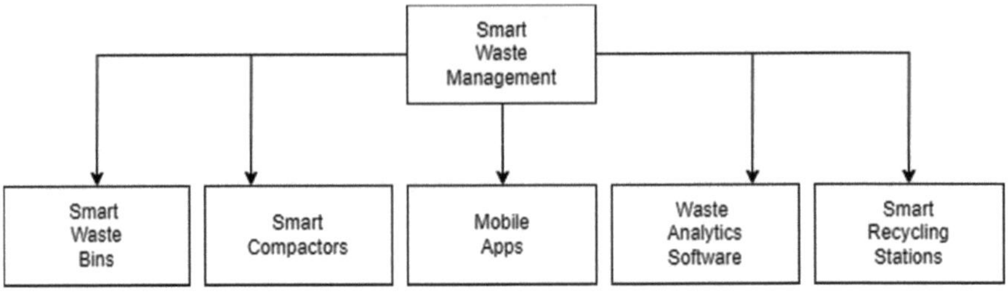

Fig. 7 Devices and software in smart waste management

smart waste management by businesses in the food, agricultural, transportation, and energy sectors in a number of case studies (Fig. 7).

Smart Waste Bins: Smart waste management systems are essential for enhancing sustainability efforts by increasing the effectiveness of waste collection, lowering costs, and decreasing the environmental impact of waste disposal. Smart bins are a key component of these systems. Smart bins have fill-level sensors that constantly track how much rubbish is present inside the bin. Bins warn waste collection teams when they reach a specific level. Authorities in charge of waste management can reduce costs by optimizing the collection schedule, staffing, and vehicle usage. Recycling may be encouraged by having the rubbish that can be recycled separated in the bins. Smart bin electricity can also be produced using solar energy.

Smart Compactors: Smart compactors are a key component of intelligent waste management systems that seek to increase the effectiveness of garbage collection and disposal procedures while fostering sustainability and minimizing environmental effects. These modern devices make better use of technology to improve trash management in general, increase waste collection, and decrease the frequency of collections. By streamlining the waste collection schedule and routes, labor and fuel use are minimized, resulting in lower greenhouse gas emissions, which support a healthier and cleaner environment. The waste in the bins can be compressed by compactors, which increases the bin's capacity to store waste. Decisions about compactors are made after data collection on waste collection trends, fill levels, and operational effectiveness.

Mobile Apps: For all parties involved, management authorities, and users to access and engage with trash-related information, services, and features, mobile applications are essential to smart waste management. The software enables waste collection scheduling, which reduces costs, and notifies or reminds users when waste collection days are. The amount of rubbish dumped illegally can also be decreased by users who promptly report violations to the authorities, allowing the waste to be cleaned up and ending the problem. Students can receive education about trash management, recycling, and environmental sustainability through games and mobile lessons. Authorities may keep an eye on the smart bins through mobile apps to ensure that they don't overflow and that collection times are kept. People who live in locations with smart recycling stations can obtain details about the station's location, the services it offers, and notifications when it needs maintenance or is full.

Waste Analytics Software: By lowering trash, raising recycling rates, and enhancing overall waste management efficiency, waste analytics software is a potent instrument for improving waste management operations, cutting costs, and encouraging sustainability. It makes it possible for waste management companies to make data-driven decisions that benefit the environment and the people they serve. To involve citizens in trash reduction and recycling initiatives, waste analytics software can be connected with mobile applications and web portals. Users can access information on recommended recycling practices and incentives for ethical trash management. This software also forecasts future waste creation patterns using historical data. Organizations can pinpoint opportunities for improvement in waste management procedures, eventually resulting in more sustainable practices, by regularly monitoring and analyzing waste data.

Smart Recycling Stations: By diverting garbage from landfills and preserving resources, smart recycling stations make it simpler for individuals and communities to participate in recycling initiatives, lower contamination, and promote a sustainable environment. Smart recycling stations frequently have distinct bins or areas for different recyclable materials, including paper, plastic, glass, and metals. Users may easily appropriately separate their waste because of the clear labeling and color-coding. Recycling compactors are a characteristic of some stations, enhancing station capacity and decreasing the frequency of emptying. As a result, collection expenses and transportation-related carbon emissions decline. Many intelligent recycling systems use solar power to cut their energy usage. Without having to rely on grid electricity, this renewable energy source aids in keeping the stations running. Accessible recycling stations for the disabled are a feature of smart recycling systems. This includes elements like lower bins for people in wheelchairs and tactile labeling for people who are blind.

Case Study:

(1) Waste comes in a variety of forms, including solid, gaseous, and liquid waste, all of which are impacted globally by population growth, urbanization, and industrialization. Reuse, recycling, and garbage creation are all aspects of waste management. This article compares the amount of plastic garbage produced in Poland and Turkey. The author's experience, observation, and review of the literature all went into this article. Poland generates roughly 110–130 million tonnes of trash annually, but Turkey will have imported 772,832 tonnes of plastic waste by 2020. It detailed how to combat plastic pollution by reducing the use of disposable plastics, implementing refilling systems, and enacting laws that target microplastics. By 2030, it is predicted that there will be a 30% reduction in plastic pollution of the environment, a twofold increase in global plastic recovery (collection and recycling), and a switch to sustainable inputs for the plastic that remains (Ünlü et al. 2023).

(2) The recent significant rise of rubbish waste, particularly in urban areas, is a major environmental concern. At commercial and residential locations, significant amounts of both organic and inorganic garbage are produced on a regular

basis. These waste materials include biodegradable food products, hospital-related garbage, domestic waste, plastics, bottles, and waste produced from dead animals, as well as industrial waste and other commercial goods. The study's information is gathered in a real-time environment using sensors and three bins to sense the presence of dangerous gas components. Collectively, 510 cases were collected. A small society in Bhubaneswar, Odisha, adopted the suggested smart waste management methodology enabled by the intelligence of things. It was able to stop dustbin overflow and send a notice when the bins were full. When the trash cans are close to the threshold limit, it joins the network and transmits a warning message to the relevant base station where the user keeps track of the situation. A choice on whether to alert the user or not can be made based on the forecast of the random forest approach because the model was fully automated with minimal human interaction (Mishra et al. 2022).

7 Conclusion

IoT and smart devices are the two factors that are and will continue to be crucial to a sustainable environment. IoT and smart devices refer to a network of networked physical objects that are integrated with software, sensors, and other technologies that allow them to exchange data online. The management of natural resources, waste reduction, and environmental impact can all be revolutionized when IoT is paired with smart devices and advanced analytics. In the study we discussed, some of the gadgets that are engaged in a sustainable environment are ones that humans can use to support a sustainable environment. Farmers may allocate resources precisely by using smart agriculture sensors to monitor crop health, weather patterns, and soil conditions in real time. Smart fire alarms make better use of cutting-edge technologies like sensors, networking, and artificial intelligence to improve efficiency overall and increase the accuracy of fire detection. HVAC systems can be precisely controlled by a smart thermostat, which optimizes energy efficiency and lowers carbon emissions. With the ultimate goal of building effective, user-centric transportation networks that minimize congestion and improve the general quality of life in metropolitan areas, smart transportation systems place a priority on accessibility and safety. Utilizing technology like sensors, IoT devices, and data analytics,mart waste management optimizes garbage collection, recycling, and disposal procedures. Smart systems provide optimal collection routes, lowering fuel consumption and emissions, by continuously monitoring the garbage level of bins. Data security and privacy, interoperability, energy use, scalability, cost, and the environmental impact of manufacturing these devices are a few of the problems they face. Predictive analysis, energy management, environmental monitoring, and many other future scopes are just a few examples.

So overall Smart technologies and the Internet of Things (IoT) have become effective tools in the fight for a sustainable environment. These technologies make it possible to gather and analyze real-time data, promoting more intelligent resource management, cutting down on waste, and lessening the environmental impact of numerous activities.

References

Afsharinejad A, Davy A, Naftaly M (2017) Variability of terahertz transmission measured in live plant leaves. IEEE Geosci Remote Sens Lett 14(5):636–638. https://doi.org/10.1109/LGRS.2017.2667225

Alahi MEE, Sukkuea A, Tina FW, Tina A, Kurdthongmee W, Suwannarat K, Mukhopadhyay SC (2023) Integration of IoT-enabled technologies and artificial intelligence (AI) for smart city scenario: recent advancements and future trends

Alam MM, Malik H, Khan MI, Pardy T, Kuusik A, Le Moullec Y (2018) A survey on the roles of communication technologies in IoT-based personalized healthcare applications. IEEE Access 6:36611–36631. https://doi.org/10.1109/ACCESS.2018.2853148

Albino V, Berardi U, Dangelico RM (2021) Smart cities: definitions, dimensions, performance, and initiatives. J Urban Technol

Alhamayani AD, Sun Q, Hallinan KP (2021) Estimating smart Wi-Fi thermostat-enabled thermal comfort control savings for any residence. Clean Technol 3(4):743–760. https://doi.org/10.3390/cleantechnol3040044

Ali J, Khan MF (2023) A trust-based secure parking allocation for IoT-enabled sustainable smart cities

Ali S, Glass T, Parr B, Potgieter J, Alam F (2021) Low cost sensor with IoT LoRaWAN connectivity and machine learning-based calibration for air pollution monitoring. IEEE Trans Instrum Meas 70. https://doi.org/10.1109/TIM.2020.3034109

Aloui N, Almukadi W, Belghith A (2023) Towards an IOT approach for smart waste management based on context ontology: a case study. Eng, Technol Appl Sci Res 13(1):10186–10191. https://doi.org/10.48084/etasr.5530

Anthopoulos LG (2017) Understanding smart cities: a tool for smart government or an industrial trick? Public Adm Inf Technol 22:47–185

Ahmed I, Zhang Y, Jeon G, Lin W (2022) A blockchain- and artificial intelligence-enabled smart IoT framework for sustainable city

Avazov K, Mukhiddinov M, Makhmudov F, Cho YI (2022) Fire detection method in smart city environments using a deep-learning-based approach. Electronics (Switzerland) 11(1). https://doi.org/10.3390/electronics11010073

Awan KA, Din IU, Almogren A, Rodrigues JJPC (2022) AutoTrust: a privacy-enhanced trust-based intrusion detection approach for the internet of smart things

Bacco M, Berton A, Gotta A, Caviglione L (2018) IEEE 802.15.4 air-ground UAV communications in smart farming scenarios. IEEE Commun Lett 22(9):1910–1913. https://doi.org/10.1109/LCOMM.2018.2855211

Banoth S, Aavula R, Kazi, Liyakat KS, Lokhande D (2022) Multiple object detection and classification based on Pruning Using YOLO

Barns S (2018) Smart cities and urban data platforms: designing interfaces for SMART governance city, culture and society, vol 12, pp 5–12

Bibri SE, Alexandre A, Sharifi A, Krogstie J (2023) Environmentally sustainable smart cities and their converging AI, IoT, and big data technologies and solutions: an integrated approach to an extensive literature review

Bouloukakis G, Zeginis C, Papadakis N, Magoutis K, Christodoulou G, Kosyfaki C, Mamoulis N (2023) SmartCityBus—a platform for smart transportation systems. In: WSDM 2023—proceedings of the 16th ACM international conference on web search and data mining. Association for Computing Machinery, Inc., pp 1299–1300 https://doi.org/10.1145/3539597.3575781

Brancoli P, Bolton K, Eriksson M (2020) Environmental impacts of waste management and valorisation pathways for surplus bread in Sweden. Waste Manag 117:136–145. https://doi.org/10.1016/j.wasman.2020.07.043

Bulatova O (2023) Using big data in smart cities transportation systems. In: E3S web of conferences, vol 371. EDP Sciences. https://doi.org/10.1051/e3sconf/202337106009

Cheng Q, Chen Y, Liu Z (2022) A bi-level programming model for the optimal lane reservation problem. Expert Syst Appl 189. https://doi.org/10.1016/j.eswa.2021.116147

Chettri L, Bera R (2020) A comprehensive survey on internet of things (IoT) toward 5G wireless systems. IEEE Internet Things J. Institute of Electrical and Electronics Engineers Inc. https://doi.org/10.1109/JIOT.2019.2948888

Chung SH (2021) Applications of smart technologies in logistics and transport: a review. Transp Res Part E: Logist Transp Rev 153. https://doi.org/10.1016/j.tre.2021.102455

Collotta M, Pau G, Talty T, Tonguz OK (2018) Bluetooth 5: a concrete step forward toward the IoT. IEEE Commun Mag 56(7):125–131. https://doi.org/10.1109/MCOM.2018.1700053

Devanand WA, Raghunath RD, Baliram AS, Kazi KS (2019) Smart agriculture system using IoT

Evans DD, Stroup ADW (1986) Methods to calculate te response time of heat and smoke detectors installed below large unobstructed ceilings, Fire Technol

Fu T, Zhao X, Chen L, Wu WS, Zhao Q, Wang XL, Wang YZ (2019) Bioinspired color changing molecular sensor toward early fire detection based on transformation of phthalonitrile to phthalocyanine. Adv Funct Mater 29(8). https://doi.org/10.1002/adfm.201806586

Gade DS, Aithal PS (2021) Smart city waste management through ICT and IoT driven Solution. Int J Appl Eng Manag Lett 51–65. https://doi.org/10.47992/ijaeml.2581.7000.0092

Ganin AA, Mersky AC, Jin AS, Kitsak M, Keisler JM, Linkov I (2019) Resilience in intelligent transportation systems (ITS). Transp Res Part C: Emerg Technol 100:318–329. https://doi.org/10.1016/j.trc.2019.01.014

Garlińska U, Iwańska M, Śliwiński R, Florek P (2021) The role of smoke detectors and carbon monoxide detectors in the fire safety of residential buildings. Saf Fire Technol 57(1):114–133. https://doi.org/10.12845/sft.57.1.2021.8

Gente R, Busch SF, Stubling EM, Schneider LM, Hirschmann CB, Balzer JC, Koch M (2016) Quality control of sugar beet seeds with THz time-domain spectroscopy. IEEE Trans Terahertz Sci Technol 6(5):754–756. https://doi.org/10.1109/TTHZ.2016.2593985

Guevara L, Cheein FA (2020) The role of 5G technologies: challenges in smart cities and intelligent transportation systems. Sustainability (Switzerland) 12(16). https://doi.org/10.3390/su12166469

Hajer M, Swilling M, Baynes T, Bergesen J, Labbé F, Musango JK, Ramaswami A, Robinson B, Salat S, Suh S, Currie P, Fang A, Hanson AK, K, R, M, Smit S, Tabory S (2018) The weight of cities: resource requirements of future urbanization (978-92-807-3699-1). UNESCO

Hamamurad QH, Jusoh NM, Ujang U (2022) Concept and evaluating of smart city

Harikrishnan R, Sivagami P (2017) Intelligent power saving system using PIR sensors. In: Proceedings of the international conferences on electronics, communication and aerospace technology (Coimbatore, India), pp 573–77

Hassan MA, Javed R, Farhatullah Granelli F, Gen X, Rizwan M, Ullah S (2023) Intelligent transportation systems in smart city: a systematic survey. In: 2023 international conference on robotics and automation in industry, ICRAI 2023. Institute of Electrical and Electronics Engineers Inc. https://doi.org/10.1109/ICRAI57502.2023.10089543

Hong X, Wang W, Liu Q (2019) Design and realization of fire detection using computer vision technology. In: Proceedings of the 31st Chinese control and decision conference, CCDC 2019. Institute of Electrical and Electronics Engineers Inc., pp 5645–5649. https://doi.org/10.1109/CCDC.2019.8832897.

Huang K, Kockelman K, Gurumurthy KM (2022a) Innovations impacting the future of transportation: an overview of connected, automated, shared, and electric technologies. Transp Letts. https://doi.org/10.1080/19427867.2022.2070091

Huang K, Jiang C, Li P, Shan A, Wan J, Qin W (2022b) A systematic framework for urban smart transportation towards traffic management and parking. Electronic Research Archive 30(11):4191–4208. https://doi.org/10.3934/era.2022212

Jalali N, Sahu KS, Oetomo A, Morita PP (2022) Usability of smart home thermostat to evaluate the impact of weekdays and seasons on sleep patterns and indoor stay: observational study. JMIR M Health U Health 10(4). https://doi.org/10.2196/28811

Jamil A, Javaid N, Khalid MU, Iqbal MN, Rashid S, Anwar N (2019) An energy efficient scheduling of a smart home based on optimization techniques. Adv Intell Syst Comput 773:3–14

Joris L, Dupont F, Laurent P, Bellier P, Stoukatch S, Redoute JM (2019) An autonomous sigfox wireless sensor node for environmental monitoring. IEEE Sens Lett 3(7). https://doi.org/10.1109/LSENS.2019.2924058

Kadir KM, Forhad MS, Fadlullah MM, Quader N, Al-Arif MMR, Dhali MA (2011) Energy saving by automatic control of power in simple home appliances. In: IEEE 3rd International conferences on communication software and networks, ICCSN (Xi'an, China), pp 311–315

Kamruzzaman MM, Alanazi S, Alruwaili M, Alshammari N, Elaiwat S, Abu-Zanona M, Innab N, Mohammad Elzaghmouri B, Ahmed Alanazi B (2021) AI- and IoT- assisted education systems during pandemics, such as COVID-19, for smart cities

Korolev SI, Muhutdinov RM, Sidorov OY, Goman V (2020) Development of the temperature prediction algorithm for smart thermostat. In: IOP conference series: materials science and engineering, vol 966. IOP Publishing Ltd. https://doi.org/10.1088/1757-899X/966/1/012108

Kumar TU, Periasamy A (2021) IoT based smart farming (E-FARM)' S. Int J Recent Adv Multidiscip Top 2(4):85–87

Li X, Sánchez del Río Saez J, Ao X, Yusuf A, Wang DY (2022) Highly sensitive fire alarm system based on cellulose paper with low-temperature response and wireless signal conversion. Chem Eng J 431. https://doi.org/10.1016/j.cej.2021.134108

Liyakat KKS (2022) IoT-based smart transportation system for passenger comfort

Masson-Delmotte V, Zhai P, Pörtner H-O, Roberts D, Skea J, Shukla PR, Pirani A, Moufouma-Okia W, Péan C, Pidcock R (eds) (2018) IPCC, 2018: global warming of 1.5 ◦C. An IPCC special report on the impacts of global warming of 1.5 ◦C above pre-industrial levels and related global greenhouse gas emission pathways, in the context of strengthening the global response to the threat of climate change, sustainable development, and efforts to eradicate poverty; in press; IPCC: Geneva, Switzerland

Mavlutova I, Atstaja D, Grasis J, Kuzmina J, Uvarova I, Roga D (2023) Urban transportation concept and sustainable urban mobility in smart cities: a review

Memic B, Dzubur AH, Avdagic-Golub E (2022) Green IoT: sustainability environment and technologies

Miezah K, Obiri-Danso K, Kádár Z, Fei-Baffoe B, Mensah MY (2015) Municipal solid waste characterization and quantification as a measure towards effective waste management in Ghana. Waste Manag 46:15–27. https://doi.org/10.1016/j.wasman.2015.09.009

Mishra S, Jena L, Tripathy HK, Gaber T (2022) Prioritized and predictive intelligence of things enabled waste management model in smart and sustainable environment. PLoS ONE 17(8). https://doi.org/10.1371/journal.pone.0272383

Naresh M, Munaswamy P (2019) Smart agriculture system using IoT technology. Int J Recent Technol Eng (IJRTE) 7(5). ISSN: 2277-3878

Nazeer O, Javaid N, Rafique AA, Kiani S, Javaid Y, Khurshid Z (2019) Home energy management using hybrid meta-heuristic optimization technique. Adv Intell Syst Comput 773:599–609

Nicol F (2007) Comfort and energy use in buildings-Getting them right. Energy Build. https://doi.org/10.1016/j.enbuild.2007.02.001

Oliver M, Bishop T, Marchant B (2013) Precision agriculture for sustainability and environmental protection

Pal A, Kant K (2019) NFMI: connectivity for short-range IoT applications. Computer 52(2):63–67. https://doi.org/10.1109/MC.2019.2892862

Pardini K, Rodrigues JJPC, Diallo O, Das AK, de Albuquerque VHC, Kozlov SA (2020) A smart waste management solution geared towards citizens. Sensors (Switzerland) 20(8). https://doi.org/10.3390/s20082380

Pertheban TR, Oqbah SB (2022) Sustainable practices of Smart Mobility Service (SMS): a case study of public transportation system in Kuala Lumpur, Malaysia

da Ponte FRP, Gomes RL, Celestino J, Madeira ERM, Patel A (2018) IoT device programmable language customization for home automation

Popli S, Jha RK, Jain S (2019) A survey on energy efficient narrowband internet of things (NBIoT): architecture, application and challenges. IEEE Access. Institute of Electrical and Electronics Engineers Inc. https://doi.org/10.1109/ACCESS.2018.2881533

Qualey JR (2000) Fire test comparisons of smoke detector response times. Fire Technol 36(2):89–108. https://doi.org/10.1023/A:1015498224060

Quy VK, Hau NV, Anh DV, Quy NM, Ban NT, Lanza S, Muzirafuti A (2022). IoT-enabled smart agriculture: architecture, applications, and challenges. Appl Sci (Switzerland). MDPI. https://doi.org/10.3390/app12073396

Rathod NH, Gurjar SC, Mahindr DU, Sinde AB (2019) Smart waste management

Sabir BE, Youssfi M, Bouattane O, Allali H (2020) Towards a new model to secure IoT-based smart home mobile agents using blockchain technology. Eng Technol Appl Sci Res 10(2):5441–5447. https://doi.org/10.48084/etasr.3394

Sahu KS, Oetomo A, Morita PP (2020) Enabling remote patient monitoring through the use of smart thermostat data in Canada: exploratory study. JMIR M Health U Health 8(11). https://doi.org/10.2196/21016

Said Mohamed E, Belal AA, Kotb Abd-Elmabod S, El-Shirbeny MA, Gad A, Zahran MB (2021) Smart farming for improving agricultural management. Egypt J Remote Sens Space Sci. Elsevier B.V. https://doi.org/10.1016/j.ejrs.2021.08.007

Sansolis E, Dumpit KA, Feliprada CA (2021) Smart fire monitoring system for a city: design and development. Tech: RomIan J Appl Sci Technol 3(10):37–47. https://doi.org/10.47577/technium.v3i10.5094

Seri F, Arnesano M, Keane MM, Revel GM (2021) Temperature sensing optimization for home thermostat retrofit. Sensors 21(11). https://doi.org/10.3390/s21113685

Shah H, Shah S, Tanwar S, Gupta R, Kumar N (2021) Fusion of AI techniques to tackle COVID-19 pandemic: models, incidence rates, and future trends

Siemens REA23 manual (2020) [Electronic resource]—Mode of access. https://www.downloads.siemens.com/download-center/Download.aspx?pos=download&fct=getasset&id1=A6V10075730. Accessed 01 June 2020

Simo A, Dzitac S, Badea GE, Meianu D (2022) Smart agriculture: IoT-based greenhouse monitoring system. Int J Comput Commun Control 17(6). https://doi.org/10.15837/ijccc.2022.6.5039

Sosunova I, Porras J (2022) IoT-Enabled smart waste management systems for smart cities: a systematic review. IEEE Access 10:73326–73363. https://doi.org/10.1109/ACCESS.2022.3188308

Suma N, Samson SR, Saranya S, Shanmugapriya G, Subhashri R (2017) International journal on recent and innovation trends in computing and communication IOT based smart agriculture monitoring system. Academia. Accessed http://www.ijritcc.org

Sushanth G, Sujatha S (2018) IOT based smart agriculture system. In: 2018 international conference on wireless communications, signal processing and networking (WiSPNET), 2018, pp 1–4. https://doi.org/10.1109/WiSPNET.2018.8538702

Tang CS, Chiang YK, Tsou YD, Ju JH, Twu CY (2016) A converged network architecture for ICT and IoT combined applications

Tripathy PK, Tripathy AK, Agarwal A, Mohanty SP (2021) MyGreen: an IoT-enabled smart greenhouse for sustainable agriculture. IEEE Consum Electron Mag 10(4):57–62. https://doi.org/10.1109/MCE.2021.3055930

Ünlü MG, Dinç V, Płaza G (2023) Plastic waste management in the smart cities: Poland and Turkey as a case study. Sci Pap Silesian Univ Technology Organ Manag Ser 2023(166):829–849. https://doi.org/10.29119/1641-3466.2022.166.52

Vaijnath SP, Siddheshwar MP, Pradip MS, Liayakat DKKS (2022) Smart safty device for women. Int J Aquat Sci 13(1):556–560

Wang X, Zhang J, Yu Z, Mao S, Periaswamy SCG, Patton J (2019) On remote temperature sensing using commercial UHF RFID tags. IEEE Internet Things J 6(6):10715–10727. https://doi.org/10.1109/JIOT.2019.2941023

Wang C, Qiu H, Yuan J, Fang A, Zhao Y (2015) Theory and application of intelligent transportation system, Chinese people's Public Security University Press

Xu H, Li Y, Huang NJ, Yu ZR, Wang PH, Zhang ZH, Tang LC (2019) Temperature-triggered sensitive resistance transition of graphene oxide wide-ribbons wrapped sponge for fire ultrafast detecting and early warning. J Hazard Mater 363:286–294. https://doi.org/10.1016/j.jhazmat.2018.09.082

Xu F, Zhang X (2013) Test on application of flame detector for large space environment. In: Procedia engineering, vol 52. Elsevier Ltd, pp 489–494. https://doi.org/10.1016/j.proeng.2013.02.173

Zhang C, Li G, Gao F, Shi C, Zhu S (2017) The study of city smart parking mode based on internet plus. Bull Surv Mapp 11:58–63. https://doi.org/10.1080/23249935.2021.202369010.13474/J.CNKI.11-2246.2017.0348

Zhang X, Shao Y, Sun C (2018) Smart mobility over the future city. Urban Transp China 16:1–7

Zhang ZH, Zhang JW, Cao CF, Guo KY, Zhao L, Zhang GD, Tang LC (2020) Temperature-responsive resistance sensitivity controlled by L-ascorbic acid and silane co-functionalization in flame-retardant GO network for efficient fire early-warning response. Chem Eng J 386. https://doi.org/10.1016/j.cej.2019.123894

Sustainable Water Management at City Corporation Level in Bangladesh: A Comparative Analysis between SCC and BCC

Imran Hossain, A. K. M. Mahmudul Haque, S. M. Akram Ullah, Mourade Azrour, Jamal Mabrouki, and Zeyneb Kılıç

Abstract This study focuses on sustainable water management practices at the city corporation level in Bangladesh, specifically comparing the cities of Sylhet and Barisal. The study aims to examine the current state of water management, identify gaps in sustainable practices, explore challenges and opportunities associated with integrating sustainable water sources, evaluate the role of city corporations in water governance, and assess the impact of community participation on sustainable water management. Primary data were collected through surveys, analyzing water sources, ownership patterns, water safety, availability, and management practices. The findings highlight variations in water sources, ownership, safety, availability, and satisfaction between the two cities. The research concludes that effective strategies are required to address the specific challenges in each city, including improving water source ownership, enhancing water safety and quality, ensuring sufficient water availability, and promoting sustainable water management practices. By implementing effective solutions and strengthening governance frameworks, both cities can achieve sustainable water management and provide clean and safe water to their residents.

I. Hossain (✉) · A. K. M. M. Haque · S. M. A. Ullah
Department of Political Science, University of Rajshahi, Rajshahi 6205, Bangladesh
e-mail: imranbd.ru@gmail.com

M. Azrour
STI Laboratory, IDMS Team, Faculty of Sciences and Techniques, Moulay Ismail University of Meknès, Meknès, Morocco

J. Mabrouki
Laboratory of Spectroscopy, Molecular Modeling, Materials, Nanomaterial, Water and Environment, CERNE2D, Faculty of Science, Mohammed V University in Rabat, Avenue Ibn Battouta, BP1014 Rabat, Agdal, Morocco

Z. Kılıç
Engineering Faculty, Civil Engineering Hydraulic Department, Adıyaman University, Adıyaman, Türkiye

I. Hossain
Department of Political Science, Varendra University, 6204 Rajshahi, Bangladesh

M. Azrour et al. (eds.), *Sustainable and Green Technologies for Water and Environmental Management*, World Sustainability Series,
https://doi.org/10.1007/978-3-031-52419-6_16

Keywords Sustainable water management · Water sources · Water safety · Water availability · Water management practices · Access to water · Water governance

1 Introduction

Water is a crucial natural resource that sustains all forms of life on Earth. However, with increasing population growth, urbanization, and climate change, the demand for water has been increasing while the supply of freshwater is becoming limited. Moreover, the issue of poor water quality is currently one of the biggest issues in developing nations like Bangladesh. In the developing world, millions of people lack access to a sufficient and safe water supply (Kimani-Murage and Ngindu 2007). Historically, developing countries have had more trouble with water and sanitation problems than developed countries (Pandit & Kumar 2015). In poor nations, primitive technologies are mainly utilized. Most residents rely on a single water source (Mahmud et al. 2007). These technologies are often unable to provide clean and safe water for the people, leading to contaminated drinking water, which in turn leads to many health problems. Bangladesh is one of those underdeveloped nations where access to clean water and proper sanitation is currently problematic. In Bangladesh, over 90% of the population rely on unimproved water sources (Hossain et al. 2023a; Bhavnani et al. 2014). Inadequate access to clean water has a profound effect on the health and well-being of the Bangladeshi population, as millions are exposed to a range of serious water-borne illnesses such as diarrhea, cholera, dysentery, and typhoid (Hasan et al. 2019a, b). Moreover, the scarcity of clean water leads to decreased educational opportunities for children, as students are often forced to stay home in order to help fetch contaminated drinking water from long distances (Nygren et al. 2016). In order to address this critical problem, the government of Bangladesh must implement effective strategies to improve access to clean water and adequate sanitation facilities. Bangladesh has advanced in terms of water accessibility and the removal of arsenic from drinking water despite its weak economy and dense population (Loewenberg 2007a, b). The understanding of sanitation and hygiene has significantly changed over the past few years. The government has implemented several initiatives, such as water safety plans, to help address water pollution and provide access to safe drinking water (Pal et al. 2011). Bangladesh's government has taken some positive steps for improving clean water and sanitation (Alam 2021). In this context, the role of local governments in ensuring sustainable water use has become even more critical. Local governments have a crucial responsibility to manage water resources sustainably, protect water quality, and ensure equitable access to safe and clean water for their citizens (Hossain & Haque 2023). Local governments are in a unique position to address water-related issues because they are closest to the communities they serve. They have a deep understanding of the local water resources and the needs of their citizens. Through their planning, policy development, and implementation, they can ensure the sustainable management of water resources, prevent pollution, and protect public health. Furthermore, local governments can also play a critical role in promoting public awareness and engagement on water issues, leading to more sustainable practices by individuals, businesses, and other organizations in their communities (Hossain & Haque 2023). In this aspect, the local government also took some effective measures. These measures included the formation of committees and departments to carry out research and analyze the implications of technological solutions, as well as formulating policies and legal frameworks for implementing these solutions (Sarkar et al. 2022).

However, the significance of this chapter lies in the urgent need to address the growing water management concerns in Bangladesh's urban areas. The population in both SCC and BCC heavily rely on various sources of drinking water, including public supply, tube wells, and submersibles. Understanding the ownership patterns of these water sources are crucial in identifying key stakeholders and their roles in sustainable water management. Moreover, assessing the safety and quality of drinking water, along with the presence of contaminants such as iron and arsenic, is essential to determine potential health risks and guide mitigation efforts. Water availability and supply duration are critical factors in assessing the adequacy of water resources for the population. By evaluating the remoteness of water sources, water supply duration, and satisfaction levels, this research will shed light on the challenges faced by SCC and BCC in meeting the water demands of their residents. Furthermore, examining the presence of personal purification systems and the adoption of water reuse practices will provide insights into the extent of individual and community efforts towards sustainable water management. Through a comprehensive analysis of primary data collected through surveys, this chapter will contribute to the existing knowledge on water management practices in urban areas of Bangladesh. The findings of this study will inform policymakers, water management authorities, and relevant stakeholders about the strengths and weaknesses of current initiatives. Furthermore, this chapter aims to investigate the state of sustainable water management at the city corporation level in Bangladesh, focusing on a comparative analysis between two cities: Sylhet City Corporation (SCC) and Barisal City Corporation (BCC). By examining these two cities, this study seeks to understand the variations in water management practices and the effectiveness of current initiatives. Such a comparative analysis can provide valuable insights into the strengths and weaknesses of existing approaches and inform strategies for enhancing water management in urban areas.

2 Methodology of the Study

The research is very much empirical because the aim of this research is to empirically investigate the present status of environmental governance in enhancing sustainable water at the city corporation level. It combines both qualitative and quantitative approaches for the validity and reliability of the data. This study has employed a comparative analysis approach to examine the sustainable water management practices at the city corporation level in Bangladesh, specifically focusing on the Sylhet

City Corporation (SCC) and Barisal City Corporation (BCC). The comparative analysis has been allowed for a comprehensive understanding of the similarities and differences in water management initiatives, policies, and challenges between the two city corporations.

Primary data have been collected through surveys administered to key stakeholders involved in water management at the city corporation level, including government officials, water supply agencies, NGOs, and community representatives. The survey questionnaire has been designed to gather information on water sources, ownership, safety, availability, satisfaction, purification systems, reuse systems, initiatives of the authority, and supervision. Secondary data have been collected from relevant literature, reports, policy documents, and scholarly articles. This will provide a theoretical framework and background information on sustainable water management practices, existing initiatives, and policies at the city corporation level in Bangladesh.

The study has adopted a purposive sampling technique to select respondents for the survey. Key stakeholders involved in water management at the SCC and BCC have been identified and included in the survey. The sample size has been determined based on the availability of respondents and the representativeness of the selected stakeholders. The sample size of the respondents is 404. Quantitative data collected from the surveys have been analyzed using descriptive statistics, including frequencies and percentages. The data have been tabulated and presented in the form of tables, charts, and graphs to facilitate data interpretation. Qualitative data obtained from open-ended survey questions have been analyzed using thematic analysis to identify common themes and patterns related to water management practices and challenges. Moreover, the data have been analyzed using simple and suitable mathematical and statistical tools like frequency, percentage, arithmetic means, and standard deviations using the SPSS program.

The collected data from both city corporations, SCC and BCC, have been compared and analyzed to identify similarities and differences in water management practices, policies, and challenges. This comparative analysis has been provided insights into the effectiveness of existing initiatives and policies and highlight areas for improvement and best practices. This research will contribute to the existing body of knowledge by providing a comparative analysis of sustainable water management practices in two city corporations in Bangladesh. The findings will have implications for policymakers, water management agencies, and stakeholders involved in urban water management, providing valuable insights for improving initiatives, policies, and practices for sustainable water management at the city corporation level.

3 Results and Discussion

3.1 Socio-Demographic Characteristics of the Respondents

All the respondents were selected from the 18 wards of the BCC and the SCC. About 9 words have been selected from BCC for collecting primary data, namely, wards no. 3, 5, 8 (category 1), 12, 15, 16 (category 2), 23, 24, and 26 (category 3). Moreover, about 10 words have been selected from the SCC for collecting data, namely, wards no. 2, 5, 9, 11, 12, 13, 14, 15, 17, 25. Figure 1 indicates that 49.0% of respondents are male and 51.0% are female. Six different categories, "illiterate", "primary (I–V)", "secondary (VI–X)", "higher secondary (XI–XII)", "graduate", and "postgraduate" were used to indicate the level of education of the respondents in the study area. It was found from Fig. 1 that the highest number of respondents (23%) had a higher secondary level of education, while 19.0 and 13.0 % completed the graduation and postgraduate levels of education, respectively. The lowest number of respondents (8.0%) is illiterate. The age distribution of the sample indicates that respondents came from different age groups. These age groups are below 16 to 25, 26 to 35, 36 to 45, and above 46. Figure 1 shows that only 10.0% of respondents belong to the young age group of <25 years, and 34.0% belong to the middle age group. It is found from Fig. 1 that the highest number (25.0%) of respondents came from business, while only 5.0% came either from farmers or laborers. 11 and 13% are engaged in non-government jobs and government jobs, respectively.

3.2 Source of Water

The primary water supplies for urban areas in Bangladesh include groundwater pumped from aquifers as well as surface water bodies like rivers, lakes, and ponds (Alam el at. 2002). Particularly important roles in meeting domestic and industrial water needs are played by rivers including the Padma, Meghna, and Jamuna (Bhuiyan et al. 2011). Despite their abundance, surface water sources are prone to contamination from a variety of sources, such as industrial discharges, household trash, and agricultural runoff. Nevertheless, groundwater suffers from a range of water contaminations, most of which are caused by toxins, pathogens, and poisons (Hasan et al. 2019a, b). Groundwater contaminated with germs has historically been recognized Bangladesh's main supply of water (Hossain et al. 2023b). Drinking contaminated water causes acute gastrointestinal diseases in children and neonates, which have a significant mortality rate (Smith et al. 2000). According to UNICEF, 38.3% of Bangladeshis drink water that has been contaminated by germs (UNICEF 2022). Water that is safe to drink is severely lacking in Bangladesh. Bangladesh's surface and groundwater are quite salty as a result of the increasing sea levels (Tilman et al. 2002). There are many practical options for supplying clean drinking water, including deep wells, traditionally dug wells, surface water treatment, and rainfall

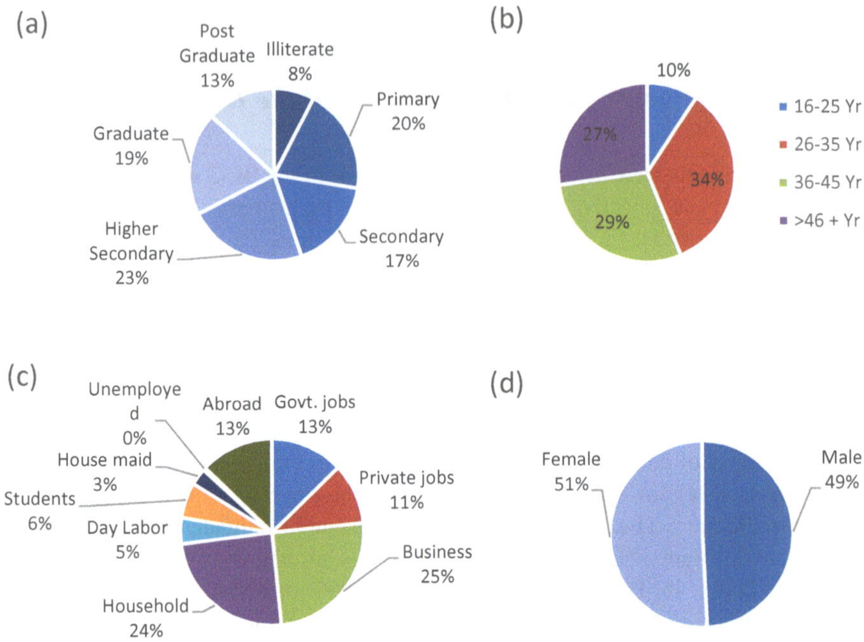

Fig. 1 Socio-demographic characteristics of the respondents. **a** Educational status of the respondents, **b** Age distribution of the respondents, **c** Occupation of the respondents and **d** Sex of the respondents

collection. Between 2000 and 2010, the government developed clean water infrastructure in Bangladesh's arsenic-affected areas (UNICEF 2018). In 2015, over 21 million people lacked access to safe water. Reducing the number of people with access to clean drinking water was important after arsenic was discovered in Bangladeshi groundwater in 1993. Over 97% of people who live in rural regions rely on groundwater as their primary source of drinking water (Hossain et al. 2023b). Figure 2 states that only 17% of Cambodians depend on groundwater as a source of water. In Timor-Leste, the highest number of people—about 93%—are also dependent on groundwater as a source of water.

However, in this comparative research of sustainable water management methods in Bangladesh (BCC), the SCC and BCC are the two City Corporations that are being looked at. This analysis's goals are to highlight the advantages and disadvantages of the two cities' current water management strategies and identify potential areas for improvement.

The study indicates that the majority (53.6%; n = 103) of urban residents in Sylhet City Corporation depend on the government for their drinking water needs. This suggests that the government's efforts to provide a reliable water supply are crucial to ensuring the well-being and health of the population. However, it is worth noting that a small percentage of people (7.8%; n = 15) still rely on tube wells,

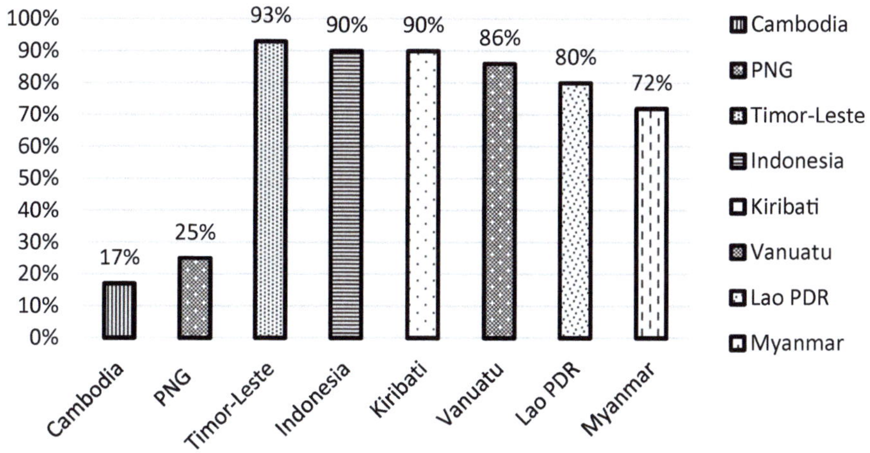

Fig. 2 Proportion of households using groundwater as their primary source of drinking water (Carrard et al. 2019)

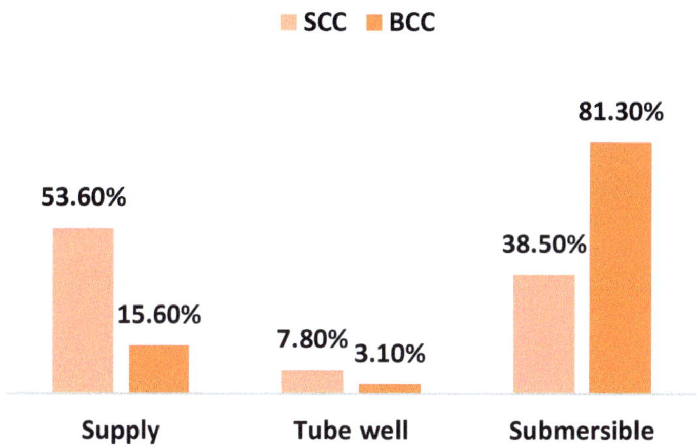

Fig. 3 Source of water

highlighting the need for further investigation into the reasons behind this choice and the quality of water obtained from such sources. Additionally, the significant percentage of respondents relying on submersible sources emphasizes on the importance of exploring and utilizing alternative water sources to meet the growing demand in urban areas (Fig. 3). Given that SCC's supply system serves as the majority of the population's main source of drinking water, these findings emphasize on the value of maintaining and enhancing it. To guarantee that those who depend on tube wells have access to clean drinking water, it is also critical to address issues with tube wells, such as possible contamination and maintenance. Considering this, a respondent from RCC argued,

I can only use the supply water for cleaning and bathing purposes; I cannot drink it. I never feel comfortable using supply water because of contamination. Even though it contains a little amount of iron, I utilize tube wells for drinking water. However, it is even more secure than supplying water. I always collect safe drinking water from a long distance because the water from the distance source is more secured. Therefore, it would be quite useful for us if the City Governance Institutions (CGI) plays a significant role in the purification of supply water (FGD, BCC, Bangladesh).

3.3 Ownership of the Water Source

On the recommendation of the World Bank and following extensive consultation with all relevant stakeholders, including NGOs and civil society groups, the National Water Policy (NWP) was taken in 1999 (Dewan 2012). Water and sanitation services are provided in Bangladesh by many different ministries. With the help of its Local Government Division, the Ministry of Local Government, Rural Development, and Cooperatives is in charge of overall sector regulation and supervision (Das Gupta et al. 2005). Additionally, the Department of Public Health Engineering (DPHE) works with City Corporation to create water supply infrastructure throughout the whole city in Dhaka, Khulna, Chittagong, and Rajshahi. In the three major cities, water and sanitation are provided by local corporations with some autonomy. Water is provided for residential, commercial, and industrial use as well as sewerage and drainage systems in the cities of Dhaka, Chittagong, Khulna, and Rajshahi by the semi-autonomous authorities as well as Dhaka Water Supply and Sewerage Authority (DWASA), Chittagong Water Supply and Sewerage Authority (CWASA), Khulna Water Supply and Sewerage Authority (KWASA), and Rajshahi Water Supply and Sewerage Authority (RWASA). Along with governmental organizations, non-governmental organizations (NGOs) and the private sector are acknowledged within the institutional sector framework of the NWMP and participate in the provision of services. Furthermore, the Bangladesh Water Development Board (BWDB) has significant control over how water resources are managed and regulated across the country, notably with regard to surface water bodies.

However, the study also demonstrates that about 15.6% ($n = 30$) of the respondents in BCC rely on water sources provided by the government. 3.1% ($n = 6$) of the population obtains its drinking water from tube wells. The submersible pump is used by about 156 people, or 81.3% of people, for collecting their drinking water (Fig. 3). The high percentage of people relying on submersible sources for drinking water highlights the importance of addressing any issues with tube wells. It is crucial to ensure that these tube wells are not contaminated, as this could pose a health risk to those who depend on them. Additionally, regular maintenance of tube wells is necessary to ensure their continued functionality and to prevent any potential disruptions in the water supply. To ensure that those who rely on submersible sources have access to safe drinking water in Barisal City Corporation, it is crucial to look into the sustainability and impact of these sources on groundwater resources.

According to the above analysis, inhabitants of SCC and BCC both substantially rely on the supply water system supplied by the government to manage their drinking water. SCC has a higher proportion of submersible sources for drinking water than BCC does. In contrast, more respondents at BCC than at SCC reported getting their drinking water through the supply system. This implies that the water sources used by the two city corporations are different. It might be claimed that whether residents of Barisal City Corporation have access to clean drinking water depends on the supply system and the presence of submersible sources. It is obvious that both cities require targeted actions to promote access to safe drinking water because different percentages of respondents in SCC and BCC obtain their water from various sources. During FGD, a respondent from BCC stated,

The quality of the water supply is not excellent. I therefore only use it for these purposes—cooking, bathing, and others. Even, we frequently receive rubbish and filth through the supply of water. In addition, I have been experiencing a terrible smell from the supply water for the past few days, but the administration has never tried to identify or address the issue. I also try to contact the city government, but I'm unsuccessful. I drink submersible water because the quality of the water source is so poor (FGD, BCC, Bangladesh).

According to this study, SCC and BCC have different ownership habits when they come to drinking water sources. In SCC, there are more government-owned businesses than in BCC (54.2% vs. 25.5%). In addition, SCC has a higher proportion of respondents (40.6%) who own their drinking water sources compared to BCC (24.0%). In contrast, a higher percentage of respondents in BCC (50.5%) and a lower percentage (5.2%) of respondents in SCC rely on NGOs or other organizations for their sources of drinking water (Fig. 4). These ownership patterns imply that the roles that various organizations have played in ensuring that everyone has access to clean drinking water. Government ownership might point to a higher degree of centralized

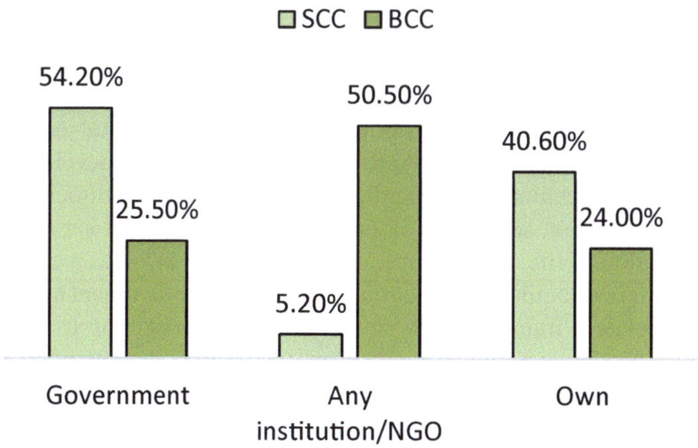

Fig. 4 Ownership of the water source

management and control, whereas private ownership might represent more decentralized approaches. The dependence on NGOs or other organizations in BCC may signify a cooperative strategy involving external entities in water management.

3.4 Safety of Water

Water pollution is a significant concern in Bangladesh, with widespread arsenic contamination in its water sources (Smith et al. 2000). This contamination has led to serious health problems for local communities and poses a substantial challenge for the country. The consequences of contaminated water are diverse, encompassing diseases such as diarrhea, skin disorders, and other severe health complications (Hasan et al. 2019a, b). Globally, water-related illnesses, including cholera, typhoid, polio, ascariasis, cryptosporidiosis, and diarrheal diseases, claim around 3.4 million lives annually (UNEP 2016). Consuming water tainted with excrement puts individual at risk of cholera, diarrhea, typhoid, and polio. Distressingly, in Bangladesh, diarrhea-related illnesses cause the death of over 100,000 children below the age of five each year (Rana 2009). Additionally, groundwater carries bacteria and other harmful elements (Henry et al. 1990). While water drawn from underground sources is comparatively less contaminated, the situation worsens in proximity to urban centers. Due to existing pollution, surface water is not considered suitable for direct consumption (Yokota et al. 2001). The quality of tube well water, a common alternative, remains subpar (Luby et al. 2008). A concerning factor is arsenic contamination, a prevalent issue. Arsenic pollution, impacting over 70 million individuals globally, stands as a significant contributor to groundwater contamination (UNESCO 2009). Despite 97% water accessibility, water purity is far from assured in Bangladesh. Notably, an estimated 57 million people in the country consume water with arsenic levels exceeding the WHO's recommended limit of 10 g/L (Robinson et al. 2011). Addressing water contamination is a scientific endeavor (Loewenberg 2007a, b). Furthermore, the presence of iron, arsenic, and other hazardous elements has been identified in water supplied by SCC and BCC. Ensuring safe drinking water is a formidable challenge. In this aspect, water safety is a fundamental aspect of public health and environmental stability. Across both developed and developing nations, escalating water contamination jeopardizes the well-being of billions (Javier et al. 2017). Access to clean, safe water is indispensable for human survival, hygiene, and overall quality of life. The assurance of water safety involves a comprehensive grasp of potential contaminants, pollution origins, effective treatment methodologies, and robust regulatory frameworks. Waterborne diseases, resulting from consuming tainted water, persist as a global concern, particularly in regions with inadequate water infrastructure.

According to the study, 59.4% of respondents ($n = 114$) in SCC think their drinking water is safe, while 7.8% ($n = 15$) feel it is neither safe nor harmful. The majority, 28.6% ($n = 55$), do not believe their water is safe to drink (Table 1). This lack of confidence in the safety of drinking water raises concerns about the

Table 1 Safety of drinking water

Variables		SCC		BCC	
		f	(%)	f	(%)
Safety of drinking water	Very safe	8	4.2	10	5.2
	Safe	114	59.4	143	74.5
	Neutral	15	7.8	8	4.2
	Unsafe	55	28.6	26	13.5
	Very unsafe	0	0.0	5	2.6

quality of the water supply in SCC. The study's findings indicate that even though the majority of SCC respondents think their drinking water is secure, a sizeable proportion still think it is unsafe. It is important for local authorities to address these concerns and take measures to ensure that the water is safe for consumption. This may involve conducting regular water quality tests, implementing stricter regulations, and providing transparent information to the community about the steps taken to maintain the safety of the drinking water. During KIIs, a government official from the SCC argued,

The safety of our city's water is a matter of utmost concern for us. The data underscores the critical need for immediate action to address any existing water safety issues. We acknowledge that there might be concerns among our residents about the safety of their drinking water. Rest assured, we are taking these concerns seriously. We have already initiated a comprehensive review of our water treatment and distribution processes. We are collaborating closely with water experts to ensure that our water treatment plants are operating optimally and that regular testing is conducted to identify any potential contaminants. Our commitment to transparency means that we will communicate openly with the public about any steps we are taking to improve water safety. We are fully aware that safe water is not a luxury; it's a necessity, and we are dedicated to ensuring that every drop that flows into our homes is of the highest quality (KII, SCC, Bangladesh).

Only 4.2% (n = 8) of respondents in BCC claimed their drinking water was neither safe nor harmful, compared to 74.5% (n = 143) who thought it was. However, a considerable proportion, 13.5% (n = 26), do think their water is harmful to drink (Table 1). This implies that a substantial number of respondents in the BCC are still apprehensive, despite the fact that the majority of respondents in the BCC feel confident in the safety of their drinking water. The competent authorities must address these concerns and take the required steps to ensure the safety and quality of the community's drinking water. The public should continue to be informed about the actions being taken to treat and monitor the water supply, as well as to resolve any possible issues. The overall results of the poll emphasize the need to maintain high.

According to the table mentioned, a large proportion of respondents in both SCC and BCC think their drinking water is safe. However, a higher proportion of respondents in SCC than in BCC feel that their drinking water is hazardous. Authorities must address the concerns of SCC people and take the necessary steps to ensure that their drinking water is safe. This can mean carrying out water quality tests more frequently, enforcing stronger regulations, and informing the public about water safety. By resolving these issues, the community may have greater confidence in the security of their drinking water as well as their overall health and well-being.

According to the study, people in SCC and BCC are not especially concerned by the taste of their water. Nevertheless, concerns remain concerning the safety of drinking water, notably in SCC, where a sizable proportion of respondents reported having health issues related to their water. To overcome these challenges, local authorities must implement stronger restrictions and educate the public about water safety. Increasing residents' trust in the safety of their drinking water will benefit their general health and well-being. In order to ensure that the drinking water in SCC and BCC is safe for consumption, the authorities need to handle this matter seriously. The citizens' health should come first, and whatever measures are required to ensure that they have access to clean, safe drinking water should be adopted.

However, a considerable proportion of those surveyed in both city corporations do mention the presence of iron, with SCC claiming a higher prevalence of iron presence than BCC. The issue of iron contamination in the water supply of both SCC and BCC must be addressed by the government. This can be facilitated by routine water source monitoring and treatment. Additionally, efforts should be launched to educate the people on the need to drink clean water and the potential health concerns associated with doing so. Making sure that everyone has access to clean, safe drinking water should remain a top goal for the SCC and BCC. None of the SCC respondents noted that their water supply contained arsenic. Arsenic in their drinking water was reported by 1.0% ($n = 2$) of BCC survey participants. The analysis shows that there is very little arsenic in the drinking water of SCC and BCC, and SCC has no known cases of arsenic contamination. The majority of respondents in the SCC (89.6%) and BCC (98.4%), according to the analysis, do not believe the taste of their drinking water to be a big worry (Table 2). This shows that the vast majority of people in these city corporations generally like the way the water tastes.

3.5 *Accessibility of Water*

The study on the accessibility and sustainability of drinking water services in the two city corporations is revealed by the data on the distance from the sources of the drinking water and the presence of a sufficient water supply in SCC and BCC. It is possible to gain a thorough understanding of the difficulties and advancements in achieving sustainable water management by analyzing these factors in relation to Sustainable Development Goal 6 (SDG 6), which ensures access to water and sanitation for everyone. The study of the lack of access to adequate water supplies and the distance from the sources of drinking water in SCC and BCC emphasizes on the need for ongoing efforts to meet SDG 6.

The study suggests that a sizable portion of BCC's inhabitants has restricted access to nearby water sources, which may affect their water supply and quality. SCC and

Table 2 Safety taste of water

Variables		SCC		BCC	
		f	(%)	f	(%)
Safety taste of the drinking water	Yes	20	10.4	3	1.6
	No	172	89.6	189	98.4
Result of the taste	Good	8	40.0	0	0.0
	Bad	12	60.0	3	100.0
Presence of iron	No iron	4	2.1	44	22.9
	A little bit	86	44.8	130	67.7
	More or less	71	37.0	17	8.9
	Highly	30	15.6	1	0.5
	Severely high	1	0.5	0	0.0
Presence of arsenic in drinking water	Yes	0	0.0	2	1.0
	No	192	100	190	99.0

BCC have significantly different drinking water source accessibility. Residents of the BCC face more difficulties because of the distance to their water sources than residents of the SCC, who typically have easier access to their water sources. Residents' access to and satisfaction with drinking water in both areas may be significantly impacted by these accessibility disparities. During KIIs, a government official stated,

The proximity of a reliable water source is more than just a convenience; it's a lifeline. Living in an area where water is readily accessible makes daily life so much easier. In my city, Sylhet, the water source is conveniently close for the majority, ensuring that we don't have to worry about long treks or waiting in line just to meet our basic water needs. However, it's crucial to remember that this isn't the reality for everyone. There are places where accessing clean water involves significant effort and time. We need to work towards a scenario where every citizen, regardless of where they live, can enjoy the ease of access to safe water. It's not just a matter of convenience; it's a matter of equity and human rights (KII, SCC, Bangladesh).

The achievement of SDG 6 depends on having services for affordable and accessible drinking water. Even though SCC shows a more favorable situation in terms of remoteness, BCC must address the issue of remote water sources to guarantee fair access to clean and safe drinking water for all residents. According to the analysis, a higher proportion of respondents in SCC (96.9%) confirmed that they had access to enough water, indicating a situation with a sufficient supply of water. In contrast, the situation at BCC is more evenly distributed, with 51.6% of respondents stating that there is an adequate supply of water and 48.4% stating that there is an insufficient supply (Table 3). A crucial element of SDG 6 is a sufficient supply of water. The higher proportion of respondents in SCC who reported having access to enough water is consistent with SDG 6 targets. However, BCC has a tough time providing for the population's water supply needs. This emphasizes how crucial it is to deal with issues of water scarcity, upgrade infrastructure, and implement efficient water management techniques in BCC to guarantee a sufficient and dependable water supply for all residents.

In BCC, emphasis should be placed on enhancing infrastructure and extending the reach of water supply networks, especially in “very remote” areas. To increase accessibility and availability of clean water, this may entail building extra water sources, pipelines, and storage facilities. Promoting water-wise practices and putting water conservation measures into place can help SCC and BCC. To lessen the strain on the available water sources, this can include public awareness campaigns, incentives for water-saving technologies, and promotion of rainwater harvesting and other sustainable water management practices.

According to the study, more respondents in SCC (58.3%) than BCC (12.5%) reported "always" receiving water supply. This could be an indication of the SCC taking a more proactive role in water management. It suggests that the SCC has made significant progress in supplying a consistent and reliable water supply to its residents. In contrast, BCC has a difficult time because so few survey participants reported having a steady supply of water. This reveals potential gaps in the current level of participation and highlights the demand for more advancements in BCC’s sustainable water management practices. The results show that both city corporations face difficulties in effectively integrating sustainable water sources and ensuring a reliable water supply (Table 4). Given that a greater percentage of respondents reported that the water supply was consistent, SCC, however, has a better opportunity for improvement.

A higher percentage of respondents reporting continuous water supply is a sign that SCC has improved its water governance, according to the data. This may show

Table 3 Easily accessible water sources

Variables		SCC		BCC	
		f	(%)	f	(%)
Remoteness of the drinking water source	Very remote	1	0.5	19	9.9
	Close	191	99.5	173	90.1x
Receiving enough water	Yes	186	96.9	99	51.6
	No	6	3.1	93	48.4

Table 4 Duration of water supply

Variables		Sylhet		Barisal	
		f	(%)	f	(%)
Duration of water supply	Always	112	58.3	24	12.5
	Most of the time	17	8.9	18	56.3
	Sometimes	63	32.8	38	19.8
	Rarely	0	0.0	22	11.5
	Don’t know	0	0.0	0	0.0

how well SCC's policies and mechanisms for water governance are working to provide its citizens with a steady supply of water. Contrarily, BCC must address the problems with water governance in order to maintain sound water governance and increase the reliability of the water supply.

3.6 Satisfaction with Water Supply

Both in SCC and BCC, a substantial percentage of respondents said they were happy with the water supply. A sizable section of the population in both municipal corporations is content with the current water services, as seen by the similar percentages of respondents (51.6%) from SCC and BCC who said they were "satisfied" with their water supply (Table 5). This could mean efforts have been made to address water management difficulties and provide a suitable water supply to a sizable section of the population in both cities. Overall, the survey's results indicate that a large proportion of people in SCC and BCC are happy with their water supply. This suggests that the respective city corporations have tried to address problems with water management and provide their residents with satisfactory water services. There may still be opportunity for improvement to ensure that all residents have access to dependable and safe water services.

Insights may be gained from comparing the satisfaction levels with water delivery between Sylhet City Corporation (SCC) and Barisal City Corporation (BCC). Compared to BCC, SCC claims a slightly lower proportion of respondents who are "highly" happy. Even if the majority of SCC residents say they are happy, there is still room for improvement. Contrarily, BCC records a marginally higher percentage of residents expressing high satisfaction, suggesting a larger number of people are happy with their water supply. The dissatisfaction percentages, however, highlight important distinctions. SCC shows a significant proportion of respondents who are dissatisfied, highlighting the need for improved sustainable water management practices. Similar to BCC, a sizeable portion of inhabitants still express some level of unhappiness despite BCC having a lower dissatisfaction rate.

The significant proportion of unhappy respondents in both SCC and BCC suggests that there are gaps in sustainable water management techniques. These gaps may

Table 5 Satisfaction with water supply

Variables		SCC		BCC	
		f	(%)	f	(%)
Satisfaction with water supply	Highly satisfied	18	9.4	24	12.5
	Satisfied	99	51.6	99	51.6
	Somewhat	2	1.0	16	8.3
	Unsatisfied	73	38.0	34	17.7
	Completely unsatisfied	0	0.0	19	9.9

include issues with water quality, dependability, accessibility, and equitable distribution, among other factors. It is crucial to find these gaps in order to address the worries of unsatisfied homeowners and enhance general water management procedures. Even if a sizable portion of inhabitants in both SCC and BCC express pleasure, there is certainly potential for improvement. Both city corporations must place a high priority on enhancing the consistency and quality of their water supply, addressing the concerns of disgruntled citizens, and attempting to boost population satisfaction. This calls for the ongoing examination, observation, and use of sustainable water management strategies. Stakeholder participation that is active is crucial for bridging gaps and increasing satisfaction.

3.7 *Water Reuse and Conservation System*

Table 6 shows that, compared to BCC (32.3%), a significantly higher percentage of residents in SCC (72.9%) own personal purification systems. This suggests that a larger proportion of people in SCC are using purification systems to ensure the safety and caliber of their drinking water on an individual basis. Contrarily, a sizable portion of BCC residents lack personal purification systems, which may limit their access to clean drinking water (Table 6).

According to the study, neither SCC nor BCC have a system in place for reusing water. This demonstrates a squandered opportunity for both cities to pursue environmentally friendly activities like recycling and water reuse. Systems for reusing water can be used to increase water availability and promote efficient water management practices. Authorities in both cities must assess if it is feasible to install water reuse systems and develop rules and guidelines to promote such operations. Furthermore, no efforts are currently being undertaken by the authorities in SCC or BCC to reuse

Table 6 Water reuse and conservation

		SCC		BCC	
		f	(%)	f	(%)
Personal purification system	Yes	140	72.9	62	32.3
	No	52	27.1	130	67.7
Owning personal water reservoir	Yes	185	96.4	46	24.0
	No	7	3.6	146	76.0
System for reusing water	Yes	0	0.0	0	0.0
	No	192	100.0	192	100.0
Initiative of the authority regarding water reuse	Yes	0	0.0	0	0.0
	No	192	100.0	192	100.0
Authorities supervision	Yes	0	0.0	0	0.0
	No	192	100.0	192	100.0

water. This lack of ambition draws attention to a problem with efficient, long-term water management strategies. Authorities place a high priority on the promotion and execution of water reuse activities because they can help with resource conservation and environmental sustainability. The development of water reuse programs should be supported and encouraged by both city corporations as a key priority. The study also demonstrates that neither SCC nor BCC officials are responsible for monitoring water management procedures. This lack of control could result in inefficiencies, inadequate infrastructure upkeep, and significant issues with water quality. In order to ensure adherence to standards, swiftly address any difficulties, and maintain accountability in water management practices, it is crucial to set up the proper authorities to supervise and regulate.

3.8 Major Challenges of City Corporations for Ensuring Sustainable Water Supply

There are several issues and restrictions related to water. Even though 70% of the earth is covered by water, just 3% of it is drinkable. Additionally, 23% of this 3% are found in Lake Baikal in Russian Siberia, which is typically frozen and inaccessible to residents of the region. The remaining 77% of the 3% (or roughly 2.2% of the Earth's surface) that is covered by water must be shared by 7 billion people. The argument is that water, a resource that humans require for survival, is incredibly limited. Water is in great demand, yet there is no way to create enough to satisfy it. Our only options for surviving are water conservation and sharing. Because of our inadequate performance in this area, billions of people face water shortages. Bangladesh has a water issue despite having a lot of consumable fresh water available. Even though 40 million people in Bangladesh still lack access to clean drinking water, access to safe water is regarded as a basic human right. In Bangladesh, there are 24,000 km of rivers that carry usable water. However, due to human activity, the water in that river is gradually becoming poisonous (Ali 2019). There is no suitable method for water conservation and reuse in this area. The BCC and SCC have not made the necessary efforts to reuse and conserve water in this regard. The city governance institutions of each city corporation have proposed some successful surface water reuse measures. Although no fruitful result can be observed, water reuse and conservation are great challenges for city governance institutions (CGI).

The key finding reveals that SCC faces challenges in providing enough water to meet the demand of its residents. The duration of the water supply indicates that sometimes the supply is limited, which can lead to water scarcity and inadequate access for the population. Insufficient Water Supply Infrastructure: The key finding reveals that SCC faces challenges in providing enough water to meet the demand of its residents. The duration of the water supply indicates that sometimes the supply is limited, which can lead to water scarcity and inadequate access for the population.

Table 7 Major challenges for ensuring sustainable water supply

Study area	Major challenges	Source
Delhi, India	• Water scarcity • Pollution • Groundwater overdraft • Slums without clean water access and sewage	Hosagrahar (2011)
Shanghai, China	• Pollution • Salt water intrusion • Influence of major hydraulic projects • Flooding	Chen and Zhou (2014)
Mumbai, India	• Water scarcity • Flooding • Slums without clean water and sewage • Pollution	Uitto and Biswas (2000)
New York, Newark United States	• Water quality • Watershed protection • Flooding	Palmer et al. (2008)
Karachi, Pakistan	• Supply problem • Contamination • Revenue recovery • Industrial pollution • Climate change	Engel et al. (2011)
Istanbul, Turkey	• Pollution • Climate change • Sanitation	Varis and Vakkilainen (2001)
Manila, Philippines	• Water supply • Sewerage system • Water privatization	Esguerra (2003)
Paris, France	• Securing water supply • Water quality	Strat (2010)
Moscow, Russia	• Water leakages • Infrastructure renovation • Pollution	Helmer and Ivanildo (1997)
SCC & BCC, Bangladesh	• Difficulties of controlling water waste • Water scarcity • Water contamination • Lack of cooperation between water supply agencies • Poor Governance • Remote water sources • Improper water reuse and conservation	(This Study)

Source Li et al. (2015)

In the case of BCC, a significant proportion of the population expresses dissatisfaction with the water supply. This highlights the challenge of providing safe and sufficient water access to the residents. It requires investments in infrastructure, such as water treatment plants and distribution networks, to improve the quality and availability of water.

The study also revealed that a substantial percentage of the water sources in BCC are classified as "very remote." This poses a unique challenge as it affects the accessibility and reliability of the water supply. Developing and maintaining infrastructure to transport water from remote sources to the city becomes crucial to ensuring a consistent and adequate water supply. Globally, people are facing various kinds of challenges related to water issues, such as water scarcity, climate change, flooding, water pollution, etc. (Table 7).

In general, combating water scarcity necessitates a multidimensional strategy that includes technological innovation, personal initiative, and governmental engagement. Together, we can make sure that everyone has access to clean, safe water for a number of generations. In conclusion, it is necessary to implement efficient methods and solutions in order to manage water scarcity and guarantee that everyone has access to safe water. The creation and upkeep of infrastructure to carry water from remote sources to the city is one such approach, especially in regions like BCC where a substantial portion of water sources are categorized as "very remote." This will make it easier to guarantee a steady and sufficient water supply, even in difficult-to-reach locations.

4 Proposed Sustainable Water Management Plan

In addressing the critical issue of sustainable water management in both Barisal City Corporation (BCC) and Sylhet City Corporation (SCC), a multi-faceted approach is essential to ensure the long-term viability of water resources. The heavy reliance on groundwater for various purposes necessitates innovative strategies to balance consumption and conservation.

Recognizing the significant role of groundwater in meeting water demands, BCC and SCC should initiate a transition towards diversifying their water sources. While groundwater remains essential, exploring alternative sources like surface water can alleviate the burden on groundwater reserves. Both cities are endowed with the presence of the Kirtankhola and Surma rivers, providing an opportunity to harness surface water for urban water supply. Developing robust infrastructure to collect, treat, and distribute this resource can significantly supplement the existing water sources and alleviate overdependence on groundwater.

Implementing rainwater harvesting systems presents another viable solution for augmenting the water supply. BCC and SCC can encourage both public institutions and homeowners to adopt these systems, thereby harnessing rainwater for non-potable purposes. By integrating advanced technologies and efficient distribution networks, the collected rainwater can serve as a valuable supplement to the available water sources, particularly during the rainy season (Nipun et al. 2022; Ashraful and Islam 2015).

To ensure the sustainable use of groundwater, BCC and SCC should focus on regulating submersible pumps that contribute significantly to groundwater depletion. The installation of water meters can empower households to monitor their usage and contribute to responsible water consumption. The utilization of Android technology to manage submersible pumps holds the potential to enhance monitoring and control. By setting limits on extraction and closely monitoring pump operations, authorities can safeguard groundwater resources from overexploitation (Paul et al. 2015; Ratnadewi et al. 2018).

Public awareness campaigns play a pivotal role in fostering a culture of responsible water use. BCC and SCC should actively engage with residents to educate them about the significance of groundwater conservation and sustainable management. Promoting the adoption of surface water and emphasizing the role of rainwater harvesting systems can empower communities to actively participate in water conservation efforts (Stavenhagen et al. 2018).

In conclusion, BCC and SCC must adopt a holistic and multi-pronged strategy to address the challenges of sustainable water management. Diversifying water sources by incorporating surface water and rainwater harvesting, coupled with efficient control mechanisms for submersible pumps, will play a crucial role in ensuring a balanced and resilient water supply system. The active involvement of both the local authorities and the community is paramount to achieving sustainable water management and securing the future water needs of these cities.

5 Conclusion and Recommendations

Several conclusions can be reached about sustainable water management at the city corporation level in Bangladesh based on the analysis and discussion of the research findings. The ownership of water sources varies between SCC and BCC. While SCC mostly uses sources controlled by the government for cooking and other uses, BCC uses sources owned more frequently by NGOs or other entities. Water quality and safety issues affect both SCC and BCC. While most people in both cities believe the water is safe, a sizeable percentage of them have complaints about the flavor and iron content. Both cities face issues with the supply and availability of water. In comparison to BCC, SCC receives more water, and the length of the water supply varies between the two cities. Greater availability for residents is indicated by the higher proportion of uninterrupted water delivery at SCC. Both in SCC and BCC, water management procedures need to be improved. Sustainable water management in both cities is severely hampered by low levels of community involvement, a lack of government effort towards water reuse, and insufficient supervision.

The following suggestions are put forth to improve sustainable water management at the municipal corporation level in Bangladesh based on the findings of the study:

To provide sufficient and dependable access to water for all inhabitants, SCC and BCC should both invest in enhancing and expanding water delivery infrastructure. This entails creating treatment facilities, expanding water storage capacity, and enhancing distribution networks.

The city corporations should improve the oversight processes and monitoring systems in place for water management. Authorities should actively encourage community engagement while promoting and enforcing sustainable practices like water reuse and home purification systems.

For sustainable water management, there must be greater community involvement. To encourage a sense of accountability and ownership among people, SCC and BCC should launch awareness campaigns, educational programs, and community-driven projects. Long-term sustainability will be aided by encouraging water conservation techniques and including community members in decision-making processes.

Regular testing and monitoring of water quality should be given top priority by both city corporations in order to address issues with taste and pollutant levels. Residents can access safe and hygienic drinking water by implementing appropriate treatment procedures and maintaining water delivery networks.

To get enough financing and resources for water management efforts, SCC, BCC, and relevant stakeholders should work together. The implementation of sustainable water practices, capacity-building initiatives, and infrastructure development should all receive enough financial assistance.

By putting these suggestions into practice, SCC and BCC may significantly advance their efforts to achieve sustainable water management, guarantee access to safe water, and enhance the general standard of living for their citizens. To assess progress and make necessary adjustments to satisfy the changing water management needs of the cities, further study, monitoring, and evaluation are required.

References

Alam MGM, Allinson G, Stagnitti F, Tanaka A, Westbrooke M (2002) Arsenic contamination in Bangladesh groundwater: a major environmental and social disaster. Int J Environ Health Res 12(3):235–253

Alam S (2021) SDGs in Bangladesh: implementation challenges and way forward. Data Sci SDGs: Challenges Oppor Rities 1–14

Ali T (2019) Waste of water, way to disaster. The Daily Star. https://www.thedailystar.net/supplements/news/waste-water-way-disaster-1718767

Ashraful AM, Islam MM (2015) A study on rain water harvesting and comparative cost analysis with withdrawal of underground water around Gazipur City. J Environ Sci Nat Resour 8(1):91–94

Bhavnani D, Goldstick JE, Cevallos W, Trueba G, Eisenberg JN (2014) Impact of rainfall on diarrheal disease risk associated with unimproved water and sanitation. Am J Trop Med Hyg 90(4):705

Bhuiyan MA, Rakib MA, Dampare SB, Ganyaglo S, Suzuki S (2011) Surface water quality assessment in the central part of Bangladesh using multivariate analysis. KSCE J Civ Eng 15:995–1003

Carrard N, Foster T, Willetts J (2019) Groundwater as a source of drinking water in southeast Asia and the Pacific: a multi-country review of current reliance and resource concerns. Water 11(8):1605

Chen K, Zhou JL (2014) Occurrence and behavior of antibiotics in water and sediments from the Huangpu River, Shanghai, China. Chemosphere 95:604–612. https://doi.org/10.1016/j.chemosphere.2013.09.119

Das Gupta A, Singh Babel M, Albert X, Mark O (2005) Water sector of Bangladesh in the context of integrated water resources management: a review. Int J Water Resour Dev 21(2):385–398

Dewan C (2012) Literature review: SWOT analysis of Bangladesh's key water policy documents

Engel K, Jokiel D, Kraljevic A, Geiger M, Smith K (2011) Big cities. Big water. Big challenges: water in an urbanizing world. http://www.wwf.se/source.php/1390895/Big_Cities_Big Water_Big_Challenges_2011.pdf. Accessed April 2015

Esguerra J (2003) The corporate muddle of Manila's water concessions. http://www.wateraid.org/~/media/Publications/privatesector-participation-manila.pdf. Accessed April 2015

Hasan MK, Shahriar A, Jim KU (2019a) Water pollution in Bangladesh and its impact on public health. Heliyon 5(8):e02145. https://doi.org/10.1016/j.heliyon.2019.e02145

Hasan MK, Shahriar A, Jim KU (2019b) Water pollution in Bangladesh and its impact on public health. Heliyon 5(8):e02145. https://doi.org/10.1016/j.heliyon.2019.e02145

Helmer R, Ivanildo H (1997) Water pollution control: a guide to the use of water quality management principles. E & FN Spon, London. ISBN: 0419229108

Henry FJ, Patwary Y, Huttly SRA, Aziz KMA (1990) Bacterial contamination of weaning foods and drinking water in rural Bangladesh. Epidemiol Infect 104(1):79–85. https://doi.org/10.1017/s0950268800054558

Hosagrahar J (2011) Landscapes of water in Delhi: negotiating global norms and local cultures. In: Sorensen A, Okata J (eds) Megacities: urban form, governance, and sustainability. Springer, Tokyo, Japan, pp 111–132. https://doi.org/10.1007/978-4-431-99267-7. ISBN: 978-4-431-99266-0

Hossain I, Haque AM (2023) Role of responsible institution in urban water governance and sanitation management: a focus on metropolitan areas of Bangladesh. Research Square. https://doi.org/10.21203/rs.3.rs-3064713/v1

Hossain I, Haque AKM, Ullah SM (2023a) Role of city governance institutions in ensuring sustainable water: a study on Gazipur City of Bangladesh. SSRN Electron J. https://doi.org/10.2139/ssrn.4480760

Hossain I, Haque AM, Ullah SA (2023b) Assessment of domestic water usage and wastage in urban Bangladesh: a study of Rajshahi City corporation. J Indones Sustain Dev Plan 4(2):109–121

Javier MS, Sara MZ, Hugh T (2017) Water pollution from agriculture: a global review. In: The food and agriculture organization of the united nations Rome and the international water management institute on behalf of the water land and ecosystems research program Colombo

Kimani-Murage EW, Ngindu AM (2007) Quality of water the slum dwellers use: the case of a Kenyan slum. J Urban Health 84:829–838

Li E, Endter-Wada J, Li S (2015) Characterizing and contextualizing the water challenges of megacities. JAWRA J Am Water Resour Assoc 51(3):589–613

Loewenberg S (2007a) Scientists tackle water contamination in Bangladesh. Lancet 370(9586):471–472. https://doi.org/10.1016/s0140-6736(07)61214-8

Loewenberg S (2007b) Scientists tackle water contamination in Bangladesh. Lancet 370(9586):471–472. https://doi.org/10.1016/s0140-6736(07)61214-8

Luby S, Gupta S, Sheikh M, Johnston R, Ram P, Islam M (2008) Tubewell water quality and predictors of contamination in three flood-prone areas in Bangladesh. J Appl Microbiol 105(4):1002–1008. https://doi.org/10.1111/j.1365-2672.2008.03826.x

Mahmud SG, Shamsuddin SAJ, Ahmed MF, Davison A, Deere D, Howard G (2007) Development and implementation of water safety plans for small water supplies in Bangladesh: benefits and lessons learned. J Water Health 5(4):585–597

Nipun MWH, Ashik-Ur-Rahman M, Rikta SY, Parven A, Pal I (2022) Rooftop rainwater harvesting for sustainable water usage in residential buildings for climate resilient city building: case study of Rajshahi, Bangladesh. Int J Disaster Resil Built Environ (ahead-of-print).

Nygren BL, Vulule J, Breiman RF, Jaron P, Nasrin D, Ayers TL, Ombok M, Lash RR, Moke F, Laserson K, Levine MM, Rajasingham A, Mintz ED, Derado G, Nataro JP, Awuor AO, Kotloff KL, Farag TH, Omore R, O'Reilly CE (2016) The relationship between distance to water source and moderate-to-severe diarrhea in the global enterics multi-center study in Kenya, 2008–2011. Am J Trop Med Hyg 94(5):1143–1149. https://doi.org/10.4269/ajtmh.15-0393

Pal SK, Adeloye AJ, Babel MS, Das Gupta A (2011) Evaluation of the effectiveness of water management policies in Bangladesh. Int J Water Resour Dev 27(2):401–417. https://doi.org/10.1080/07900627.2011.564973

Palmer PM, Wilson LR, O'Keefe P, Sheridan R, King T, Chen C (2008) Sources of pharmaceutical pollution in the New York City Watershed. Sci Total Environ 394(1):90–102. https://doi.org/10.1016/j.scitotenv.2008.01.011

Pandit AB, Kumar JK (2015) Clean water for developing countries. Annu Rev Chem Biomol Eng 6(1):217–246. https://doi.org/10.1146/annurev-chembioeng-061114-123432

Paul S, Das M, Sau A, Patra S (2015) Android based smart water pump controller with water level detection technique. Int J Adv Res Comput Commun Eng 4(12):2278–1021

Rana AKMM (2009) Effect of water, sanitation and hygiene intervention in reducing self-reported waterborne diseases in rural Bangladesh. RED research report: online

Ratnadewi R, Nurdiyanto H, Najmurrokhman A, Prabowo C, Idmayanti R, Eteruddin H, Rahim R (2018) Control and notification automatic water pump with Arduino and SMS gateway. In: IOP conference series: materials science and engineering, vol 407. IOP Publishing, p 012160

Robinson C, Von Broemssen M, Bhattacharya P, Häller S, Bivén A, Hossain M, Thunvik R (2011) Dynamics of arsenic adsorption in the targeted arsenic-safe aquifers in Matlab, south-eastern Bangladesh: insight from experimental studies. Appl Geochem 26(4):624–635

Sarkar MSK, Okitasari M, Ahsan MR, Al-Amin AQ (2022) Localisation of sustainable development goals (SDGs) in Bangladesh: an inclusive framework under local governments. Sustainability 14(17):10817

Smith AH, Lingas EO, Rahman M (2000) Contamination of drinking-water by arsenic in Bangladesh: a public health emergency. Bull World Health Organ 78(9):1093–1103

Stavenhagen M, Buurman J, Tortajada C (2018) Saving water in cities: assessing policies for residential water demand management in four cities in Europe. Cities 79:187–195

Strat AL (2010) Paris: local authorities regain control of water management. The transnational institute (TNI). http://www.tni.org/sites/www.tni.org/files/Paris; Chapter by Anne Le Strat En_final.pdf. Accessed April 2015

Tilman D, Cassman KG, Matson PA, Naylor R, Polasky S (2002) Agricultural sustainability and intensive production practices. Nature 418(6898):671–677. https://doi.org/10.1038/nature01014

Uitto JI, Biswas AK (2000) Water for urban areas: challenges and perspectives. United Nations University Press, Tokyo, Japan

UNEP (2016) A snapshot of the world's water quality: towards a global assessment. United Nations Environment Programme. https://uneplive.unep.org/media/docs/assessments/unep_wwqa_report_web.pdf

UNESCO (2009) The united nations world water development report 3 water in a changing world. UNESCO Publishing Group, Paris

UNICEF (2018) Water, sanitation and hygiene annual report 2018. UNICEF, New York

UNICEF (2022) Water, sanitation and hygiene annual report 2022. UNICEF, New York

Varis O, Vakkilainen P (2001) China's 8 challenges to water resources management in the first quarter of the 21st century. Geomorphology 41(2):93–104

Yokota H, Tanabe K, Sezaki M, Akiyoshi Y, Miyata T, Kawahara K, Tsushima S, Hironaka H, Takafuji H, Rahman M, Ahmad S, Sayed M, Faruquee M (2001) Arsenic contamination of ground and pond water and water purification system using pond water in Bangladesh. Eng Geol 60(1–4):323–331. https://doi.org/10.1016/s0013-7952(00)00112-5

How Can Green Energy Be Improved by Integrating IoT into the Microalgae Process?

Khadija EL-Moustaqim, Jamal Mabrouki, Mourade Azrour, and Driss Hmouni

Abstract Green energy is a key factor in replacing fossil fuels, which have harmful effects on the environment. And in reducing air pollution, benefits not just the planet, but human and animal health too. The growing reliance on the Internet of Things (IoT) has contributed to the evolution of the energy sector. Today, green energies benefit through the Internet of Things (IoT), which allows industries to optimize the use of data and enhance efficiency and sustainability. To enhance all technical and economic aspects of the green and sustainable energy production process. Integrating microalgae into the Internet of Things is leading to the creation of new innovative technologies. Today, developing and adopting digitization technologies in the algae sector has become paramount to achieving the desired production levels. This chapter discusses the use of advanced process control strategies, improved digitization, and data analysis in microalgae cultivation processes.

Keywords Green energy · Internet of Things · Microalgae · Production · Environment

K. EL-Moustaqim · D. Hmouni
Improvement and Valuation of Plant Resources, Ibn Tofaïl University—KENITRA-University Campus, 14000 Kenitra, Morocco

J. Mabrouki (✉)
Laboratory of Spectroscopy, Molecular Modelling, Materials, Nanomaterials, Water and Environment, CERN2D, Mohammed V University, AV Ibn Battouta, 10106 Agdal, Rabat, Morocco
e-mail: j.mabrouki@um5r.ac.ma

M. Azrour
STI Laboratory, IDMS Team, Moulay Ismail University of Meknès, Meknes, Morocco

M. Azrour et al. (eds.), *Sustainable and Green Technologies for Water and Environmental Management*, World Sustainability Series,
https://doi.org/10.1007/978-3-031-52419-6_17

1 Introduction

Today, a rapidly growing world population is leading to a severe lack of access to natural resources and significant demands on our ability to preserve the environment. The agricultural sector, in particular feeding our population, has also come under closer scrutiny (Hou et al. 2022; Kholssi et al. 2022). We are increasingly aware of our impact on the environment and the importance of switching to renewable energy sources (Giannino et al. 2018; Atoche et al. 2023; Farias Neves et al. 2019). Green energy refers to the type of energy that brings the greatest number of environmental benefits and is derived mainly derived directly from renewable energy sources such as solar, wind, geothermal, and biomass (Nanjundan and Jaisingh 2023). As shown in Fig. 1, the various energy sources that can produce green energy are represented as follows.

The environment can benefit from green energy by replacing the damaging effects associated with fossil fuels through more eco-friendly options (Guldhe et al. 2019). In the context of releasing much less greenhouse gases and few or no atmospheric contaminants, green energy plays a very important role in keeping these buildings clean, when considered over their entire lifecycle (Mutaf et al. 2023; Arunakumara 2011; Putri et al. 2019). For many years, microalgae, as aquatic micro-organisms, have been used for nutrition, beauty care and pharmaceutical products to obtain valuable metabolites like proteins, vitamins, pigments, fatty acids, and intracellular and extracellular polysaccharides (Mutaf et al. 2023). The microalgae biomass may be

Fig. 1 Representation of energy sources that can produce green energy

utilized for produce several different forms of biofuel and is one of the most effective and versatile technologies for practical and commercial use (Elahi et al. 2020). Through photosynthesis, microalgae are unicellular organisms which transform the sun's from energy to chemical energy (Zeng et al. 2021). These cells can contain a wide range of added-value bioactive substances, like proteins, lipids, carbohydrates, fatty acids, pigments, and vitamins (Levasseur et al. 2020). However, the development of microalgal biomass production started in earnest in the middle of the last century (Dębowski et al. 2020; Hawrot-Paw et al. 2020). The development of the IoT (Internet of Things) has enabled energy companies to evolve (Mabrouki et al. 2021, 2022; Pozzobon et al. 2020). IoT applications for green energy production use sensors mounted on production, transmission, and distribution equipment (Kagita et al. 2020; Shanmugapriya 2023). Alternative energy sources offer several advantages above traditional energy sources. As the demand for cleaner energy rises and respect for the environment becomes the norm, the development of Internet of Things (IoT) technology solutions for energy management continues unabated (Atoche et al. 2023; Hui et al. 2020; Chamara et al. 2022; Mavani et al. 2022; Chavan et al. 2018; Fabris et al. 2020; Mishra et al. 2022). Today, green energies are benefiting from IoT technology, which offers manufacturers within this sector the opportunity to make the most of their data and enhance the efficiency and safety of their facilities (Vinuesa et al. 2020; Divino 2030). Today's technologies, such as the Internet of Things (IoT), open up a huge range of potential applications in the energy sector, for example in the provision, transmission, and distribution of energy, as well as in energy demand (Tamilselvan and Thangaraj 2020). IoT technology helps boost energy efficiency, develop renewable energy production and reduce environmental pollution caused by excessive energy use (Joshi et al. 2022; Khatua et al. 2020; Hossein Motlagh et al. 2020).

This article discusses the current research on the deployment within energy systems of IoT technology, generally, and in the context of intelligent grids, in particular while focusing on the integration of the microalgae process with the aim of producing green energy. In addition, we examine the enabling technologies in IoT and microalgae, including the conversion of biomass into biofuels as a new energy pathway. Finally, we examine the deployment challenges of IoT in the energy industry, covering privacy and security, as well as the impact of IoT on microalgae biomass.

2 IoT Technology in Energy System

The development of IoT (Internet of Things), also referred to an industrial Internet or Internet of Everything, represents an enabling technology that has given rise to the worldwide interaction of machine networks. IoT technology is playing a key role in future technologies, as major companies are increasingly turning their attention to it (Elahi et al. 2018; Elahi 2021). The importance of IoT from an industrial point of view becomes clear when devices connected to the global network of machines can interact with each other for customer support systems, intelligence applications for

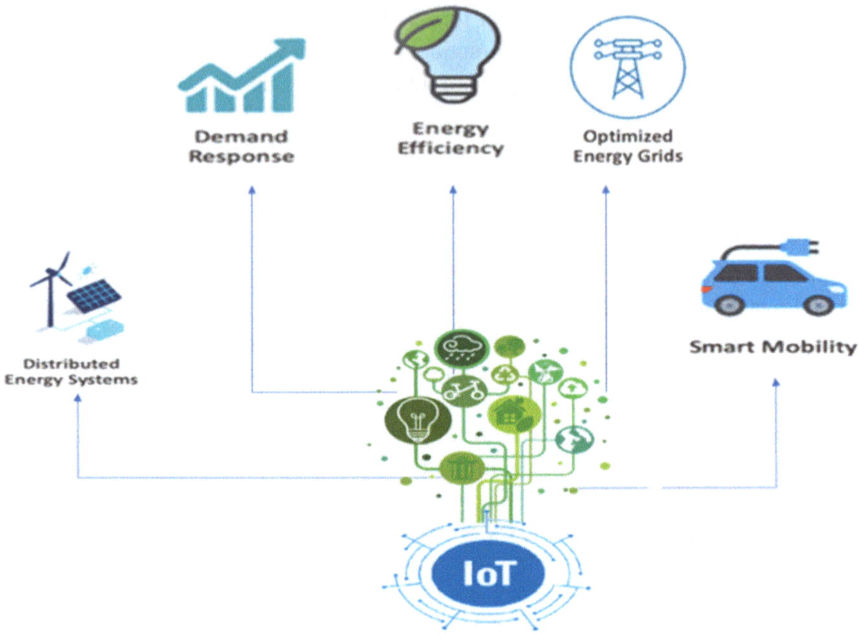

Fig. 2 Examples of the application of IoT in an advanced, intelligent system of energy integration

businesses, inventory systems for vendors based on requirements, and also to enable analytics operations to be performed for the enterprise (Elahi et al. 2020). As a result, companies are fast adopting such technology to obtain new technological solutions and cope with the pressure of competition (Michaels et al. 2020; Pierro et al. 2020).

The mechanism for disseminating IoT in renewable energies involves applying information and communication technologies to the entire energy chain, thanks to various embedded systems, sensors and actuators (Nayak et al. 2018). The viability of renewable energy sources, as illustrated in Fig. 2, and advances in sensor and communication systems technologies have led to the development of modern, sustainable energy systems. Creating a smart network involves combining smart sensors and controllers, giving the energy system the ability to link up and share energy data derived from renewable sources with an information system and end-users as part of a communication network. Thanks to IoT (Internet of Things), this data can be monitored in real-time.

2.1 *Using IoT for Energy Generation*

In the 1990s, automated process control and measurement systems have become increasingly important for the energy business (Ramamurthy and Jain 2017; Prabha and Sundaram 2018). Thanks to equipment and process monitoring and control functionalities, the energy sector has benefited from IoT advances by mitigating the risk of lost production or power outages (Bhardwaj 2015). The main challenges associated with older power plants include reliability, efficiency, environmental impact and maintenance issues. Along with maintenance issues, old equipment in the power sector can lead to a significant degree of energy waste as well as a certain reliability deficit (Utilities and Energy 2023; Immelt 2015). IoT-connected devices can detect any anomaly in operation or irregular drop in energy effectiveness, triggering an alert on how maintenance should be carried out (Tamilselvan and Thangaraj 2020; Connolly et al. 2016). This increases system reliability and efficiency, and reduces management costs (Tan et al. 2017; Anjana and Shaji 2018; Rajawat et al. 2022).

Figure 3 shows the devices that make up an IoT platform. Other peripherals that count as secondary components for IoT are likely to take the shape of captors, activators, or other equipment that participates in the data acquisition or processing cycle.

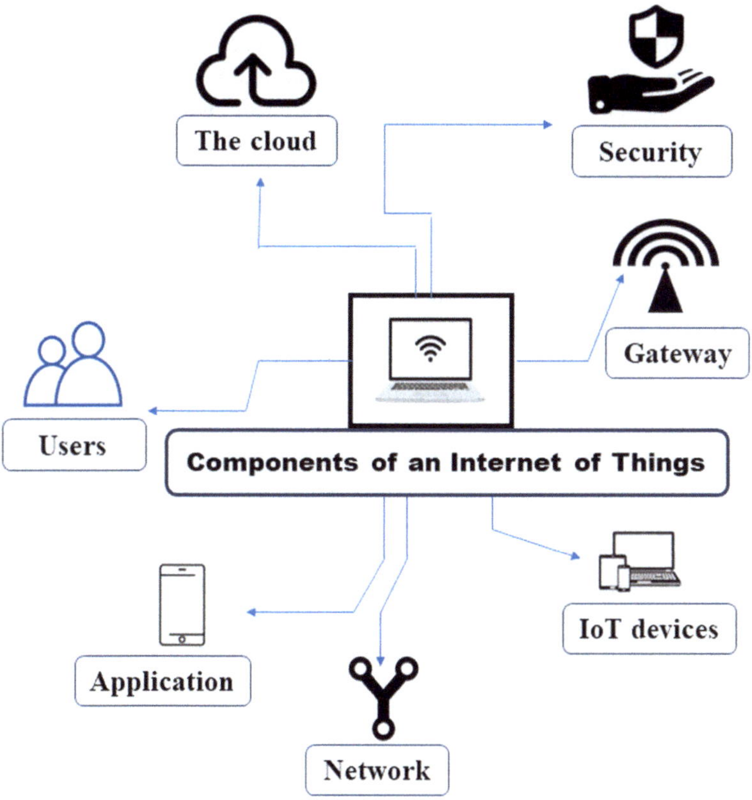

Fig. 3 The major components of IoT

3 Converting Biomass into Biofuels as a New Energy Source

With the help of sunlight, which is considered to be very useful energy in photosynthesis, microscopic unicellular, eukaryotic and autotrophic microorganisms are known for their consumption of CO_2 from the atmosphere. They are highly responsive and are classified according to their habitat and morphology (Aravind et al. 2020). Microalgae use artificial light as an energy factor to promote the development of their growth and carry out metabolic activities. Artificial light also helps to reduce anthropogenic carbon emissions into the atmosphere, thereby helping to combat climate change (Koyande et al. 2019). Table 1 below shows the transformation of biomass and biomolecules from microalgae into various biofuels.

Biomass is converted into various forms of biofuel. Biodiesel is produced from the lipids of microalgae, in particular TAG, which is converted to fatty acid methyl esters (FAME)/biodiesel, as well as glycerol by transesterification, where several acids and enzyme groups or alcohols are exploited as catalysts to ensure high-quality biodiesel production under low temperature and pressure, contributing to a biomass conversion rate of up to 98%, always in the presence of a catalyzer that may increase process costs (Chen et al. 2011; Soydemir et al. 2016; Pandit and Fulekar 2019; Zhu et al. 2019). For bioethanol production, the carbohydrates contained in the microalgae biomass must undergo saccharification into glucose before they can be used as a raw material (Abdulla et al. 2020). The fermentation process can be carried out in the presence of enzymes which have a positive impact on fermentation time but cause an increase in production cost, giving rise to the advances that have been made in the fermentation of microalgal biomass to improve the efficiency of production and reduce processing costs on a laboratory scale (Hernández et al. 2015; Kumar et al. 2016). Microalgal bio-photolysis, either direct or indirect, is used mainly to protect biohydrogen and to photoferment microalgal biomass (Ren et al. 2018). Indirect photolysis involves splitting water to form the oxygen (O_2) and H_2 in the cells of the microalgae, while the resulting H_2 is liberated directly to the atmosphere, ensuring sustainable biohydrogen production through fermentation (Goswami et al. 2021; Wang and Yin 2018). Nutrients such as nitrogen, phosphorus and sulfur are considered major inhibitors of biohydrogen production, since the level of formation increases in their absence through the reduction of O_2 (Shastik et al. 2019). The transformation of moist biomass in contact with O_2 at mild temperatures results in the production of biochar and pyrolysis (Kumar et al. 2020), and that this product has a wide range of uses as a bioenergy source bioenergy, including charcoal fuel, synthesis gas production, biofertilizers and also bioadsorbents, with the aim of reducing the cost of biochar production (Yu et al. 2018; Lee et al. 2020).

Table 1 Transformation of biomass and microalgae biomolecules into various biofuels

Biofuels	*Microalgae used*	The transformation method	Performances	References
Biodiesel	*The Chlorella vulgaris*	Trans-esterification with Sulphuric acid and tetrahydrofuran	95% (FAME)	Lam and Lee (2014)
	A microalgae	Pre-esterification with CaO catalyst and Alkali transesterification	99%	Dong et al. (2015)
	A Isochrysis galbana	Two-step esterification-transesterification	99.7% (FAME)	Silitonga et al. (2017)
	A microalgae biomass	In situ trans-esterification with an acid catalyst	90%wt	Im et al. (2014)
Bioethanol	*Chlorella vulgaris FSPE*	SHF with Saccharomyces cerevisiae	1.7 (g/L)	Ho et al. (2013)
	Chlamydomonas sp.	Fermentation with Saccharomyces cerevisiae	61 (g/L)	Teo et al. (2014a)
	A mixed microalga	Fermentation with Saccharomyces cerevisiae	12.87 (g/L)	Teo et al. (2014b)
	Chlamydomonas mexicana	Immobilized yeast with SSF	9.7 (g/L)	Lakshmikandan et al. (2020)
Biohydrogen	*Chlorella pyrenoidosa*	Sulphur deprivation conditions	45 (mL/L)	Shastik et al. (2019)
	Chlorella sp. biomass	Photo-fermentation using PNSB Rhodobacter sphaeroidesTISTR 1952	125.0 (mL/gVS)	Phanduang et al. (2019)
	Chlorella vulgaris FSPE	Dark fermentation using Clostridium butyricum CGS5	2.87 (mmol)	Ho et al. 2013)
	Chlorella sorokiniana	Dark fermentation using Thermoanaerobacterium sp.	958 vS (dm3/kg)	Roy et al. 2014)
Biomethane and biogas	*Chlamydomonas sp.*	Process of anaerobic digestion	Biogas: 765 $\pm$ 20 mL/gVS/d Biomethane: 478 $\pm$ 15 (mL/gVS/d)	Klassen et al. (2020)
	Arthrospira platensis	Process of anaerobic digestion	The Biomethane: 4 g VS/L/d	Herrmann et al. (2016)
	Chlorella vulgaris	Process of anaerobic digestion	The Biogas: 238 (mL/gVS) The Biomethane: 99 mL/gVS	Doğan-Subaşı and Demirer (2016)
	A microalgal biomass	Anaerobic digestion	The Biomethane: 377 mL	Martı́n Juárez et al. (2018)

4 The Effect of Microalgae on Biofuel Production

Before the biofuel product can be marketed, consideration must be given to the yield parameter, which relies on the level of photosynthetic effectiveness of the microalgae stock, process and biomass conversion to biofuel (Chernova et al. 2014). Also, we can deduct from the literature that efficiency may depend on the extraction of the oil from microalgal biomass, and depending on the solvent used, it is possible to have a yield of 95% of the lipids, making biofuel production more efficient (Zewdie and Ali 2020). However, processing efficiency is proportional to biodiesel production efficiency (Tan et al. 2020). Starch and carbohydrates occupy a very important place in bioethanol production as they are contained in microalgal biomass, and glucose and lipids through hydrolysis are thus important tools for efficiency and yield (Kumar et al. 2016). Thermochemical conversion has been shown to have highest transformation efficiency relative to biologically and chemically based methods, and that yields require high temperatures and pressures, giving permission to focus on developing research aimed at improving biofuel yields and even reducing energy requirements (Serrà et al. 2020; Chang et al. 2019; Nzayisenga et al. 2018).

5 The Impact of IoT on Microalgae Biomass

Converting a biomass resource to fuel energy is carried out using modern biological and ecological processes and some original alternatives engineering technologies, with the aim of achieve efficient, clean and universal conversion of biomass resources (Khoshnevisan et al. 2021). In fact, nutrients such as carbon dioxide are consumed by aquatic micro-organisms such as microalgae, which is very interesting for ensuring their photosynthesis as well as for forming proteins, lipids and carbohydrates (Mat Aron et al. 2020). Figure 4 shows a few examples of the various aspects of introducing IoT technology into microalgal biomass production.

The biorefinery can improve through sensors based on IoT systems to have biomass harvesting which reduces operating costs as well as monitoring microalgae growth and production in real terms and even minimizing waste production (Tao et al. 2018; Uhlemann et al. 2017). Indeed, cultivated microalgae have the capacity to produce biofuels. The IoT has a highly recognized advantage: it enables control, data sharing and continuous monitoring (Rahmat et al. 2020). Green energies are beneficial for the environment by respecting the replacement of fossil fuels into more convenient and less harmful options, since the IoT can associate various information sensing devices to the internet, and especially in the classification of microalgae and in quantify biomass levels using economical methods that would be beneficial for biorefinery (Atoche et al. 2023; Mavani et al. 2022; Acién et al. 2014). The microalgae process remains the most usable technique for energy production and is environmentally friendly, thanks to CO_2 absorption and sequestration.

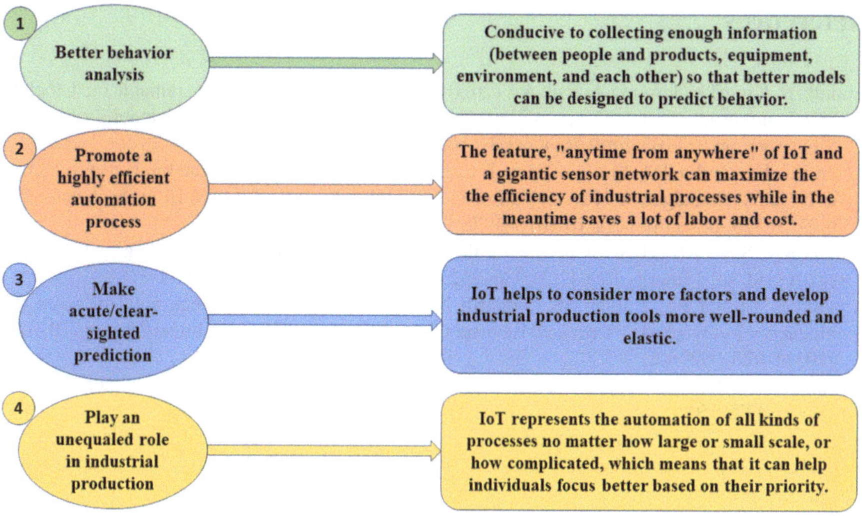

Fig. 4 An example of different aspects of IoT on microalgae biomass

6 Conclusion

The demand and need for fossil fuels remains a major preoccupation in today's world, as the population continues to grow, while research continues into environmentally-friendly, less costly solutions that deliver a very high biomass yield. The IoT can make this task easier, given its usefulness in technological progress. The diversity of microalgae is such that a vast amount of data and information is available, especially for conventional conversion of microalgae and their biorefining, which are considered expensive and difficult processes. However, the optimization of microalgae biorefining can be overcome by using advanced IoT-based statistical tools. The thermochemical method has become more advantageous in the conversion of microalgae metabolites for improving the biorefinery process by integrating green technologies, which enables wet biomass to be exploited, dispensing with drying steps. The use of microalgae in biorefineries to produce biofuels using IoT-integrated co-products can represent a move in the direction of managing the energy challenge, or even minimizing the demand for energy from fossil fuels opening up a new way of improving the planet's environmental preservation and economic situation.

References

Abdulla R, King TK, Jambo SA, Faik AA (2020) Microalgae chlorella as a sustainable feedstock for bioethanol production. In: Yaser AZ (ed) Green engineering for campus sustainability. Springer, Singapore, pp 81–103. https://doi.org/10.1007/978-981-13-7260-5_7

Acién FG, Fernández JM, Molina-Grima E (2014) Economics of microalgae biomass production, in biofuels from algae. In: Pandey A, Lee D-J, Chisti Y, Soccol CR (eds). Elsevier, Amsterdam, pp 313–325. https://doi.org/10.1016/B978-0-444-59558-4.00014-0

Anjana KR, Shaji RS (2018) A review on the features and technologies for energy efficiency of smart grid. Int J Energy Res 42(3):936–952. https://doi.org/10.1002/er.3852

Aravind S, Kumar PS, Kumar NS, Siddarth N (2020) Conversion of green algal biomass into bioenergy by pyrolysis. A review. Environ Chem Lett 18(3):829–849. https://doi.org/10.1007/s10311-020-00990-2

Arunakumara KKIU (2011) Microalgae: the future of green energy. Microal Biotechnol Microbiol Energy 227–250

Atoche AC et al (2023) A study on the renewable power generation capacity of microalgae microbial fuel cells for powering GIoT sensor nodes. J Power Sources 580:233185. https://doi.org/10.1016/j.jpowsour.2023.233185

Bhardwaj A (2015) Utilizing the internet of things and analytics for smart energy management. TATA Consult Adm Mumbai India

Chamara N, Islam MD, (Frank) Bai G, Shi Y, Ge Y (2022) Ag-IoT for crop and environment monitoring: past, present, and future. Agric Syst 203:103497. https://doi.org/10.1016/j.agsy.2022.103497

Chang H et al (2019) Microalgal lipids production and nutrients recovery from landfill leachate using membrane photobioreactor. Bioresour Technol 277:18–26. https://doi.org/10.1016/j.biortech.2019.01.027

Chavan MS, Patil MVP, Chavan S, Sana S, Shinde C (2018) Design and implementation of IOT based real time monitoring system for aquaculture using raspberry pi. Int J Recent Innov Trends Comput Commun 6(3):159–161

Chen C-Y, Yeh K-L, Aisyah R, Lee D-J, Chang J-S (2011) Cultivation, photobioreactor design and harvesting of microalgae for biodiesel production: a critical review. Bioresour Technol 102(1):71–81. https://doi.org/10.1016/j.biortech.2010.06.159

Chernova NI, Kiseleva SV, Popel' OS (2014) Efficiency of the biodiesel production from microalgae. Therm Eng 61(6):399–405. https://doi.org/10.1134/S0040601514060019

Connolly D, Lund H, Mathiesen BV (2016) Smart energy Europe: the technical and economic impact of one potential 100% renewable energy scenario for the European Union. Renew Sustain Energy Rev 60:1634–1653. https://doi.org/10.1016/j.rser.2016.02.025

de Farias Neves F, Demarco M, Tribuzi G (2019) Drying and quality of microalgal powders for human alimentation, in Microalgae-from physiology to application. IntechOpen

Dębowski M et al (2020) The cultivation of lipid-rich microalgae biomass as anaerobic digestate valorization technology—a pilot-scale study. Processes 8(5). https://doi.org/10.3390/pr8050517

Divino S (2021) Artificial intelligence, law and the 2030 agenda for sustainable development. Rev Direitos Sociais E Políticas Públicas UNIFAFIBE, vol 9, p 671. https://doi.org/10.25245/rdspp.v9i1.800

Doğan-Subaşı E, Demirer GN (2016) Anaerobic digestion of microalgal (Chlorella vulgaris) biomass as a source of biogas and biofertilizer. Environ Prog Sustain Energy 35(4):936–941. https://doi.org/10.1002/ep.12294

Dong T et al (2015) Two-step microalgal biodiesel production using acidic catalyst generated from pyrolysis-derived bio-char. Energy Convers Manage 105:1389–1396. https://doi.org/10.1016/j.enconman.2015.06.072

Elahi H (2021) The investigation on structural health monitoring of aerospace structures via piezoelectric aeroelastic energy harvesting. Microsyst Technol 27(7):2605–2613. https://doi.org/10.1007/s00542-020-05017-y

Elahi H, Eugeni M, Gaudenzi P (2018) A review on mechanisms for piezoelectric-based energy harvesters. Energies 11(7). https://doi.org/10.3390/en11071850
Elahi H, Munir K, Eugeni M, Atek S, Gaudenzi P (2020) Energy harvesting towards self-powered IoT devices. Energies 13(21):5528
Elahi H et al (2020) A review on applications of piezoelectric materials in aerospace industry. Integr Ferroelectr 211(1):25–44. https://doi.org/10.1080/10584587.2020.1803672
Fabris M et al (2023) Emerging technologies in algal biotechnology: toward the establishment of a sustainable, algae-based bioeconomy. Front Plant Sci 11. https://doi.org/10.3389/fpls.2020.00279
Giannino F, Esposito S, Diano M, Cuomo S, Toraldo G (2018) A predictive decision support system (DSS) for a microalgae production plant based on Internet of Things paradigm. Concurr Comput Pract Exp 30(15):e4476. https://doi.org/10.1002/cpe.4476
Goswami RK, Mehariya S, Obulisamy PK, Verma P (2021) Advanced microalgae-based renewable biohydrogen production systems: a review. Bioresour Technol 320:124301. https://doi.org/10.1016/j.biortech.2020.124301
Guldhe A, Singh B, Mutanda T, Permaul K (2019) Microalgae as a potential tool for biorefinery: a review. Renew Sustain Energy Rev 104:324–342
Hawrot-Paw M, Koniuszy A, Gałczyńska M (2020) Sustainable production of monoraphidium microalgae biomass as a source of bioenergy. Energies 13(22). https://doi.org/10.3390/en13225975
Hernández D, Riaño B, Coca M, García-González MC (2015) Saccharification of carbohydrates in microalgal biomass by physical, chemical and enzymatic pre-treatments as a previous step for bioethanol production. Chem Eng J 262:939–945. https://doi.org/10.1016/j.cej.2014.10.049
Herrmann C, Kalita N, Wall D, Xia A, Murphy JD (2016) Optimised biogas production from microalgae through co-digestion with carbon-rich co-substrates. Bioresour Technol 214:328–337. https://doi.org/10.1016/j.biortech.2016.04.119
Ho S-H, Huang S-W, Chen C-Y, Hasunuma T, Kondo A, Chang JS (2013) Bioethanol production using carbohydrate-rich microalgae biomass as feedstock. Bioresour Technol 135:191–198. https://doi.org/10.1016/j.biortech.2012.10.015
Hossein Motlagh N, Mohammadrezaei M, Hunt J, Zakeri B (2020) Internet of things (IoT) and the energy sector. Energies 13(2). https://doi.org/10.3390/en13020494
Hou J et al (2022) An assessment of biochar as a potential amendment to enhance plant nutrient uptake. Environ Res 214:113909. https://doi.org/10.1016/j.envres.2022.113909
Hui H, Ding Y, Shi Q, Li F, Song Y, Yan J (2020) 5G network-based Internet of Things for demand response in smart grid: a survey on application potential. Appl Energy 257:113972. https://doi.org/10.1016/j.apenergy.2019.113972
Im H, Lee H, Park MS, Yang J-W, Lee JW (2014) Concurrent extraction and reaction for the production of biodiesel from wet microalgae. Bioresour Technol 152:534–537. https://doi.org/10.1016/j.biortech.2013.11.023
Immelt JR (2015) The future of electricity is digital. In: Technical Report, General Electric
Joshi D et al (2022) An optimized open pit mine application for limestone quarry production scheduling to maximize net present value. Mathematics 10(21). https://doi.org/10.3390/math10214140
Kagita MK, Thilakarathne N, Rajput DS, Lanka DS (2020) A detail study of security and privacy issues of internet of things. https://doi.org/10.48550/arXiv.2009.06341
Khatua PK, Ramachandaramurthy VK, Kasinathan P, Yong JY, Pasupuleti J, Rajagopalan A (2020) Application and assessment of internet of things toward the sustainability of energy systems: challenges and issues. Sustain Cities Soc 53:101957. https://doi.org/10.1016/j.scs.2019.101957
Kholssi R et al (2022) Green agriculture: a review of the application of micro-and macroalgae and their impact on crop production on soil quality. J Soil Sci Plant Nutr 22(4):4627–4641
Khoshnevisan B et al (2021) A critical review on livestock manure biorefinery technologies: sustainability, challenges, and future perspectives. Renew Sustain Energy Rev 135:110033. https://doi.org/10.1016/j.rser.2020.110033

Klassen V, Blifernez-Klassen O, Bax J, Kruse O (2020) Wastewater-borne microalga Chlamydomonas sp.: A robust chassis for efficient biomass and biomethane production applying low-N cultivation strategy. Bioresour Technol 315:123825

Koyande AK, Chew KW, Rambabu K, Tao Y, Chu D-T, Show PL (2019) Microalgae: a potential alternative to health supplementation for humans. Food Sci Hum Wellness 8(1):16–24. https://doi.org/10.1016/j.fshw.2019.03.001

Kumar VB, Pulidindi IN, Kinel-Tahan Y, Yehoshua Y, Gedanken A (2016) Evaluation of the potential of chlorella vulgaris for bioethanol production. Energy Fuels 30(4):3161–3166. https://doi.org/10.1021/acs.energyfuels.6b00253

Kumar M, Sun Y, Rathour R, Pandey A, Thakur IS, Tsang DCW (2020) Algae as potential feedstock for the production of biofuels and value-added products: opportunities and challenges. Sci Total Environ 716:137116. https://doi.org/10.1016/j.scitotenv.2020.137116

Lakshmikandan M, Murugesan AG, Wang S, Abomohra AEF, Jovita PA, Kiruthiga S (2020) Sustainable biomass production under CO2 conditions and effective wet microalgae lipid extraction for biodiesel production. J Clean Prod 247:119398. https://doi.org/10.1016/j.jclepro.2019.119398

Lam MK, Lee KT (2014) Cultivation of Chlorella vulgaris in a pilot-scale sequential-baffled column photobioreactor for biomass and biodiesel production. Energy Convers Manage 88:399–410. https://doi.org/10.1016/j.enconman.2014.08.063

Lee T, Nam I-H, Jung S, Park Y-K, Kwon EE (2020) Synthesis of nickel/biochar composite from pyrolysis of Microcystis aeruginosa and its practical use for syngas production. Bioresour Technol 300:122712. https://doi.org/10.1016/j.biortech.2019.122712

Levasseur W, Perré P, Pozzobon V (2020) A review of high value-added molecules production by microalgae in light of the classification. Biotechnol Adv 41:107545. https://doi.org/10.1016/j.biotechadv.2020.107545

Mabrouki J, Azrour M, Fattah G, Dhiba D, El Hajjaji S (2021) Intelligent monitoring system for biogas detection based on the Internet of Things: Mohammedia, Morocco city landfill case. Big Data Min Anal 4(1):10–17

Mabrouki J et al (2022) Smart system for monitoring and controlling of agricultural production by the IoT. In: IoT and smart devices for sustainable environment. Springer, pp 103–115

Martín Juárez J, Riol Pastor E, Fernández Sevilla JM, Muñoz Torre R, García-Encina PA, Bolado Rodríguez S (2018) Effect of pretreatments on biogas production from microalgae biomass grown in pig manure treatment plants. Bioresour Technol 257:30–38. https://doi.org/10.1016/j.biortech.2018.02.063

Mat Aron NS, Khoo KS, Chew KW, Show PL, Chen W-H, Nguyen THP (2020) Sustainability of the four generations of biofuels—a review. Int J Energy Res 44(12):9266–9282. https://doi.org/10.1002/er.5557

Mavani NR, Ali JM, Othman S, Hussain MA, Hashim H, Rahman NA (2022) Application of artificial intelligence in food industry—a guideline. Food Eng Rev 14(1):134–175. https://doi.org/10.1007/s12393-021-09290-z

Michaels H et al (2020) Dye-sensitized solar cells under ambient light powering machine learning: towards autonomous smart sensors for the internet of things. Chem Sci 11(11):2895–2906. https://doi.org/10.1039/C9SC06145B

Mishra R, Naik BKR, Raut RD, Kumar M (2022) Internet of Things (IoT) adoption challenges in renewable energy: A case study from a developing economy. J Clean Prod 371:133595. https://doi.org/10.1016/j.jclepro.2022.133595

Mutaf T, Bİlgİn GC, Öncel S, Elİbol M (2023) Metal nanopartiküllerin mikroalgler aracılığı ile yeşil sentezi. Ege J Fish Aquat Sci 40(1). https://doi.org/10.12714/egejfas.40.1.12

Mutaf T, Çalışkan G, Öncel S, Elibol M (2023) Green synthesis of metal nanoparticles by microalgae. Ege J Fish Aquat Sci 40:81–89. https://doi.org/10.12714/egejfas.40.1.12

Nanjundan P, Jaisingh W (2023) IOT contribution in construct of green energy. In: Vijayalakshmi S, Balusamy SB, Dhanaraj RK (eds) AI-powered IoT in the energy industry: digital technology and sustainable energy systems. Power Systems, pp 119–145. Springer International Publishing, Cham. https://doi.org/10.1007/978-3-031-15044-9_6

Nayak M, Dhanarajan G, Dineshkumar R, Sen R (2018) Artificial intelligence driven process optimization for cleaner production of biomass with co-valorization of wastewater and flue gas in an algal biorefinery. J Clean Prod 201:1092–1100. https://doi.org/10.1016/j.jclepro.2018.08.048
Nzayisenga JC, Eriksson K, Sellstedt A (2018) Mixotrophic and heterotrophic production of lipids and carbohydrates by a locally isolated microalga using wastewater as a growth medium. Bioresour Technol 257:260–265. https://doi.org/10.1016/j.biortech.2018.02.085
Pandit PR, Fulekar MH (2019) Biodiesel production from microalgal biomass using CaO catalyst synthesized from natural waste material. Renew Energy 136:837–845. https://doi.org/10.1016/j.renene.2019.01.047
Phanduang O, Lunprom S, Salakkam A, Liao Q, Reungsang A (2019) Improvement in energy recovery from Chlorella sp. biomass by integrated dark-photo biohydrogen production and dark fermentation-anaerobic digestion processes. Int J Hydrog Energy 44(43):23899–23911. https://doi.org/10.1016/j.ijhydene.2019.07.103
Pierro M, Moser D, Perez R, Cornaro C (2020) The value of PV power forecast and the paradox of the "single pricing" scheme: the Italian case study. Energies 13(15). https://doi.org/10.3390/en13153945
Pozzobon V, Levasseur W, Viau E, Michiels E, Clément T, Perré P (2020) Machine learning processing of microalgae flow cytometry readings: illustrated with Chlorella vulgaris viability assays. J Appl Phycol 32(5):2967–2976. https://doi.org/10.1007/s10811-020-02180-7
Prabha G, Sundaram K (2018) Design of feasible energy generation using solar panel and control using an IoT. Int J Eng Technol 7:191–196. https://doi.org/10.14419/ijet.v7i2.24.12029
Putri A, Pratama AO, Rohdiana A, Saraswati RR (2019) Smart microalgae photobioreactor helped by solar energy as eco-green technology to reduce CO_2 emissions in Jakarta, Indonesia, Zenodo, 1 février (2019). https://doi.org/10.5281/zenodo.2541282
Rahmat A, Jaya I, Hestirianoto T, Jusadi D, Kawaroe M (2020) Design a photobioreactor for microalgae cultivation with the IOTs (internet of things) system. Omni-Akuatika 16(1). https://doi.org/10.20884/1.oa.2020.16.1.791
Rajawat AS, Barhanpurkar K, Shaw RN, Ghosh A (2022) IoT in renewable energy generation for conservation of energy using artificial intelligence. In: Shaw RN, Ghosh A, Mekhilef S, Balas VE (eds) Applications of AI and IOT in renewable energy. Academic Press, pp 89–105. https://doi.org/10.1016/B978-0-323-91699-8.00005-X
Ramamurthy A, Jain P (2017) The internet of things in the power sector opportunities in Asia and the Pacific
Ren H-Y, Liu B-F, Kong F, Zhao L, Ma J, Ren N-Q (2018) Favorable energy conversion efficiency of coupling dark fermentation and microalgae production from food wastes. Energy Convers Manage 166:156–162. https://doi.org/10.1016/j.enconman.2018.04.032
Roy S, Kumar K, Ghosh S, Das D (2014) Thermophilic biohydrogen production using pre-treated algal biomass as substrate. Biomass Bioenergy 61:157–166. https://doi.org/10.1016/j.biombioe.2013.12.006
Serrà A, Artal R, García-Amorós J, Gómez E, Philippe L (2020) Circular zero-residue process using microalgae for efficient water decontamination, biofuel production, and carbon dioxide fixation. Chem Eng J 388:124278. https://doi.org/10.1016/j.cej.2020.124278
Shanmugapriya I (2023) A survey on energy management evolution and techniques for green IoT environment. In: Bindhu V, Tavares JMRS, Vuppalapati C (eds) Proceedings of fourth international conference on communication, computing and electronics systems. Lecture notes in electrical engineering. Springer Nature, Singapore, pp 155–165. https://doi.org/10.1007/978-981-19-7753-4_13
Shastik E, Li L, Liu J (2019) methods for hydrogen production by marine microalga Chlorella pyrenoidosa in natural seawater. Int J Hydrog Energy 44(29):14707–14714. https://doi.org/10.1016/j.ijhydene.2019.04.178

Silitonga AS, Masjuki HH, Ong HC, Mahlia TMI, Kusumo F (2017) Optimization of extraction of lipid from Isochrysis galbana microalgae species for biodiesel synthesis. Energy Sources Part Recov Util Environ Eff 39(11):1167–1175. https://doi.org/10.1080/15567036.2017.1310957

Soydemir G, Keris-Sen UD, Sen U, Gurol MD (2016) Biodiesel production potential of mixed microalgal culture grown in domestic wastewater. Bioprocess Biosyst Eng 39(1):45–51. https://doi.org/10.1007/s00449-015-1487-3

Tamilselvan K, Thangaraj P (2020) Pods—a novel intelligent energy efficient and dynamic frequency scalings for multi-core embedded architectures in an IoT environment. Microprocess Microsyst 72:102907. https://doi.org/10.1016/j.micpro.2019.102907

Tan YS, Ng YT, Low JSC (2017) Internet-of-things enabled real-time monitoring of energy efficiency on manufacturing shop floors. Proc CIRP 61:376–381. https://doi.org/10.1016/j.procir.2016.11.242

Tan KM, Kassim MA, Ng ZJ, Lalung J (2020) Isolation and characterization of novel acidophilic microalgae from abandoned mining site area for carbohydrate biosynthesis and its kinetic growth study in photobioreactor. IOP Conf Ser Mater Sci Eng 012011. IOP Publishing

Tao F, Cheng J, Qi Q, Zhang M, Zhang H, Sui F (2018) Digital twin-driven product design, manufacturing and service with big data. Int J Adv Manuf Technol 94(9):3563–3576. https://doi.org/10.1007/s00170-017-0233-1

Teo SH, Islam A, Yusaf T, Taufiq-Yap YH (2014) Transesterification of Nannochloropsis oculata microalga's oil to biodiesel using calcium methoxide catalyst. Energy 78:63–71. https://doi.org/10.1016/j.energy.2014.07.045

Teo CL, Atta M, Bukhari A, Taisir M, Yusuf AM, Idris A (2014) Enhancing growth and lipid production of marine microalgae for biodiesel production via the use of different LED wavelengths. Bioresour Technol 162:38–44. https://doi.org/10.1016/j.biortech.2014.03.113

Uhlemann THJ, Lehmann C, Steinhilper R (2017) The digital twin: realizing the cyber-physical production system for industry 4.0. Proc CIRP 61:335–340. https://doi.org/10.1016/j.procir.2016.11.152

Utilities and Energy, Sigfox 0G Technology. https://www.sigfox.com/use-cases/utilities-energy/ (consulté le 1 août 2023)

Vinuesa R et al (2020) The role of artificial intelligence in achieving the Sustainable Development Goals. Nat Commun 11(1). https://doi.org/10.1038/s41467-019-14108-y

Wang J, Yin Y (2018) Fermentative hydrogen production using various biomass-based materials as feedstock. Renew Sustain Energy Rev 92:284–306. https://doi.org/10.1016/j.rser.2018.04.033

Yu KL, Show PL, Ong HC, Ling TC, Chen W-H, Salleh MAM (2018) Biochar production from microalgae cultivation through pyrolysis as a sustainable carbon sequestration and biorefinery approach. Clean Technol Environ Policy 20(9):2047–2055. https://doi.org/10.1007/s10098-018-1521-7

Zeng J, Wang Z, Chen G (2021) Biological characteristics of energy conversion in carbon fixation by microalgae. Renew Sustain Energy Rev 152:111661. https://doi.org/10.1016/j.rser.2021.111661

Zewdie DT, Ali AY (2020) Cultivation of microalgae for biofuel production: coupling with sugarcane-processing factories. Energy Sustain Soc 10(1):27. https://doi.org/10.1186/s13705-020-00262-5

Zhu L et al (2019) Effects of nitrogen source heterogeneity on nutrient removal and biodiesel production of mono- and mix-cultured microalgae. Energy Convers Manage 201:112144. https://doi.org/10.1016/j.enconman.2019.112144

Evaluation of the Resistance of 10 New Citrus Rootstocks to Root Rot Caused by *Phytophthora Parasitica*

Dalal Boudoudou, Allal Douira, and Hamid Benyahia

Abstract In Morocco, the sour orange (*Citrus aurantium L.*) is the most anciently used rootstock, it remains dominant. Due to its better affinity with most commercial varieties, its adaptation to a wide range of soils and mainly due to its resistance to root rot caused by *Phytophthora* spp. However, it gives susceptible associations to tristeza. In vivo, several methods were used for the screening of citrus rootstocks in relation to *Phytophthora* spp. The objective of our experimentation is the development of an in vitro screening test of rootstocks caused by *Phytophthora parasitica.* Plants of Citrumelo (57. 98. 502), Poncirus Trifoliata, Citron troyer, Citron Troyer c 35 B 6 A11, Cleopatra mandarin tree x CC 30 575, Cleopatra mandarin tree xP.T 30 584, Cleopatra mandarin tree x cc B2 30 576, Mandarin tree changsa B2, Mandariniecleop X CC 30,577 and a South African Hybrid aged 2 months were watered daily and fertilized every 15 days using a nutrient solution. Inoculation was done by soaking the root systems in a spore suspension of *P. parasitica* for 24 h. Inoculation with *P. parasitica* caused root rot (Rot-tips) with a reduction in feeder fiber production and colonization of root systems that varied between rootstocks. Indeed, the Mandarin hybrids Cleopatra x cc 30 575 and the Mandariner Cleopatra x cc B2 30 576, were the most susceptible rootstocks to *P. parasitica* while the hybrids Citrange Troyes, Citrange Troyer C 35 B 6 A11 and the hybrid of South Africa are the most resistant rootstocks. Variability of rootstocks colonization by *P. parasitica* is thought to be variability in the production ability of 6, 7-dimethoxycoumarin (DMC) in root tissues of rootstocks.

Keywords Rootstocks · *P. parasitica* · Root rot · 6.7 dimethoxycoumarine

D. Boudoudou (✉) · A. Douira
Department of Biology, Laboratry of Plant Science, Biotechnology and Crop Protection, Ibn Tofail University, BP.242, Kenitra, Morocco
e-mail: dboudoudou@yahoo.fr

H. Benyahia
Unit of Plant Breeding and Germoplasm Conservation, Laboratry of Citrus Rootstocks Breeding, National Institute of Agricultural Research (INRA), Regional Center of Kenitra, BP 14000, Kenitra, Morocco

M. Azrour et al. (eds.), *Sustainable and Green Technologies for Water and Environmental Management*, World Sustainability Series,
https://doi.org/10.1007/978-3-031-52419-6_18

1 Introduction

Among the fungal diseases that threaten Moroccan citrus orchards related to *Phytophthora* spp., are the most important (Benyahia 1998; Boudoudou et al. 2015; Dalal et al. 2016; Serrhini 1986; Vanderweyen 1974). These pathogens can cause root rot, parasitic gummosis on the trunk and brown rot on the fruit (Timmer et al. 2000).

The use of a rootstock in citrus orchards has several advantages. In fact, the rootstock exhibits resistance to salinity, low temperatures and diseases such as those caused by *Phytophthora spp* (Agostini et al. 1991; Najwa et al. 2016; Boava et al. 2011; Graham 1995; Jim et al. 2011; Graham and Feichtenberger 2015; Graham et al. 2019; Sulistyowati et al. 1990).

Due to the lack of a specific selection program, the rootstocks retained in organic farming are the same as those used in conventional agriculture (Bénaouf 2005). The ideal rootstock does not exist, so it is necessary to retain the best compromise (Benyahia 2007; Bénaouf 2005). In nursery production of plants, rootstocks are obtained by multiplying seedlings (Bénaouf 2005). The selection of rootstocks, apart from agronomic characteristics and problems of incompatibility with the graft, is mainly based on sanitary criteria and in particular the tolerance of the rootstock to the Tristeza and the Resistance or tolerance to Gummosis due to *Phytophthora* (Ouissam 2014). The selection is geared towards the adaptability of new rootstocks to heavy soil conditions (Bénaouf 2005). The choice of the producer must take into account several criteria: The pedoclimatic constraints, the sanitary aspects, the type of cultivar and the chosen cultural behavior (Ouissam 2014). The sour orange is the most anciently used rootstock, it remains dominant in Morocco. However, it gives susceptible associations to tristeza (Benyahia 2007), it is now replaced by other tolerant rootstocks (Benyahia et al. 2011).

The national institute for agronomic research of Morocco currently has a collection of more than 200 rootstocks, the majority of them are not aware of the behaviour towards *Phytophthora spp* (Benyahia et al. 2011). Several methods have been used for the screening of citrus rootstocks in relation with *Phytophthora spp* (Afek and Sztejnberg 1990; Najwa et al. 2016; Carpenter and Furr 1962). These methods consist in inoculating plants in vivo under greenhouse or in the field, which is a laborious requiring both time and material (Benyahia et al. 2011).

Although these techniques have been successful, it is necessary to define a faster, reliable and reproducible methodology allowing mass screening of citrus rootstocks in relation to *Phytophthora spp* (Benyahia 2007). It is in this context that we have designed this work, which is a part of a program to improve the rootstocks. Our aim is to develop an in vitro screening test for rootstocks against *Phytophthora parasitica.*

Table 1 List of used rootstocks to evaluate their resistance against *Phytophthora parasitica*

Cultivars	Origin
Citrumelo (57. 98. 502)	USA
PoncirusTrifoliata	USA
Citrangetroyer	Morocco
CitrangeTroyer c 35 B 6 A11	France
Mandarinier cléopatre x cc 30 575	France
Mandarinier cléopatrexP.T 30 584	France
Mandarinier cléopatre x cc B2 30 576	France
Mandarinier changsa B2	France
Mandariniecléop X C.C 30,577	France
Hybride (Afrique du sud)	South Africa

2 Materials and Methods

2.1 Preparation of Plants

The seeds of 10 rootstocks (Table1) are from the collection of rootstocks of the EL Menzeh station of the Kenitra Regional Center for Agricultural Research. The semi was made in basins containing a sterile sand peat mixture.

After emergence, the plantlets from each rootstock were transplanted into 500-L plastic pots containing a mixture of sterile/sable peat. They were then placed under plastic tunnel, at temperature variations between (25 and 30 °C). The plantlets were daily watered and fertilized every 15 days using a nutrient solution, consisting of 100 L of water, 300 g of ammonium nitrate (33.5%), 30 g of Triple superphosphate (45%) and 25 g of potassium sulphate (50%).

After 2 months, 6 homogeneous plants for each rootstock were selected for the inoculation operation (Fig. 1).

2.2 Preparation of the Inoculum

The *Phytophthora parasitica* strain used for inoculation of rootstock roots was isolated from Elmenzeh citrus soils. The isolate (I9) was grown in petri dishes containing the PDA medium; proved to be the most aggressive. The cultures were incubated at 28 °C in the dark for 6 days before sampling, in the colony's growth front, ten agar discs, each 6 cm in diameter. These discs containing the mycelium were transferred to petri dishes containing 8 mL of soil extract. This extract came from the supernatant recovered after filtration of the addition of 15 g of non-sterile soil and one liter of distilled water, followed by agitation for 8 H and then settling for 16 h. The culture dishes were then incubated at 24 °C for 5 days to stimulate

Fig. 1 Plantlets of the used various rootstocks

a significant sporangia production. The release of the zoospores was then activated by transfer of the dishes at 4 °C for 15 min. Immediately after this heat shock, the contents of each dish were poured into containers and the final concentration of the sporangia was adjusted to 9 sporangia/mL using a Malassez cell.

For inoculation, the root systems of various dug up rootstocks were soaked for 24 h in containers containing the zoospores suspension. In parallel, un-inoculated controls were soaked only in distilled water. Then, the plants were separately transplanted in pots containing sand and peat and they were distributed according to the initial experimental device. Watering was carried out with the Hoagland solution at 50% for 1 month. The plants were then dug up and their root system was washed under running water until all the soil and peat particles were eliminated (Fig. 2).

2.3 Expression of Results

One month after inoculation, the plants were dug up. The root system was washed by tap water until all soil and peat particles were eliminated. The evaluation of *Phytophthora parasitica* attacks on the root system of plants of the various rootstocks was based on the estimation of:

- **Severity of root attack**, based on the scale of Zitko et al. (1991);
- **Number of feeder fibres or rot-tips**, by counting feed fibres at the level of each root system;
- **Root colonization of citrus rootstocks by P.** ***parasitica***, according to a technique used by Affailal (1987); Benyahia et al. 2004): for each rootstock, a sample of

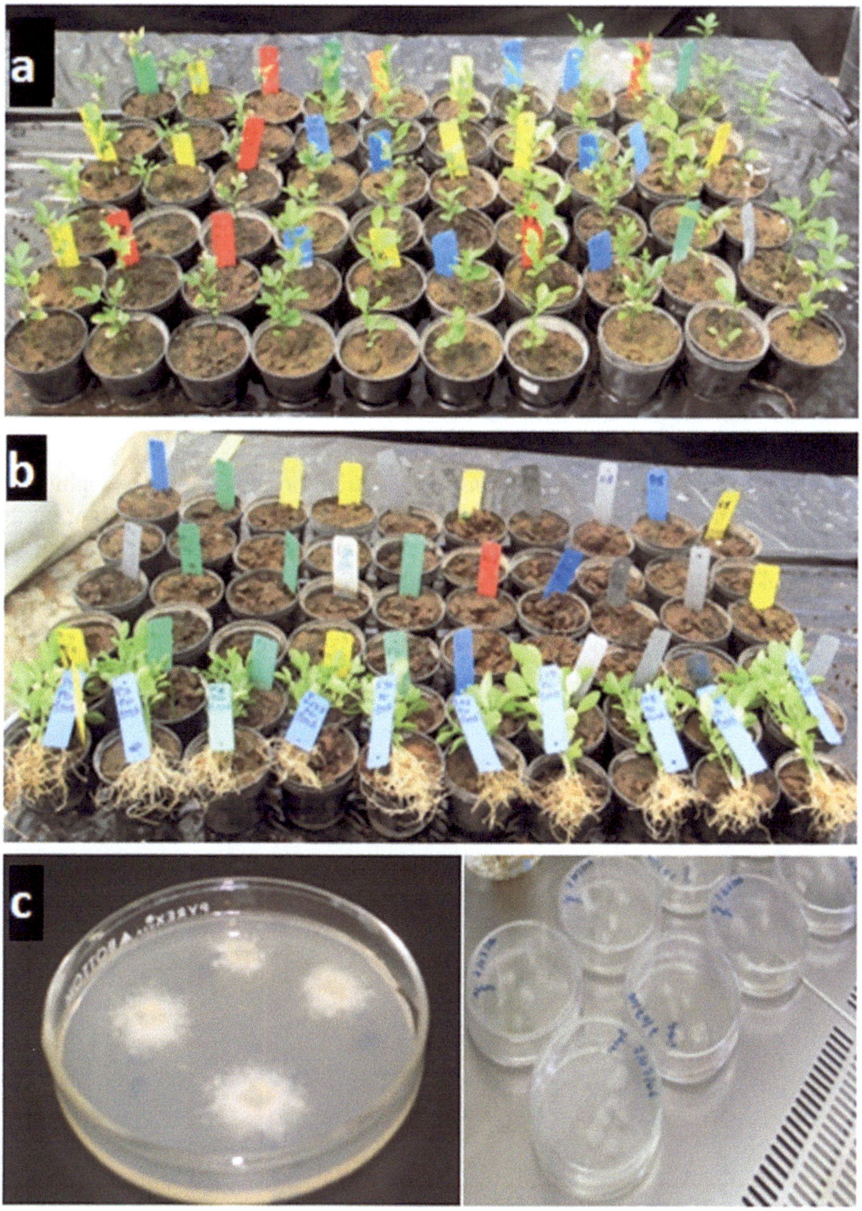

Fig. 2 Stages of inoculation of rootstocks by *P. parasitica* (**a** Breeding of the plantlets; **b** grubbing and washing of root systems; **c** preparation of *P. parasitica* inoculum)

2 g of fresh roots from each tested rootstocks were crushed for 2 min in a blender containing 20 mL of sterile distilled water. An extract of one mL of this root maceration was spread in a petri dish containing the selective medium BARPHY 72, developed by Benyahia (1998). This is indeed a selective medium for citrus pathogenic *Phytophthora.*

Three petri dishes were prepared for each sampled plant. These boxes were incubated for 4 days in the dark and at 28 °C before the *P. parasitica* colonies were counted. Assuming that a propagule was the origin of a colony's observation, the results of the analysis could be expressed as a quantity of germs/g of fresh tissue.

3 Results

Inoculation of rootstocks by the *P. parasitica* strain promoted the development of root rot symptoms in different rootstocks with varying intensity depending on the rootstock. These symptoms correspond to browning of root with alteration of the epidermal cells of the root and extremities) (Fig. 3). Incubation of root fragments taken from plants inoculated with *P. parasitica* into petri dishes containing BARPHY 72 medium (Benyahia 1998) revealed *P. parasitica* development after 48 h at 28 °C in darkness. On the other hand, fragments from uninoculated plants did not give rise to *P. parasitica.*).

3.1 Severity of Root Rot

The attack of *P. parasitica* on the different rootstocks was variable. Indeed, some rootstocks were attacked more than others (Fig. 4). Analysis of variance of the root

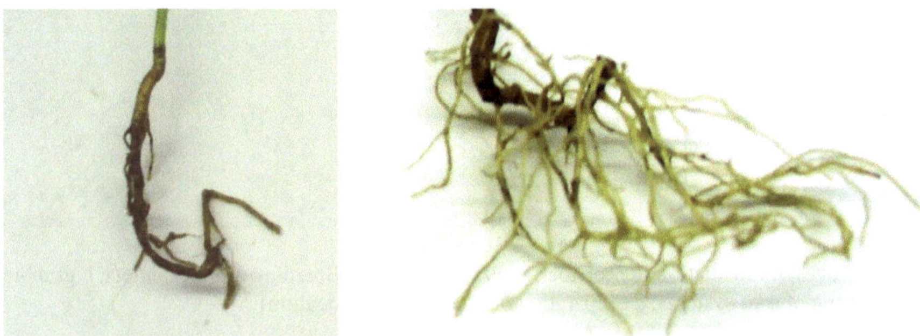

(a)Inoculated by *P. parasitica* **(b) non-inoculated by *P.parasitica***

Fig. 3 Symptoms of the citrus rootstocks root inoculated with *P. parasitica*at (**a**) and non-inoculated (**b**)

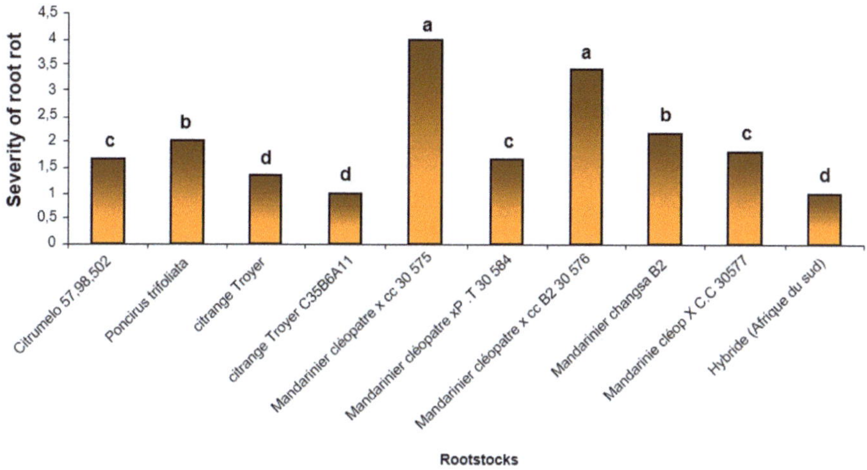

Fig. 4 Severity of root rot registered in different rootstocks after inoculation with *P. parasitica* (The rootstocks followed by the same letter does not differ significantly to the single of 5 (Duncan's test))

Table 2 Synthetic analysis of variance of the root rot severity

Source	DF	F	Pr > F
Porte-greffe	9	17,26	<0.0001
Bloc	5	0,80	0,5571

rot severity variable showed a very highly significant difference between the used rootstocks (table 2).

Comparing averages of root rot severity resulted in different statistically distinct groups (Fig. 5):

- **Group 1**: This group contains two rootstocks, Mandarin Cleopatra x cc 30 575 and Mandarin Cleopatra x cc B2 30 576. These two rootstocks showed the highest attack severity of 3.5 to 4.
- **Group 2**: This group also contains two rootstocks namely the Poncirus trifoliata and Mandarin Changsa. These two rootstocks were moderately attacked in relation to the rootstocks of the first group.
- **Group 3**: Three rootstocks appear in this group: Mandarin Cleop X C. C 30,577, Mandarin Cleopatra XP. T 30 584 and Citrumelo (57. 98.502). These rootstocks showed resistance to *P. parasitica* in terms of root rot severity. The severities, which were registered in this group, do not reach a value of 2.

Finally, the **last group** contains the rootstocks, which showed a very significant resistance to *P. parasitica*. These are Citrange Troyer, Citrange Troyes c 35 B 6 A11 and the hybrid of South Africa. The severity registered in this group is in the vicinity of the value 1.

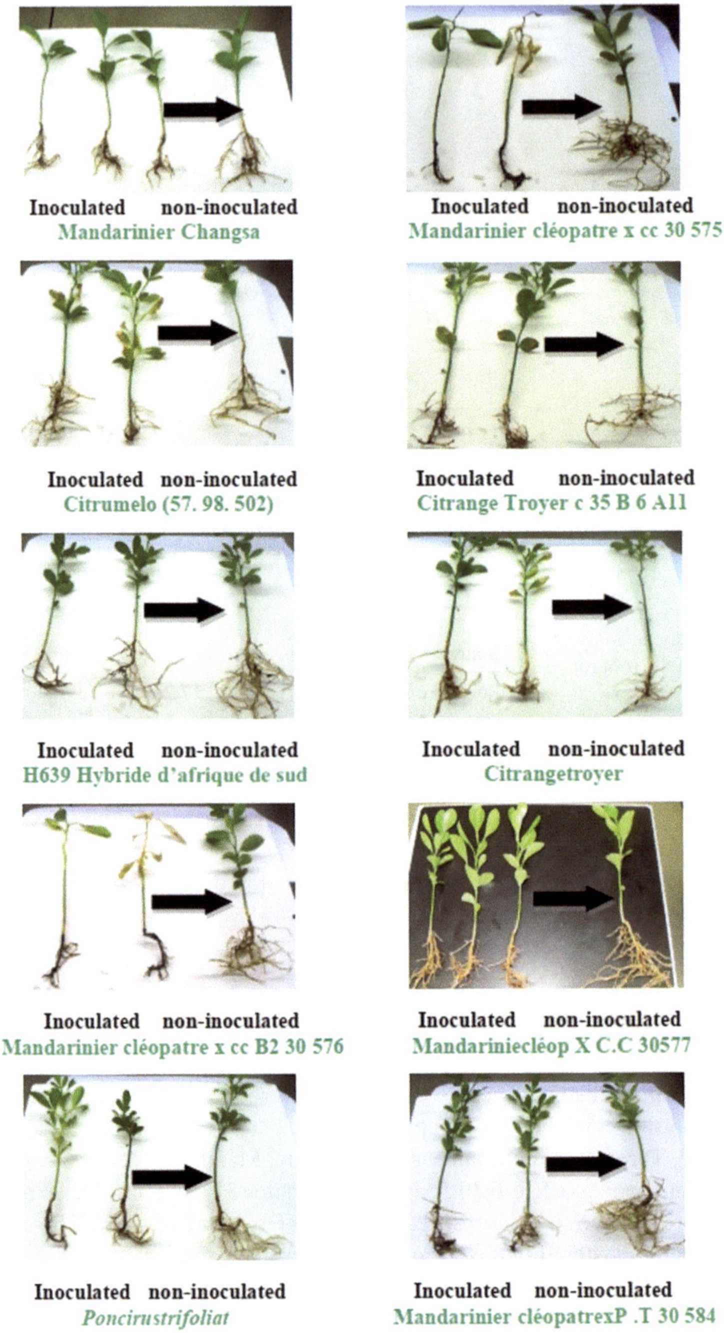

Fig. 5 Symptoms of *P. parasitica*-induced root rot in different rootstocks

Table 3 Synthetic analysis of variance of root rot severity

Source	DF	F	Pr > F
Porte-greffe	9	100,21	<0.0001
Bloc	5	1,67	0,1624

3.2 Colonization of Rootstock Roots by P. Parasitica

Root colonization of rootstocks by *P. parasitica* estimated in number of propagule per gram of root fresh weight, varied from one rootstock to another. Indeed, the analysis of variance of the variable number of propagule per gram of root fresh weight showed the existence of a very highly significant difference between the used rootstocks (Table 3).

The comparison of the root colonization means of rootstocks in terms of number of propagule per gram of root (Figs. 5, 6 and 7) shows the different rootstock groups, which differ statistically.

- **Group 1**: This group contains a single rootstock which was the Mandarin Cleopatra x cc 30 575. This rootstock most colonized by *P. parasitica* under conditions of our experimentation.
- **Group2**: In this group there were three rootstocks namely Citrange Troyer, Mandarin Cleopatra x P. T 30 584 and Mandarin Cleopatra x cc B2 30 576. The rootstocks of this group were moderately colonized by *P. parasitica* followed by the group containing the rootstocks Citrumelo (57. 98.502) Poncirus Trifoliata and Citrange Troyer C 35 B 6 A11.
- The last group contains the rootstocks that showed the lowest degree of colonization. This group contains the Mandarin Changsa B2, Mandarin cleop X c. C 30,577 and the Hybrid (South Africa)

3.3 Number of Lactating Fibres 'Rot-Tips'

Inoculation of rootstocks by *P. parasitica* significantly affected the regeneration ability of feeder fibers by used rootstocks, but with varying degrees depending on rootstocks. The analysis of variance of the feeder fibers number showed in fact a difference in the regeneration ability of these organs between the used rootstocks (Table 4).

Figure 8 shows the existence of different groups of rootstocks. The rootstock group showing the highest regeneration ability of root fibres is represented by Mandarin Cleop X C. C 30,577 Hybrid (South Africa). Both rootstocks recorded the highest number of root fibers.

It is also apparent from this figure that the rootstocks Citrange Troyes C 35 B 6 A11 and Mandariner Cleopatra × P. T 30 584 are rootstocks with the lowest regeneration ability of root fibres.

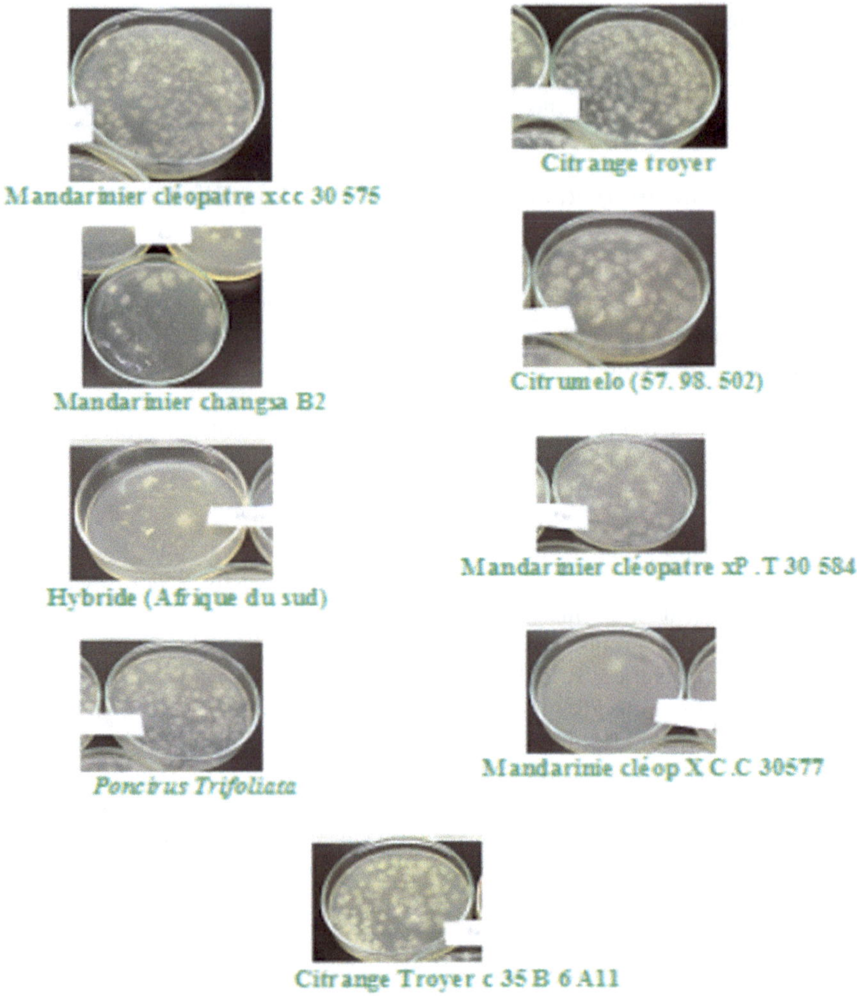

Fig. 6 Colony of *P. parasitica* from different rootstocks after inoculation

4 Discussion and Conclusion

In this study, the inoculation of rootstocks by *P. parasitica* caused on one hand root rot (rottips) with a reduction in the production of feeder fibers. On the other hand, a colonization of the root systems by *P Parasitica* which varies between rootstocks.

Observed Root rot symptoms in inoculated plants correspond to intense root browning with epidermal cell alteration.

The root rot severity varies between the used rootstocks. Indeed, the hybrids Mandarin Cleopatra × cc 30 575 and the Mandarin Cleopatra × cc B2 30 576, proved to be the most susceptible rootstocks to *P. Parasitica* while the hybrids Citrange

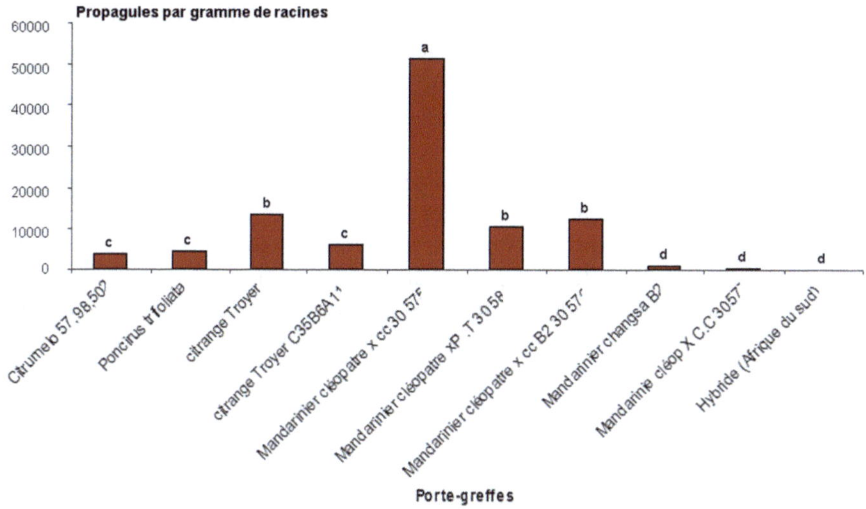

Fig. 7 Colonization of the roots of the various rootstocks by *P. Parasitica* (The rootstocks followed by the same letter does not differ significantly to the single of 5 Duncan's test)

Table 4 Synthetic analysis of variance of the feeder fibres number (rot-tips)

Source	DF	F	Pr > F
Rootstock	9	7,94	<0.0001
Bloc	5	0,77	0,5769

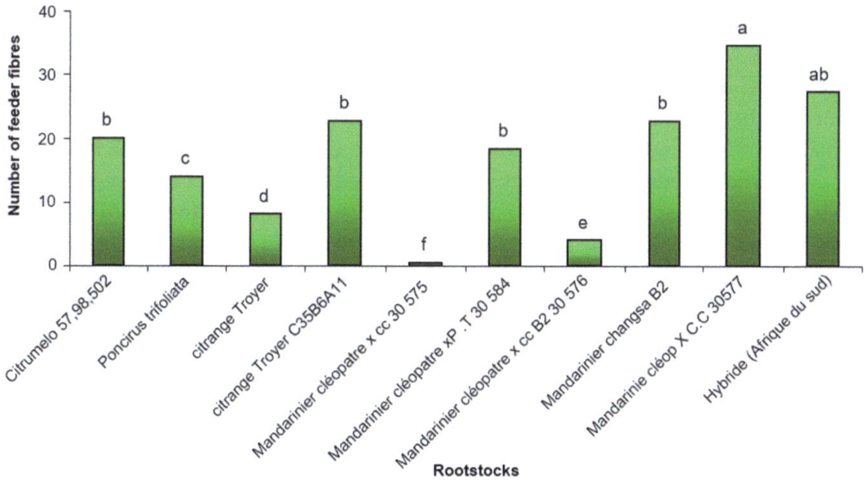

Fig. 8 Number of feeder fibres in different rootstocks inoculated with *P. Parasitica* (The rootstocks followed by the same letter does not differ significantly to the single of 5 (Duncan's test))

Troyes, Citrange Troyer C 35 B 6 A11 and the hybrid of South Africa are the most resistant rootstocks to *P. Parasitica* in terms of root rot based on the results obtained in this study. The resistance of Citrange Troyer to *Phytophthora* attacks, including root rot, has been reported by several authors (Graham 1990; Graham et al. 1998; Whiteside et al. 1988). The Citrange Troyer rootstock was used in this experiment as a reference rootstock in order to estimate the degree of rootstocks resistance to *Phytophthora* without undue risk of error. According to the results obtained the rootstocks Citrange Troyes C 35 B 6 A11, the hybrid of South Africa has the same degree of resistance to *P. Parasitica* as that of the Citrange Troyer. The Poncirus trifoliate, although this rootstock is considered a source of resistance to *Phytophthora* (Graham 1990; Graham et al. 1998) and is the parent of the Citrange Troyer showed a weaker resistance than the latter. This result could be explained by heterogeneity of the used plant material. In this sense, Benyahia (1993) had noted the existence of variability in susceptibility to root rot in different sour orange clones.

In this study, we also showed variability in root colonization of rootstocks by *P. parasitica.* According to the obtained results, the rootstocks Mandarin changsa B2, Mandarin cleop X c. C 30,577 and the hybrid (South Africa) are the least colonized rootstocks by *P. Parasitica.* It is recognized that in species of the *Citrus* genus, the resistance of rootstocks to trunk rot caused by *P Citrophthora*would be closely related to the accumulation capacity of a phytoalexin: 6.7 dimethoxycoumarine (DMC) (Sulistyowati et al. 1990; Sulistyowati and Keane 1992; Sulistyowati 1993). One of the explanatory assumptions of the variability of rootstock colonization by *P. Parasitica* would then be variability in the production ability of DMC in rootstock root tissues.

In terms of the regeneration ability of root fibers, this study shows that this biological activity varies between the used rootstocks. Indeed, the rootstocks Mandarin cléop X c. C 30,577 Hybrid (South Africa) showed the highest regeneration ability of feeder fibers. Graham (1990) reported that resistant rootstocks could be infected with *P. Parasitica* under artificial conditions and even in orchards but with low intensity compared to sensitive rootstocks. He has stated that only tolerant rootstocks can persist towards *P. parasitica* and that the tolerance corresponds to the ability of the rootstocks to regenerate the refined fiber caused by *P. parasitica* infection. Indeed, according to Zitko et al. (1991), *P. parasitica* specifically attacks the root fibers of citrus rootstocks. According to our obtained results: the rootstocks having shown the highest ability to regenerate feeder fibers are the most tolerant to *P. parasitica.*

5 Conclusion

The evaluation of the used rootstocks resistance the against *P. parasitica*, this study made it possible to highlight certain promoter rootstocks in terms of resistance to this pathogen, this is the case of Mandarin tree cleop X C.C 30,577 and Hybrid (South Africa) rootstocks which showed the highest ability to regenerate feeder fibres.

It would therefore be interesting to inoculate these rootstocks at trunk level to assess their degree of resistance to gummosis of the trunk, in fact, according to Graham (1995), there is no a correlation between resistance to root rot due to *P. parasitica* and gummosis of the trunk induced by *P.citrophthora* in citrus rootstocks.

Furthermore, the search for biochemical markers of resistance or sensitivity of citrus rootstocks to *Phytophthora* is desirable with a view to mass screening of rootstocks on the one hand and on the other hand overcome the problems of contradictory results.

References

Afek U, Sztejnberg A (1990) A rapid method for evaluating citrus seedlings for resistance to foot rot caused by *Phytophthora citrophthora*. Plant Dis 74:66–68

Affailal A (1987) Manifestation de la verticilliose sur les tomates sensibles et résistantes : effet de la salinité sur le développement des deux races de *Verticillium dahliae Kleb.* et sur la réaction des plantes à l'agent pathogène. Fac Sci Rabat thèse, Maroc, 127p

Agostini JP, Timmer LW, Castle WS, Mitchell DJ (1991) Effect of citrus rootstocks on soil populations of *Phytophthora parasitica*. Plant Dis 75:296–300

Bénaouf G (2005) (CIVAM BIO CORSE) Corse Pôle agronomique—20230 San SAN GIULIANO. Produire des agrumes en agriculture biologique

Benyahia H (1993) Evaluation de la résistance de 18 clones du bigaradier à la pourriture racinaire due à *Phytophthora citrophthora* et *Phytophthora parasitica*. Mémoire présentée à la faculté des Sciences Semlalia pour l'obtention du certificat des études approfondies en physiologie végétale, 70p

Benyahia H (1998) Effet de la salinité sur le développement des maladies à *Phytophthora* des agrumes au Maroc. Thèse de 3ème cycle, Université Cadi Ayyad, Faculté des Sciences Semlalia, Marrakech, 170 pp

Benyahia H (2007) Amélioration de la résistance des porte-greffes d'agrumes vis-à-vis des contraintes biotiques et abiotiques, thèse de Doctorat. Université Mohamed Ben Abdellah, Faculté des Sciences Dhar El Mahraz

Benyahia H, Ait HM, M., Jrifi A. and Lamsettef Y. (2004) Effet de la salinité de l'eau d'irrigation sur la colonisation des racines des porte-greffes d'agrumes par *Phytophthora parasitica*. Fruits 59(2004):101–108

Benyahia H, Beniken L, Omari FE, Benazzouze A, Handaji N, Msatef Y, Olitrault P (2011) Evaluation of the resistance of few citrus rootstocks to alkalinity by applying a fast test of screening. Afr J Agric Res 6(4):780–784

Boava LP, Cristofani-Yaly M, Stuart RM, Machado MA (2011) Expression of defenserelated genes in response to mechanical wounding and *Phytophthora parasitica* infection in *Poncirus trifoliata* and *Citrus sunki*. Physiol Mol Plant Pathol 77:1–7

Boudoudou D, Fadli A, Talha A, Bourachdi Y, Douira A, Benyahia H (2015) Effect of seasonal and *citrus* rootstocks on inoculum density of *Phytophthora* sp. in*Citrus* orchard in a heavy soil of the Gharb region of Morocco. Biolife 3(2):367–377. https://doi.org/10.17812/blj2015.32.4

Carpenter JW, Furr JR (1962) Evaluation of tolerance to root rot caused by *Phytophthora parasitica* in seedlings of citrus and related genera. Phytopathology 52:1277–1285

Dalal B, Abdelhak T, Anas F, Allal D, Hamid B (2016) Influence of citrus rootstocks on soil populations of *phytophthora* sp. In the gharb region in morocco. Int J Recent Sci Res 7(11):14230–14236

Graham JH (1990) Evaluation of tolerance of citrus roostocks to phytophthora root rots in chlamydospore-infested soil. Am Phytopathol Soc 74(10):743–746

Graham JH (1995) Root regeneration and tolerance of citrus roots tocks to root rot caused by *Phytophtora nicatianae*. Phytopatholgy 85:111–117

Graham J, Feichtenberger E (2015) Citrus Phytophthora diseases: management challenges and successes. J Citrus Pathol 2(1):1–11

Graham JH, Timmer LW, Drouillard DL, Peever TL (1998) Characterization of *Ph tophthora spp.* causing outbreaks of citrus brown rot in Florida. Phytopathology 88:724–729

Graham JH, Timmer LW, Dewdney MM (2019) Florida citrus pest management guide: phytophthora foot rot and root rot. The Plant Pathology Department, UF/IFAS Extension, IFAS publication, Florida, pp 156. 7 p

Jim G, Irey M, Taylor J (2011) Phytophthora damage to roots A potential contributor to decline of HLB affected Trees. Citrus Ind

Najwa B, Metoui N, Naïma Boughalleb-M'Hamdi. (2016) Screening fortolerance of different citrus rootstocks against zoospores of *Phytophthora nicotianae* in infested soil. J Phytopathol Pest Manage 3(3):63–75

Ouissam K (2014) Erosion génétiquedes espèces agrumicolesdans la wilaya de Skikda: Contraintes de production, Université 20 Août 1955 Skikda

Serrhini MN (1986) La résistance de *Phytophthora citrophthora* au metalaxyl. Université catholique de Louvain Belgique, Thèse de Doctorat, p 150

Sulistyowati L (1993) Effect of salinity on development of root infections caused by *Phytophthora citrophthora* in Citrus rootstocks growing in hydroponics. Agrivita 16(1):13–19

Sulistyowati L, andKeane, P. J. (1992) Effect of soil salinity and water content on stem rot caused by *Phytophthora citrophthora* and accumulation of phytoalexines in Citrus rootstocks. Phytopathology 82:771–777

Sulistyowati L, Keane PJ, Anderson JW (1990) Accumulation of the phytoalexin,6,7-dimethoxycoumarin, in roots and stems of Citrus seedlings following inoculation with *Phytophthoracitrophthora*. Physiol Mol Plant Pathol 37:451–461

Timmer, LW,, Garnsey, SM., Graham, JH. (2000) Compendium of citrus diseases, 2nd edn. APS Press Inc., St. Paul, MN, USA

Vanderweyen A (1974) La gommose à Phytophthora des agrumes au Maroc. Les agents pathogènes et la recherche de porte-greffe résistants. Al-Awamia 51:83–127

Whiteside JO, Garnsey SM, Timmer LW, eds, (1988). Compendium of Citrus Diseases. St Paul, MN, USA: APS Pres

Zitko SE, Timmer LW, Sandler HA (1991) Isolation of Phy-tophthora palmivora pathogenic to citrus in Florida. Plant Dis 75:532–535

Deep Learning-Based Tracking System for Detecting Toxic Elements in Water to Protect Public Health

Jamal Mabrouki, Mohammed Benchrifa, Karima Azoulay, Imane Bencheikh, Mourade Azrour, and Hajar Raji

Abstract Microorganisms are to blame for contaminating water resources, toxic substances or industrial waste. It can be found in rivers, aquifers and brackish bodies of water, as well as in rain water, dew, salt water, snow and arctic pack oceans. Bio sensing techniques based on cell growth use the full range of existing cytotoxic response to external inputs, via a to indicate the toxicity of aquatic targets. These bio-detection techniques can provide an effective means of indicating the water's toxicity to the safety of humans and the environment and health of marine organisms. The substantial increase in the level of pollutants entering fresh water requires rapid and accurate methods to monitoring the aqueous environment and detection of water aquatic aquatic poisoning. In this project, we review the recent evolution of current water biosensing techniques based on cell culture quality assessment, explain their major properties and capabilities, and the prospects for development in the coming years.

Keywords System · Water · Toxicity · Biodetection · Deep learning

J. Mabrouki (✉) · M. Benchrifa · K. Azoulay · I. Bencheikh
Laboratory of Spectroscopy, Molecular Modeling, Materials, Nanomaterial, Water and Environment, CERNE2D, Faculty of Science, Mohammed V University in Rabat, Avenue Ibn Battouta, BP1014 Rabat, Agdal, Morocco
e-mail: j.mabrouki@um5r.ac.ma

M. Azrour
STI Laboratory, IDMS Team, Faculty of Sciences and Techniques, Moulay Ismail University of Meknès, Meknes, Morocco

H. Raji
Laboratory of Studies and Research in Economics and Management Sciences, Superior School of Technology, Sultan Moulay Slimane University, University Campus M'ghila BP:591, 23000 Beni Mellal, Morocco

M. Azrour et al. (eds.), *Sustainable and Green Technologies for Water and Environmental Management*, World Sustainability Series,
https://doi.org/10.1007/978-3-031-52419-6_19

1 Introduction

Pollution of the environment can take many shapes and forms, from chemical to biological and thermic, and the types of water or bodies of land concerned can be fresh, salty or brackish, ground or surface water. It could even be a question of rain or dew, snow or arctic ice. It can also concern interfaces, estuaries and other wetlands, and in particular the sediments they support, which can first absorb (Cabelli et al. 1983) and then discharge certain contaminants. Generally deposits can be heavily impacted by past contamination. Combating such pollution is made all the more difficult by the fact that what is at the bottom of water, or diluted in the water, is often impossible to see, that certain toxic elements infiltrated into the soil frequently take a long time to take effect, and that many pollutants act through complex synergies. For instance, an accidental spill of 4,000 L of tetrachloromethane near Erstein in 1970 (Hartmann et al. 1993), which went almost undetected at the time, led twenty years later to the contamination of the water table and a well dug seven kilometers further downstream. Nitrates can reach certain water tables only after decades (Hill 2019).

The primary problem for heavy gases in the medium environment is their persistence. In contrast to the organics, they are neither degraded by chemical or physical means nor by organic compounds. Although the chemical nature of metals can be modified by oxidation or reduction, their elemental nature remains the same and excludes any possibility of thermal decomposition or microbiological degradation. Consequently, they tend to accumulate in soils or sediments, where they can develop a very complex chemistry due to the abiotic (pH, organic matter, redox potential) and biotic (microbial activity, plant roots) influences of the environment (White and Engelen 1997). Indeed, they can be adsorbed and desorbed, modified in their chemical form and speciation, complexed with various ligands, precipitated and redissolved. One consequence of the accumulation, persistence and behaviour of metals in the environment is that complex processes are required to remediate contaminated sites (Stratton and Smith 1988).

The development of an inexpensive, flexible, field-usable heavy metal detection and quantification system that would provide information on the level of contamination at a site would therefore be a valuable contribution to this field. During the last two decades, several biological tests have been developed to evaluate the toxicity and bioavailability of different compounds. These biotests or bioassays use a wide variety of organisms such as bacteria, daphnia, mussels or trout (Horsburgh et al. 2002). These approaches are effective in determining the bioavailable concentration of toxic compounds, but specific distinction of contaminants is difficult.

The innovative Microtox-OS. A custom design has been devised for use with microbial probes marked Lux and has been tested for toxin control in treatment plants and for the estimation of hazardous substances at environmentally impacted sites involving the surveillance of gross water supply and groundwater well tunnels (Killham and Yeomans 2001). In addition to the benefits of having automated monitoring for specific toxicity, on-line testing also enables the sampling of large and difficult with the help of analyses of kinetics (dose–response) and logistical (time-response) of the multisensor output (Turner 1999). The major limitation of automated, open cell, on-line testing for toxicity is the transport system for the photosensor, which can tend to be voluminous and heavy. These are currently largely storage and reactivation based, and typically employ a large-scale operating room based carousel to assure that the storage of cells is maintained reanimated sensors are timely deposited at the test site proper. The restriction of biosensor shipment should be surmounted if the materials are to be shipped over long time periods from a unique source. Slow delivery matrixes, in which the lyophilised photosensors are fixed, potentially allow for these distribution patterns. A range of various immobilisation templates have been investigated for the conservation/protection of microbes, including the alginate (Cassidy et al. 1996) and the polymerised polyvinyl alcool (PVA) (Chen and Kundu 1994; Lozinsky and Plieva 1998). PVA seems very interesting for the cryoimobilization of biosensors for microbial toxicity assays (Rainina et al. 1996). While bacterial biosensor capture methods for use in toxicity assays have been disseminated to a large extent, the use of the capturing techniques and later slow release of microbial probes for on-line tests has not been fully studied. To combine the advantages of biological tests with the specificity and sensitivity of a physicochemical analysis, biotechnology research efforts have focused on the development of bacterial biosensors. These are genetically modified living organisms coupled to a signal transduction system allowing the detection of bioavailable quantities of a specific pollutant. Biosensors are constructed so that the organism can sense environmental stress and translate it into an easily measurable signal (Zaanen et al. 1994). Their main interest is that they allow the assessment of the bioavailability of the pollutant, with a high specificity (Heitzer et al. 1992), data vital for analysis of the state of the environment and the implementation of remediation methods.

The possibility of a durable supply of loose biosensor units would provide many benefits for tox screens, including increased instrument design options, allowing various types of tox tests to be carried out based on the type of test sample and the user's information needs. The object of this research was to examine the future development of VPA for releasing lux-based self-sensors in on-line toxicity testing. In addition, the potentials of on-line toxicity testing for impregnation of air pollutants were investigated, with the aim of merging these materials into new toxicity determining instruments.

2 Literature Review

The issue of free heavy gases in the natural environment is of anthropogenic and human activity. Natural atmospheric sources consist of particulate emission due to volcanic activities, forest fires, or leeched soil particles (Nriagu 1989). The main source of trace minerals in aquatic organisms is from municipal effluents (notably As, Cr, Cu, Mn and Ni), coal-burning generating plants (As, Hg and Se), from metals casting plants (Cd, Cr, Mo, Ni, Pb, Se and Zn in some cases) and from sludge from treatment works (As, Mn and Pb). The disposal of toxic chemicals in the Aquatic Ecology Program results in the deposition of these metals in the water column of the lake and in the sediments of the sea. At this level, they can be precipitated and immobilized, but there is also the possibility of mobilization and diffusion towards the water table (Förstner and Salomons 1991). The two major sources in soils are coal combustion ash deposits and the general discharge of commercial and industrial wastes. The waste from municipal sources is a main factor for Cu, Hg, Pb or Zn, with some contributions from Cd and Pb via the airport. The large volumes associated with the food sector can significantly affect the trace element content in some soils, particularly agricultural soils, on which these wastes are dumped. Heavy metals in sewage plant sludge are not a risk overall, but in some cases the sludge is highly loaded with metals, so that discharge to land is problematic; this is perhaps the most influential single point of metal exposure in the area (Riether 2001).

In this section, we will outline the key drivers of biovailability of local heavy chemicals. De Haan and Koudijs (1994) developed the word "bioavailability" to define the extent to which a stimulant or pollutant can be readily taken up by a biological system: this can be quantified by the effect of the substance on the organism. For metals, they are considered to be bioavailable if they are in the form of free ions, and therefore generally soluble and mobile (Roane et al. 2001). In the course of its activity, however, the composition of metals is affected by various factors. As we shall see, these factors are interdependent, making it difficult to postulate. He origin of life probably developed around hydrothermal-oceanic sources, with a high proportion of heavy metals such as Zn, Cu, Hg or Pb (Rasmussen and Rasmussen 2000). Thus, the main challenge of the first cells was to maintain an appropriate intracellular concentration of essential trace elements and to exclude toxic metals at the same time. This premature ecological response has been a major factor in the development of heavy metal health and the detoxification process (Gatei et al. 2000). Thus, microscopic organisms have developed resistant systems to handle the toxic metal-driven stress. These systems, which are based on chromosomal or plasmid genes or on transposons, confer resistance to almost all the elements of the periodic table (Silver and Misra 1988).

3 Propose System

The internet of objects is the linking of objects, environments and physical places to physical world. The term refers to a growing number of objects connected to the Internet, enabling communications between our supposedly physical assets and those that exist digitally. These forms of connection make it possible to collect new masses of data on the network and consequently new knowledge and forms of knowledge. The intelligent system developed in this research as a laboratory is specialized in physic-chemical, microbiological and biological analyses on all types of products related to the environment. The research department has worked on different types of collectors. This sensor, impermeable and robust to a hostile environment once deployed in the water, is able to measure the potential of chemical elements present in the water and to transmit these values directly by wireless radio link to the analysis station remote from the water point. The sensors consist of a galvanostat made of various precision electrical components and a radio transmitter module. This wireless radio network for the development of connected objects is the best known and distinguishes itself from its direct competitor (Fig. 1).

The main Arduino Uno platform consists of an ATMega328 clocked at 16 MHz. This is the most straightforward and least expensive Arduino board. Connectors on the outer edges of the board are used to connect a series of additional modules (Fig. 2a) (Kumar et al. 2016).

This sensor is ideal for observing the evolution of water quality remotely, over time, over long periods of time, without human intervention. Its deployment and use is simple: once attached to the bank of the water point to be analyzed, the program implanted in the card can retrieve the data from the receiver which transfers them to the sending device in order to send the data to the basic station which has the receptive module.

These data are then received via the receiver module and are then transmitted by a simple Bluetooth link (Fig. 2b) or wifi module (Fig. 2c) to a Pc or a Smartphone with an interface dedicated to this system that generates live: curves, data tables

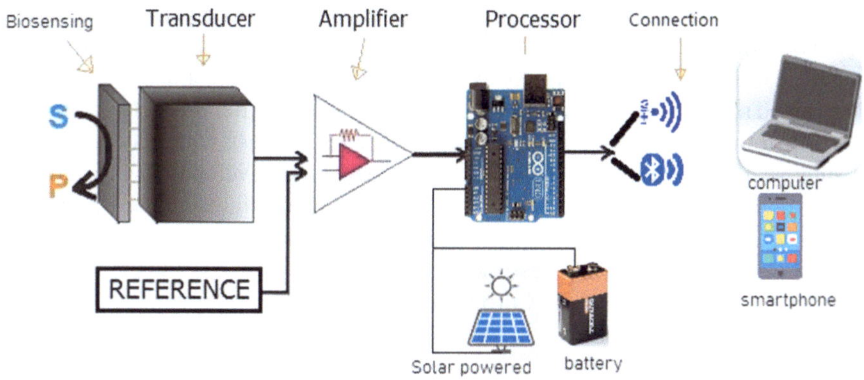

Fig. 1 The diagram of proposed system

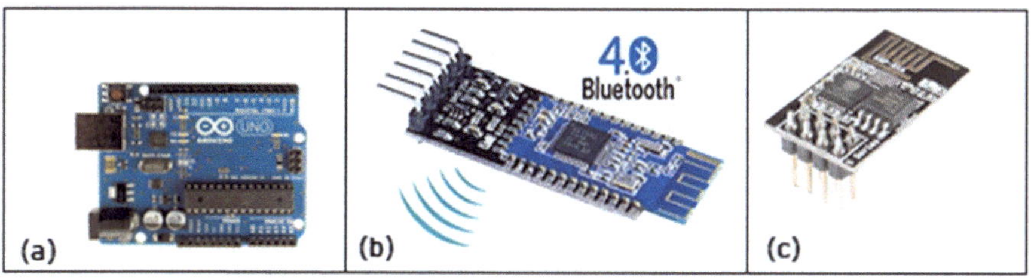

Fig. 2 **a** Arduino card; **b** Bluetooth HM-10 module, **c** Wi-Fi module esp8266

and other data according to the user's choice. Thanks to this system it ensures the communication of pollution episode to the port director in case of abnormal variations observed by the laboratory (Mabrouki et al. 2021).

4 Materials Used

4.1 The Biosensor

The Bacteria are an attractive biological medium for the design of a photosensor. Bacteria live in almost all natural environments, are easy to manipulate and constitute a largely unexplored reservoir of enzymes. One of the key advantages of utilizing bacteria is that they can be generated quickly, in large amounts and at little cost. Their response to any change in the environment is rapid, making them good indicators of environmental disturbance (Briscoe et al. 1996). The selected biosensor for the present work consisted. The full-length lux CDABE module is present in Escherichia coli HB101 pUCD607 (Rattray et al. 1990). Once cells had reached the end of the log phase, they were lyophilized in 1 ml aliquots and kept at −20.8 C until needed (Sousa Santos 1998).

By mixing 1 ml of PVA-based substrate with the contents of one vial of lyophilized E. coli solution, the microsensor was immobilised in the chosen slow-release matrix. In order to shorten the drying period of the matrix after the inclusion of the cells, the PVA-based matrix was dried at room temperature for 3/4 h. After properly combining, PVA and biosensor were placed in a desiccator for at least 16 h. The sensitivity of microorganisms to various pollutants has led to the development of a number of microbial tests that are not specific to the particular pollutant. The best known examples are the Microtox test, which is based on a naturally luminescent bacterium (Phosphorus spp.) whose luminescence changes in the presence of toxic substances (Michel et al. 1994), or the SOS Chromo test system, which is based on the response to SOS-responsive genes in E. coli, used to assess genotoxicity of chemicals (Allard et al. 1982).

In desalinated water with 1 M HNO3 conditioning and maintained at 3/5 8C, parent Zn compounds were created from sulphate. In deionized water, new raw materials for 3, 2-bromo-2-nitro-1, 3-propanediol (Bronopol) and DCP were created. Dilute the stock solutions with deionized water to make the Zn, DCP, or Bronopol solutions. If necessary, raise the pH of all the standards to 5.5 by adding HCl or NaOH. By equilibrating 20 ml deionized water and 10 ml benzene or benzene for 2 h at 25 °C in a general-purpose flask, you can create benzene and toluene stock solutions using the method proposed by Boyd (1997); Mabrouki et al. 2019). The concentrations of the higher standard were determined by gas chromatography after the benzene and toluene dilutions from the higher standards were prepared in water with a pH of 5.5 right before use.

These are genes that confer bioluminescence. The luc gene is the code for eukaryotic luc iferasis, which is required for the formation of the firefly Photinus or pyralis lights (Revilla et al. 1999). The reaction that eukaryotic luc iferase catalysis involves the substratum luciferin, a heterocylic carboxylic acid that is optimized in the presence of ATP and oxygen. This conversion results in light emission at 560 nm (Vela et al. 1995). In the case of the luc gene, the luminescence reaction is dependent on ATP supply and thus on cell physiology. Bacterial luciferase, encoded by the luxAB genes, is notably native to marine bacteria such as Vibrio harveyi and Fischeri (Meighen 1991). In the presence of oxygen and the reduced form of flavin mononucleotide (FMNH2), bacterial luciferase oxidizes an aliphatic aldehyde to an organic acid with concomitant light emission at 490 nm. Oxygen and FMNH2 are available in most aerobic bacterial cells, whereas aldehyde is synthesized in naturally luminescent bacteria by the products of the lux CDE genes, encoding the fatty acid reductase complex (Vuuren et al. 1989). Thus, by using the complete lux operon (luxCDABE) as a reporter, no additional external substrate is actually required (Meighen 1994).

Instantly, one-liter dual effluent samples were heated to kinetic impression The effluent samples came from a metallizing plant, a hard wood plant and a burners plant The effluent samples were stored at 4.8 C prior to the analysis If required for testing toxicity, lyophilized microsensors were heated for 1 instant within 10 ml of MKCl 0 1 for 25.8 C If necessary for testing, lyophilized photosensors were reheated for 1 instant within 10 ml of MKCl 0 1 for 25 8 C. The proposal in heat pipetted for volumes to 100 ml in the cuvettes due luminometer bucket 900 ml to sample solutions and mixed on eternal pipetting for 10 s Intermittency in heat then pipetted into the cuvettes Intermittency in heat added to the successive test cuvettes for 15 s interval Brightness in heat measured beforehand a retrospective at 15 min to the test solutions with the help of a Granit bigophone luminometer Biotests were carried out for the ambient isolation which varied to 18 for 20.8 C Each test resolution in heat analyzed for double copies.

4.2 Detection of Toxic Effects of Chemicals

A combination of dose- and time-response data make up the poisonous fingerprint. For a variety of contamination strengths, luminescence measurements were taken at a number of time points. A 3D map of the biosensor's response to the potential pollutant (toxicity fingerprint) is produced by combining the dose- and time-response data. Using a condensed online tool, the toxicity fingerprinting of three benchmark dangerous chemicals and three environmental samples was performed. For one hour at 25 °C, multisensor suspension in 100 ml of 0.1 M KCl.

De-ionized water forms a sample or a range of standard concentrations. Prior to the kinetic analysis, sample aliquots are metered into glass containers (18 ml for Zn, Bronopol, and the ambient samples, and 72 ml for 3,5-DCP). The sample is mixed with the biosensor cell suspension at the beginning of the analysis (2 ml for Zn, Bronopol, and ambient Samples, and 8 ml for 3,5-DCP), after which the mixture is instantly pumped into the in-line system. The sample was pumped at a rate of 4.8 ml per minute, and the online monitor was programmed to read luminescence readings at 20-s intervals. There is a 75 s lag between adding the biosensor to the sample and the first reading of luminescence due to the time it takes for the solution to enter the photomultiplier tube of the online instrument. A deionized water blank was run in between each sample categorization as part of the same process, which was repeated for each sample classification. In order to evaluate each setup for a longer (15 min.) period of time, the maximum sample volume was used for the DCP fingerprint.

Each fingerprint point's values are expressed as a percentage of the control luminescence. The control luminescence value has been determined as the average of the present luminescence in the previous and following deionized water blanks for each exposure time. Surfer software was used to map the three-dimensional kinetic responses (time and dosage response) for each sample.

5 Results and Discussion

A variety of potential environmental contaminants in water are sensitive to the E. coli HB101. The EC50 values obtained during on-line testing are similar with those obtained during batching testing. Zinc and bronopol considerably increased the sensitivity of the online test method (P B/0.001). Figure 3 depicts the characteristic curves for each contamination based on three conventional dose–response analyses and time-response curves. The biosensor responded to Zn with a quick reduction of luminescence over time and a loss of brightness at all doses tested.

However, the 3,5-DCP response demonstrated a marked enhancement of photoemission at the lowest tested dose (2.5/10 mg 11 DCP), followed by a sharp decline in photoemission levels at concentrations more than 10 mg 11. The luminescence of bronopol significantly decreased linearly over time. The three environmental

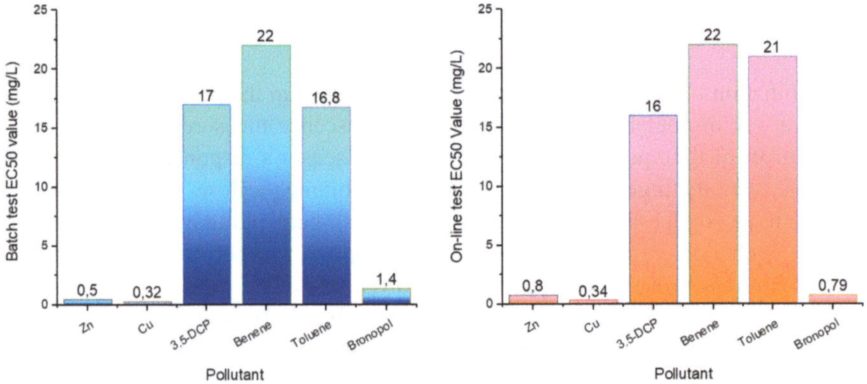

Fig. 3 Batch and online comparison of EC50 values for a range of contaminants

samples under investigation's toxicology fingerprints demonstrated the biosensor's great sensitivity to effluent and its ease in differentiating form patterns (Fig. 3).

It would seem that the PVA template used for the gradual distribution of lyophilized biosensors is a good fit for this use. Luminescence is monitored using an in-line equipment after an initial period of about 30 min. The speed of KCl administration to the array and the length of time needed to dry the Biosensor/PVA combination determine when cells will liberate. About an hour after the matrix released the cells, and gradually over a period of up to 16 h, the luminescence levels were high enough to permit toxicity assessment. Cells may be vulnerable to both zinc and DCP, according to preliminary toxicity test results utilising slow-release biosensors. According to estimates, the EC50 values for zinc and 3,5-DCP are respectively 0.25 and 30 mg/L. Both pollutants' sensitivity was of the same order of magnitude as the batch method's sensitivity (Fig. 3).

6 Conclusion

The purpose is of course to protect the health of the students, but also to anticipate the new WHO regulations. For businesses, the additional cost of these measurement systems will be offset by improved productivity due, for example, to fewer medical absences of staff, or even lengthy legal proceedings in the most serious cases of contamination, which also tarnish the image of companies. However, the installation of a monitoring system always has something to offer and the system will have to be as discreet as possible so that, on the one hand, the employee forgets about monitoring and, on the other hand, that it does not interfere with the employee's daily life at the workplace. The installation of the system for measuring personal exposure must also be followed by an accompaniment by persons capable of explaining the use of the system. As a first stage, the implementation of an environmental parameter monitoring internet of things (IoT) was tested to verify the metrics. This system provided a low-power alternative for setting up a meteorological station station system. The device was tested in an indoor setting and successfully adapted sensor data to weather and environmental conditions. The information recorded by the system will be useful for any inspections carried out in the future. The model can also be extended to the observation on pollution in new and modern urban areas. To protect the public good against pollution, the model provides effective response and ongoing observation with minimum effort.

References

Allard C, Quillard J, Rothschild C, Barthelemy M, Passeron J (1982) ORL localization of angioimmunoblastic lymphadenopathy. La Nouvelle Presse Medicale 11(38):2848–2849

Boyd T (1997) Am I black enough for you?: Popular culture from the'hood and beyond. Indiana University Press

Briscoe BJ, Evans PD, Pellilo E, Sinha SK (1996) Scratching maps for polymers. Wear 200(1–2):137–147

Cabelli VJ, Dufour AP, McCabe LJ, Levin MA (1983) A marine recreational water quality criterion consistent with indicator concepts and risk analysis. J (Water Pollut Control Fed):1306–1314

Cassidy MB, Lee H, Trevors JT (1996) Environmental applications of immobilized microbial cells: a review. J Ind Microbiol Biotechnol 16(2):79–101

Chen JL, Kundu A (1994) Rotation and gray scale transform invariant texture identification using wavelet decomposition and hidden Markov model. IEEE Trans Pattern Anal Mach Intell 16(2):208–214

De Haan A, Koudijs JC (1994) A linear relationship between ATP degradation and fatigue during high-intensity dynamic exercise in rat skeletal muscle. Exp. Physiol.: Transl. Integr. 79(5):865–868

Förstner U, Salomons W (1991) Mobilization of metals from sediments. Metals and their compounds in the environment. pp 379–398

Gatei M, Young D, Cerosaletti KM, Desai-Mehta A, Spring K, Kozlov S, Khanna K (2000) ATM-dependent phosphorylation of nibrin in response to radiation exposure. Nat Genet 25(1):115–119

Hartmann E, Görlich D, Kostka S, Otto A, Kraft R, Knespel S, Prehn S (1993) A tetrameric complex of membrane proteins in the endoplasmic reticulum. Eur J Biochem 214(2):375–381

Heitzer A, Webb OF, Thonnard JE, Sayler GS (1992) Specific and quantitative assessment of naphthalene and salicylate bioavailability by using a bioluminescent catabolic reporter bacterium. Appl Environ Microbiol 58(6):1839–1846

Hill AR (2019) Groundwater nitrate removal in riparian buffer zones: a review of research progress in the past 20 years. Biogeochemistry 143(3):347–369

Horsburgh AM, Mardlin DP, Turner NL, Henkler R, Strachan N, Glover LA, Killham K (2002) On-line microbial biosensing and fingerprinting of water pollutants. Biosens Bioelectron 17(6–7):495–501

Killham K, Yeomans C (2001) Rhizosphere carbon flow measurement and implications: from isotopes to reporter genes. Plant Soil 232(1):91–96

Kumar NS, Vuayalakshmi B, Prarthana RJ, Shankar A (2016) IOT based smart garbage alert system using Arduino UNO. In: 2016 IEEE Region 10 conference (TENCON). IEEE, pp 1028–1034

Lozinsky VI, Plieva FM (1998) Poly (vinyl alcohol) cryogels employed as matrices for cell immobilization. 3. Overview of recent research and developments. Enzym Microb Technol 23(3–4):227–242

Mabrouki J, Bencheikh I, Azoulay K, Es-soufy M, El Hajjaji S (2019) Smart monitoring system for the long-term control of aerobic leachate treatment: dumping case Mohammedia (Morocco). In: International conference on big data and networks technologies. pp 220–230

Mabrouki J, Azrour M, Fattah G, Dhiba D, El Hajjaji S (2021) Intelligent monitoring system for biogas detection based on the Internet of Things: Mohammedia, Morocco city landfill case. Big Data Min Anal 4(1):10–17

Meighen EA (1991) Molecular biology of bacterial bioluminescence. Microbiol Mol Biol Rev 55(1):123–142

Meighen EA (1994) Genetics of bacterial bioluminescence. Annu Rev Genet 28(1):117–139

Michel G, Gluckman E, Esperou-Bourdeau H, Reiffers J, Pico JL, Bordigoni P, Jouet JP (1994) Allogeneic bone marrow transplantation for children with acute myeloblastic leukemia in first complete remission: impact of conditioning regimen without total-body irradiation–a report from the Société Française de Greffe de Moelle. J Clin Oncol 12(6):1217–1222

Nriagu JO (1989) A global assessment of natural sources of atmospheric trace metals. Nature 338(6210):47–49
Paton DJ, Sands JJ, Lowings JP, Smith JE, Ibata G, Edwards S (1995) A proposed division of the pestivirus genus using monoclonal antibodies, supported by cross-neutralisation assays and genetic sequencing. Vet Res 26(2):92–109
Rainina EI, Efremenco EN, Varfolomeyev SD, Simonian AL, Wild JR (1996) The development of a new biosensor based on recombinant E. coli for the direct detection of organophosphorus neurotoxins. Biosens Bioelectron, 11(10):991–1000
Rasmussen UF, Rasmussen HN (2000) Human skeletal muscle mitochondrial capacity. Acta Physiol Scand 168(4):473–480
Rattray EA, Prosser JI, Killham K, Glover LA (1990) Luminescence-based nonextractive technique for in situ detection of Escherichia coli in soil. Appl Environ Microbiol 56(11):3368–3374
Revilla E, Delibes M, Travaini A, Palomares F (1999) Physical and popularion parameters of eurasian badgers (Meles meles L.) from mediterranean Spain
Riether KB (2001) Développement d'un biocapteur bactérien pour la détection de métaux lourds. Doctoral dissertation, Université Paul Verlaine-Metz
Roane TM, Josephson KL, Pepper IL (2001) Dual-bioaugmentation strategy to enhance remediation of cocontaminated soil. Appl Environ Microbiol 67(7):3208–3215
Silver S, Misra TK (1988) Plasmid-mediated heavy metal resistances. Annu. Rev. Microbiol. 42(1):717–743
Sousa Santos B (1998) Participatory budgeting in Porto Alegre: toward a redistributive democracy. Polit Soc 26(4):461–510
Stratton GW, Smith TM (1988) Interaction of organic solvents with the green algaChlorella pyrenoidosa. Bull Environ Contam Toxicol 40(5):736–742
Turner BR (1999) Tectonostratigraphical development of the Upper Karooforeland basin: Orogenic unloading versus thermally-induced Gondwana rifting. J Afr Earth Sc 28(1):215–238
Van Vuuren AM, Bergsma K, Frol-Kramer F, Van Beers JAC (1989) Effects of addition of cell wall degrading enzymes on the chemical composition and the in sacco degradation of grass silage. Grass Forage Sci 44(2):223–230
van Zaanen HC, van der Lelie H, Timmer JG, Fürst P, Sauerwein HP (1994) Parenteral glutamine dipeptide supplementation does not ameliorate chemotherapy-induced toxicity. Cancer 74(10):2879–2884
Vela J, Cebolla VL, Membrado L, Andrés JM (1995) Quantitative hydrocarbon group type analysis of petroleum hydroconversion products using an improved TLC-FID system. J Chromatogr Sci 33(8):417–425
White R, Engelen G (1997) Cellular automata as the basis of integrated dynamic regional modelling. Environ Plann B Plann Des 24(2):235–246

MIX
Papier aus verantwortungsvollen Quellen
Paper from responsible sources
FSC® C105338

If you have any concerns about our products,
you can contact us on
ProductSafety@springernature.com

In case Publisher is established outside the EU,
the EU authorized representative is:
Springer Nature Customer Service Center GmbH
Europaplatz 3, 69115 Heidelberg, Germany

Printed by Libri Plureos GmbH
in Hamburg, Germany